D0985697

THE GENETIC BASIS OF EVOLUTIONARY CHANGE / NUMBER XXV IN THE COLUMBIA BIOLOGICAL SERIES

R. C. LEWONTIN

THE GENETIC BASIS OF EVOLUTIONARY CHANGE

NEW YORK
COLUMBIA UNIVERSITY PRESS

R. C. Lewontin is Alexander Agassiz Professor
of Zoology at Harvard University.

Library of Congress Cataloging in Publication Data

Lewontin, Richard C 1929–

The genetic basic of evolutionary change.

(Columbia biological series no. 25)

Bibliography: p.

1. Population genetics. 2. Evolution.
3. Variation (Biology) I. Title. II. Series.
[DNLM: 1. Evolution. 2. Genetics. QH431
L678g 1973]

QH455.L48 575.1 73-19786

ISBN 0-231-03392-3

ISBN 0-231-08318-1 (pbk.)

AMS 1970 Subject Classifications 92–02, 92A10

Printed in the United States of America

Second cloth and Fourth paperback printing.

TO THEODOSIUS DOBZHANSKY

Nel mezzo del cammin di nostra vita
mi ritrovai per una selva oscura,
che la diritta via era smarrita.
Ahi quanto a dir qual era è cosa dura
questa selva selvaggia ed aspra e forte,
che nel pensier rinnova la paura.

DANTE

PREFACE

This book has grown, or better, exploded out of the Jesup Lectures that I gave at Columbia in 1969. Previous lecturers, in their presentations and in the books that resulted, had laid the foundations of modern evolutionary and systematic studies and had thereby contributed to a mystique of the Jesup Lectures that was overpowering, especially to a former Columbia student. My own consciousness of being an epigone permeated the lectures themselves and has contributed substantially to the delay in turning those lectures into a book. In the end, the delay has been beneficial, because the intervening three years have seen a great increase in theoretical and practical knowledge of the problems discussed, and because my own understanding has made some advance over the naive and disjointed views I then held.

The subject of the lectures and this book, the nature of genetic diversity among organisms, has always seemed the basic problem of evolutionary genetics. Because of the immense methodological difficulties and ambiguities, a characterization of that genetic variation seemed always to elude us. Then, with the discoveries of molecular biology about the relationship between genes and proteins, the possibility of an unambiguous characterization of genetic variation among individuals was opened. The first experiments revealed an extraordinary wealth of genetic diversity and, quite naturally, those of us involved in the work felt an immense elation in having finally given a direct answer to the major problem that had been plaguing our field. Some sense of that tension and resolution is conveyed in part II of this book.

As we tried to explain the great variation that had been observed, our original elation gave way slowly to disappointment. For no explanation is really satisfactory, and the kind of ambiguity that originally permeated the observations now pervades the theoretical explanations.

The problem may be even more serious. I have had the growing suspicion that in revealing the great variety of genic forms segregating in populations we have given the right answer to the wrong question. That is, the question was never really, How much genetic variation is there between individuals? but rather, What is the nature of genetic variation for *fitness* in a population? If, as some suspect, the genic diversity we have exposed is irrelevant to natural selection, or if, as the theory expounded in part III of this book suggests, the individual locus is not the appropriate unit of observation, then we have indeed given the right answer to the wrong question.

In fact, the methodological problem that is described in part II, and that seemed to be resolved by new methods and concepts, is really much deeper than I have suggested there. For I now realize that the question was not simply, How much genetic variation is there? nor even, How much genetic variation in fitness is there? but rather, How much genetic variation is there that can be the basis of adaptive evolution? To answer that question it is not sufficient to measure genetic variation, which we can now do, nor to measure the present variation in fitness associated with that genetic variation, which we have not done but which may be possible with a proper reorientation of theory. We require, further, that we be able to assess the *potentiality* for adaptive evolution in genetic variation that may currently be nonadaptive. But such an assessment will depend on an understanding of the relation between gene and organism that far transcends any present knowledge of development, physiology, and behavior. In fact, it demands the answer to every other question that now lies open in biology.

Progress, both observational and theoretical, has been rapid in evolutionary genetics in the last few years. As a result, in the three years that have intervened between the Jesup Lectures and the finishing of this book, a lot has been discovered. Much of it has been published, but much has not, and I am very grateful to my friends and colleagues who have told me of their work or shown it to me in manuscript form. Among these are Francisco Ayala, Avigdor Beiles, George Carmody, Brian Charlesworth, Ian Franklin, Ove Frydenberg, Moto Kimura, Ken-ichi Kojima, Terumi Mukai, Satya Prakash, Dick Richardson, Rollin Richmond, Tom Schopf, Robert Selander, Monte Slatkin, Tsuneyuki Yamazaki, and Eleutherios Zouros. I have discussed many of the concepts with them and with all my colleagues in Chicago, especially Brian Charlesworth, Ian Franklin, and Dick Levins, who have helped me to understand many things that were obscure to me. The book has been substantially improved by the enlightening comments made on the original manuscript by James Crow and Tim Prout, each of whom read it with his characteristic care and intelligence.

Above all, it has been my association with Jack Hubby that has given rise to the experimental work on which much of this book is based. It would be absurd to express my "gratitude" to him. Although his name does not appear on the title page, he is an "author" of this book as much as I am.

The text has been made much more readable and many opacities have been rendered transparent by the careful editorial work of Maria Caliandro. The tedious and unrewarding job of typing the several versions of the manuscript was carried out with her usual scrupulousness by Frances La Duke. Mary Jane Lewontin took on the maddening job of making the thousands of decisions necessary to compiling the index, and I am immensely grateful to her for it.

As is usually the case, I have undoubtedly forgotten to express my thanks to people who have made really important contributions to the work. I ask their indulgence.

Chicago
September 1972 *R. C. Lewontin*

CONTENTS

PART I / THE PROBLEM

CHAPTER 1 / The Structure of Evolutionary Genetics 3

PART II / THE FACTS

CHAPTER 2 / The Struggle to Measure Variation 19

CHAPTER 3 / Genic Variation in Natural Populations 95

CHAPTER 4 / The Genetics of Species Formation 159

PART III / THE THEORY

CHAPTER 5 / The Paradox of Variation 189

CHAPTER 6 / The Genome as the Unit of Selection 273

BIBLIOGRAPHY 319

INDEX 337

THE GENETIC BASIS OF EVOLUTIONARY CHANGE

PART I / THE PROBLEM

CHAPTER 1 / THE STRUCTURE OF EVOLUTIONARY GENETICS

When different people consider the work of a revolutionary like Darwin, they see different aspects of it as representing the "real" or "fundamental" element that separates it from the preexistent conformity of thought. To many, Darwinism means "evolution" and the commitment to an evolutionary world view, but historical evidence makes clear that Darwin only applied rigorously to the organic world what was already accepted as characteristic of the inorganic universe and of human culture (Lewontin, 1968). To others, especially students of the evolutionary process, Darwin's unique intellectual contribution was the idea of natural selection. For them, Darwinism is the theory that evolution occurs because, in a world of finite resources, some organisms will make more efficient use of those resources in producing their progeny and so will leave more descendants than their less efficient relatives.

Yet it is by no means certain, even now, what proportion of all evolutionary change arises from natural selection. Attitudes toward the importance of random events as opposed to selective ones vary from time to time and place to place. The famous Committee on Common Problems of Genetics, Palentology and Systematics, whose work led to the publication of Genetics, Paleontology and Evolution (Jepsen, Mayr, and Simpson, 1949) embodied within it divergent views on this question. In his article entitled "Speciation and Systematics," Mayr made clear that, in his view, the divergence between isolated populations is the result of their adaptation to different environments and that "geographical races are invariably to a lesser or greater degree also ecological races" (p. 291). But Muller's essay in the same volume puts emphasis on "cryptic genetic change" (p. 425) that is not reflected in phenotypic differentiation but that may result in sufficient genetic divergence between groups to result in speciation. For Mayr, natural selection is vital in the divergence between isolated populations while for Muller natural selection is always primarily a cleansing agent, rejecting "inharmonious" gene combinations, and not necessarily the causative agent in the initial divergence between incipient species. Nor is this conflict of viewpoints yet resolved. During the last few years there has been a flowering of interest in evolution by purely random processes in which natural selection plays no role at all. Kimura and Ohta suggest, for example (1971a), that *most* of the genetic divergence between species that is observable at the molecular level is nonselective, or, as proponents of this view term it, "non-Darwinian," although they do not deny that the evolution of obviously adaptive characters like the elephant's trunk and the camel's hump are the result of natural selection. If the empirical fact should be that most of the genetic change in species formation is indeed of this non-Darwinian sort, then where is the revolution that Darwin made?

The answer is that the essential nature of the Darwinian revolution was neither the introduction of evolutionism as a world view (since historically that is not the case) nor the emphasis on natural selection as the main motive force in evolution (since empirically that may not be the case), but rather the replacement of a metaphysical view of variation among organisms by a materialistic view (Lewontin, 1973). For Darwin, evolution was the conversion of the variation among individuals within an interbreeding group into variation between groups in space and time. Such a theory of evolution necessarily takes the variation between individuals as of the essence. Ernst Mayr has many times pointed out, especially in *Animal Species and Evolution* (1963), that this emphasis on individual variation as the central reality of the living world is the mark of modern evolutionary thought and distinguishes it from the typological doctrine of previous times.

In the thought of the pre-Darwinians, the Platonic and Aristotelian notion of the "ideal" or "type" to which actual objects were imperfect approximations was a central feature. Nature, and not just living nature, was understood by the pre-Darwinians only in terms of the ideal; and the failure of individual cases to match the ideal was a measure of the imperfection of nature. Such a metaphysical construct is not without importance in science, as Newton's mechanics prove so well. The first law in the *De Motu Corporum* is that

> Every body perseveres in its state of rest, or of uniform motion in a right line, unless it is compelled to change that state by forces impressed thereon.

Yet Newton points out immediately that even "the great bodies of the planets and comets" have such perturbing forces impressed upon them and that no body perseveres indefinitely in its motion. The metaphysical introduction of ideal bodies moving in ideal paths, so essential to the proper development of physics and so consonant with the habits of thought of the seventeenth century, was precisely what had to be destroyed in the creation of evolutionary biology. Darwin rejected the metaphysical object and replaced it with the material one. He called attention to the *actual* variation among *actual* organisms as the most essential and illuminating fact of nature. Rather than regarding the variation among members of the same species as an annoying distraction, as a shimmering of the air that distorts our view of the essential object, he made that variation the cornerstone of his theory. Let us remember that the *Origin of Species* begins with a discussion of variation under domestication.

The conflict between the real and the ideal was also important in a second realm that is relevant to evolution. What we know as the science of genetics is meant to explain two apparently antithetical observations—that organisms resemble their parents and differ from their parents. That is, genetics deals with both the problem of heredity and the problem of variation. It is in fact the triumph of genetics that a single theory, down to the molecular level, explains in one synthesis both the constancy of inheritance and its variation. It is the Hegelian's dream. But this synthesis was not possible until sufficient importance was attached to variation. Francis Galton attempted to construct a theory of inheritance based upon the degree of resemblance of offspring to their parents. Galton's Law of Filial Regression placed the emphasis on the fact that the offspring of extreme parents tended to "regress" back to the mean of the parental generation. The law was derived from observations of the mean height of all offspring whose parents belonged to a specific height class, but it placed no weight at all on the variation between parental pairs whose offspring were the same height, nor between sibs, nor between sibships whose parents belonged to the same height class. Galton's scheme depended entirely upon the regression of *means* on *means*.

In striking contrast, Mendel placed his emphasis on the *variations* among the offspring, rather than on any average description of them, *and derived his laws from the nature of the variations.* Thus, for Mendelism, as for Darwinism, the fact of variation and its nature are central and essential. Of course, at a second level, Mendel idealized his laws, much as Newton did, and every student knows that Mendel's F_2 ratios were suspiciously close to those 3 to 1 ideals. But this in no way vitiates the centrality of variation in Mendel's thought. Modern evolutionary genetics is then a union of two systems of knowledge, both of which took variation to be the essential fact of nature. It is not surprising, therefore, that the study of genetically determined variation within and between species should be the starting point of evolutionary investigation.

A DIGRESSION INTO FORMALISM

Dynamic Sufficiency When we say that we have an evolutionary perspective on a system or that we are interested in the evolutionary dynamics of some phenomenon, we mean that we are interested in the change of state of some universe in time. Whether we look at the evolution of societies, languages, species, geological features, or stars, there is a formal representation that is in common to all. At some time t the system is in some state E, and we are interested in the state of the system, E', at a future time, or past time, τ time units away. We must then construct laws of transformation T that will enable us to predict E' given E. Formally, we may represent this as

$$E(t) \xrightarrow{T} E'(t + \tau)$$

The laws of transformation T cannot be of an arbitrary form. Usually they contain some parameters Π, values that are not themselves a function of time or the state of the system. Second, they will contain the elapsed time τ, except in the description of equilibrium systems in which no change is taking place. They may or may not refer specifically to the absolute time t, depending on whether the system carries in its present state some history of its past. For example, if I wish to know what proportion of a population will be alive a year from now, I need to know how long ago the individuals were born, since survival probabilities change with age. But if I wish to know what proportion of my teacups will last through the next year, I do not have to know how old they are since, to a first order of approximation, their breakage probability per year is independent of their age. Finally, and essentially, the laws of transformation must contain the present state of the system E and obviously must produce as part of their output the new state E'. The laws of transformation are a machinery that must be built to process information about the current state E and to produce, as an output, the new state E'. But that means that the description of the system, E, must be chosen in such a way that laws of transformation can indeed be constructed using it.

For example, one cannot predict the future position of a space capsule from its present position alone. No set of laws can be constructed that will transform $E(t)$ into $E'(t + \tau)$ if E is the position of the capsule in three-dimensional space. It is necessary, in addition, to specify the present velocity of the capsule in three orthogonal directions, and its present acceleration in three orthogonal directions. If the state description is a function of those nine variables, then laws of transformation are possible and, as a result, capsules get to the moon and back. We will say that the state description with all nine variables is a *dynamically sufficient description* because, given that description, it is possible to find laws $T(E, \Pi, \tau, t)$:

$$E(t) \xrightarrow{T(E,\,\Pi,\,\tau,\,t)} E'(t + \tau)$$

The transformation of state in time has a geometrical interpretation. Let us take each variable used in specifying the state of the system as an axis in a Cartesian space. Then the state of the system at any time is represented as a point in that space, located by the projection onto the various axes. Evolution of the system is movement of the point through the space, tracing out a trajectory, and the laws of transformation are the equation whose solution is that trajectory. The axes of the space are the *state variables*, and the space they span is the *state space* of our system.

Looked at in this way, the problem of constructing an evolutionary theory is the problem of constructing a state space that will be dynamically sufficient, and a set of laws of transformation in that state space that will transform all the state variables. It is not always appreciated that the problem of theory building is a constant interaction between constructing laws and finding an appropriate set of descriptive state variables such that laws can be constructed. We cannot go out and describe the world in any old way we please and then sit back and demand that an explanatory and predictive theory be built on that description. The description may be dynamically insufficient. Such is the agony of community ecology. We do not really know what a sufficient description of a community is because we do not know what the laws of transformation are like, nor can we construct those laws until we have chosen a set of state variables. That is not to say that there is an insoluble contradiction. Rather, there is a process of trial and synthesis going on in community ecology, in which both state descriptions and laws are being fitted together.

Tolerance Limits In the development of a real science about a real and practical world, it is impossible and undesirable to search for an exactly sufficient description. The nature of the physical universe is such that the change of state of every part of it affects the change of state of every other part, no matter how remote. While a space capsule evolves in a nine-dimensional space for all practical purposes, an absolutely exact treatment of its motion would have to take into account the fact that it affects the motion of the earth and moon and of every other celestial body in some tiny degree. Yet no serious person would suggest that we really need to take into account the impetus given to the earth at the moment of launching. In each domain of practice we have a notion of the appropriate accuracy of prediction, and we are satisfied if our theory gives results that are somehow close enough. More exactly, corresponding to each state of a system E we establish a *tolerance set* ϵ of states that are similar enough to be regarded as indistinguishable. That does not mean that they *are* indistinguishable, but only that we do not care about the differences among states within a tolerance set. In practice, then, a sufficient dimensionality is one that allows us to describe the evolution of tolerance sets in time rather than the evolution of exact state descriptions.

In population ecology and evolutionary genetics, the tolerance limits remain matters of debate and choice. Ought we to be satisfied with theories that predict only that one community will have more species than another or that one population will be more polymorphic than another? If so, a very low dimensionality may be sufficient. Or do we really want to predict the number of breeding individuals of each species each year, or the gene frequencies at various loci, and if so, how accurately? Exactly the same domain of science may require quite different degrees of accuracy in different applications and thus use models of very different dimensionality. For example, population ecologists are generally satisfied to explain to one order of magnitude the increases and decreases in population size of the organisms studied, and for this purpose net fecundity and mortality are usually sufficient. Game and fish management, however, may require prediction of population changes to an accuracy of 10 to 20 percent, and for this purpose complete age-specific mortality and fecundity schedules are required. Finally, the human demographer needs to project human population sizes to better than 1 percent accuracy, and to do so needs fecundity and mortality figures by age, sex, socioeconomic class, education, geographical location, and so on. The building of a dynamically sufficient theory of evolutionary processes will really entail the simultaneous development of theories of different dimensionalities, each appropriate to the tolerance limits acceptable in its domain of explanation.

Empirical Sufficiency There is a second problem of theory construction in evolution, which we may term the problem of *empirical sufficiency*. The laws of transformation contain two elements that require measurement: the state variables that make up E and the parameters that make up Π. Even when a dynamically sufficient state space and a set of transformation laws have been arrived at, some of the state variables or the parameters may turn out to be, in practice, unmeasureable. Such unmeasurability is often not absolute; instead, the accuracy of measurement is low compared to the sensitivity of predictions to small perturbations in the values of the variables. To return to our space capsule example, when the capsule leaves its "parking orbit" around the earth, the smallest error in exit angle, smaller than can be controlled, will cause it to miss the moon by many miles. That is why mid-course corrections are necessary.

But empirical sufficiency is not always a matter of accuracy. It may lie much deeper. An example to which I shall return is the measurement of fitness of genotypes. One component of fitness is the probability of survival from conception to the age of reproduction. But by definition the probability of survival is an ensemble property, not the property of a single individual, who either will or will not survive. To measure this probability we then need to produce an ensemble of individuals, all of the same genotype. But if we are concerned with the alternative genotypes at a single locus, we need to randomize the rest of the genome. In sexually reproducing organisms, there is no way known to produce an ensemble of individuals that are all identical with respect to a single locus but randomized over other loci. Thus a theory of evolution that depends on the characterization of fitness of genotypes with respect to single loci is in serious trouble, trouble that cannot be cleared up by a quantitative improvement in the accuracy of measurement. The theory suffers from an epistemological paradox.

It is a remarkable feature of the sociology of science that evolutionary biologists have persistently ignored the problem of empirical sufficiency. The literature of population genetics is littered with estimates lacking standard errors and with methods for deciding between alternatives that have no sensitivity analyses or tests of hypotheses. No one has been exempt from this methodological naïveté, including myself, so that any specific discussion of the problem becomes immediately offensive to almost everyone. There are, however, a few positive landmarks in the assessment of these methodological difficulties that may be mentioned to the credit of their discoverers.

Since the invention by Muller of the *ClB* technique in 1928, many investigations have been made of the distribution of viabilities and fecundities of chromosomal homozygotes in various species of Drosophila. Yet it was not until 25 years later that Wallace and Madden (1953) and Dobzhansky and Spassky (1953) showed what proportion of the observed variance among chromosomal homozygotes and heterozygotes was, in fact, genetic, as opposed to experimental error. Since the error variance turned out be be between 59 and 99 percent of the total variance in different samples of heterozygotes and between 24 and 68 percent in homozygotes (Dobzhansky and Spassky, 1953, table 6), it is difficult to know what to make of studies without estimates of error.

An important parameter to be estimated for an understanding of the differentiation between populations is the effective breeding size, N. An elaborate theory exists for the estimation of this quantity from the allelism of lethal genes. The estimation procedure involves the reciprocal of the difference between two small numbers, a fact that in itself makes the use of the technique dubious. In an empirical test of procedure, Prout (1954) showed that in practice one cannot detect the difference between a population of 5000 and an infinite population.

Since the first experiments of L'Heritier and Teissier (1933) and of Wright and Dobzhansky (1946), adaptive values of Drosophila genotypes in population cages have been estimated from changes in genotype frequencies, but despite the use of a number of statistically sophisticated estimation techniques, standard errors of fitness estimates from population cages were not commonly calculated until Anderson, Oshima, Watanabe, Dobzhansky, and Pavlovsky did so in 1968, with rather disturbing results. Fitness estimates as disparate as 0.72 and 0.19 were not significantly different and estimates of 0.54, 0.47, and 0.59 had standard errors of 0.78, 1.15, and 1.23, respectively. Admittedly these experiments involved the estimation of nine fitness values, but this simply points up the problem of estimation for even a small number of genotypes under rigorously controlled laboratory conditions. The possibility of fitness estimation with smaller standard errors in natural populations seems remote, under the circumstances.

While dynamic sufficiency is an absolute and basic requirement for the building of an evolutionary theory, empirical sufficiency adds yet another stricture that may render a formally perfect theory useless. If one simply cannot measure the state variables or the parameters with which the theory is constructed, or if their measurement is so laden with error that no discrimination between alternative hypotheses is possible, the theory becomes a vacuous exercise in formal logic that has no points of contact with the contingent world. The theory explains nothing because it explains everything. It is my contention that a good deal of the structure of evolutionary genetics comes perilously close to being of this sort.

POPULATION GENETICS

Population genetics sets a much more modest goal than general evolutionary theory. If we take the Darwinian view that evolution is the conversion of variation between individuals into variation between populations and species in time and space, then an essential ingredient in the study of evolution is a study of the origin and dynamics of genetic variation within populations. This study, population genetics, is *an* essential ingredient, but it is not the entire soup. While population genetics has a great deal to say about changes or stability of the frequencies of genes in populations and about the rate of divergence of gene frequencies in populations partly or wholly isolated from each other, it has contributed little to our understanding of speciation and nothing to our understanding of extinction. Yet speciation and extinction are as much aspects of evolution as is the phyletic evolution that is the subject of evolutionary genetics, strictly speaking. That is not to say that speciation and extinction are not the natural extensions of changes within populations, but only that our present theories do not deal with these processes except on the most general and nonrigorous plane.

Even the partial task set for population genetics is a tremendously ambitious program. If we could succeed in providing a description of the genetic state of populations and laws of transformation of state that were both dynamically and empirically sufficient, we would create a complete theory concerning a vastly more complex domain than any yet dealt with by physics or molecular biology.

The present structure of population genetics theory may be represented as

$$G_1 \xrightarrow{T_1} P_1 \xrightarrow{T_2} P_2 \xrightarrow{T_3} G_2 \xrightarrow{T_4} G_1' \xrightarrow{T_1}$$

where G_1 and G'_1 represent a genetic description of population at times t and $t + \Delta t$, and P_1, P_2 and G_2 represent phenotypic and genotypic descriptions of states during the transformation and the laws of transformation are

T_1: a set of epigenetic laws that give the distribution of phenotypes that result from the development of various genotypes in various environments

T_2: the laws of mating, of migration, and of natural selection that transform the phenotypic array in a population *within* the span of a generation

T_3: an immense set of epigenetic relations that allow inferences about the distribution of genotypes corresponding to the distribution of phenotypes, P_2

T_4: the genetic rules of Mendel and Morgan that allow us to predict the array of genotypes in the next generation produced from gametogenesis and fertilization, given an array of parental genotypes

It would appear that both genotypes and phenotypes are state variables and that what population genetic theory does is to map a set of genotypes into a set of phenotypes, provide a transformation in the phenotype space, then map these new phenotypes back into genotypes, where a final transformation occurs to produce the genotypic array in the next generation. Figure 1 shows these transformations schematically. From this schema one would imagine that both phenotypic and genotypic state variables would enter into a sufficient dimensionality for the description of population evolution, and that the laws of population genetics would be framed in terms of both genetic and phenotypic variables. Yet an examination of population genetic theory shows a paradoxical situation in this respect. One body of theory, what we might call the "Mendelian" corpus, used almost exclusively by those interested in the genetics of natural populations, seems to be framed entirely in genetic terms. The other theoretical system, the "biometric," used almost exclusively in plant and animal breeding, appears to be framed in completely phenotypic terms. In Mendelian population genetics, for example, the expression

$$\Delta q = \frac{q(1-q)}{2} \frac{d \ln \bar{w}}{dq} \tag{1}$$

expresses the change in relative *allele frequency* Δq of an allele at a locus after one generation, in terms of the present allele frequency q

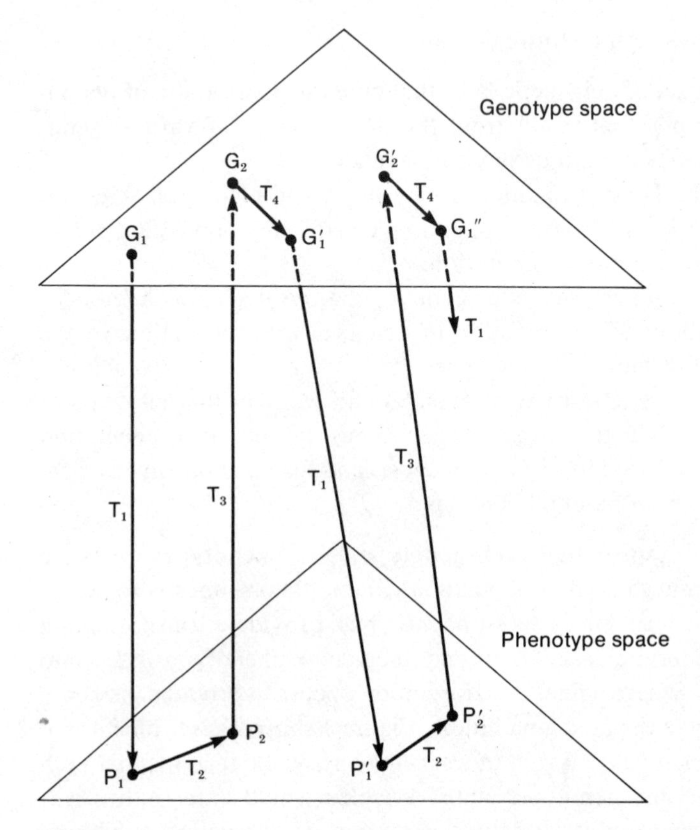

FIGURE 1

Schematic representation of the paths of transformation of population genotype from one generation to the next. G and P are the spaces of genotypic and phenotypic description. G_1, G'_1, G_2, and G'_2 are genotypic descriptions at various points in time within successive generations. P_1, P'_1, P_2, and P'_2 are phenotypic descriptions. T_1, T_2, T_3, and T_4 are laws of transformation. Details are given in the text.

and the mean fitness $\bar{w}$ of the genotypes in the population.

In biometric genetics, the prediction equation

$$\Delta P = ih^2 = i\frac{\sigma_g^2}{\sigma_p^2} \tag{2}$$

predicts the change in mean *phenotype*, ΔP, in one generation in terms of i, the difference in phenotype between the population as a whole and the selected parents, the phenotypic variance of the character in the population, σ_p^2, the additive genetic variance, σ_g^2, and the heritability of the trait, h^2.

Apparently, then, we have two parallel systems of evolutionary dynamics, one operating in the space of genotypes and bypassing the phenotypic space, and another operating entirely in the phenotypic domain. But this impression is illusory and arises from a bit of sleight-of-hand in which variables are made to appear as merely parameters that need to be experimentally determined, constants that are not themselves transformed by the evolutionary process. In equation (1) these pseudoparameters are the fitnesses associated with the individual genotypes in computing the function $\bar{w}$. Fitness, however, is a function of the phenotype, not the genotype, although in special circumstances it might turn out that a one-to-one constant correspondence existed between genotype and fitness. More usually, however, relative fitnesses of phenotypes will be a function of the phenotype composition of the population as a whole, so that the fitnesses assigned to genotypes in each generation will themselves follow a law of transformation that depends upon the genotype-phenotype relations. Although, for convenience, geneticists usually use a "constant genotypic fitness" model, this has led to a number of paradoxes that can be resolved properly only when phenotypic relations are taken fully into account. An example is the paradox of *genetic load*, which I will discuss more fully in chapter 5.

The biometric model presents even greater difficulty. The heritability h^2 and the genetic and phenotypic variances are themselves determined by the genetic variation in the population and undergo constant change in the course of the population's evolution. The laws of change of these variables cannot be framed without reference to the genetic determination of phenotype, including the degree of dominance of genes, the amount of interaction between them, and the relative frequencies of the alleles at the loci determining the character under selection. Thus, while equation (2) masquerades as a phenotypic law, genetic state variables must be added to provide a sufficient dimensionality. Because the transition between generations depends upon genetic laws, no sufficient description in terms of phenotypes alone is ever possible, and attempts like those of Slatkin (1970) can only succeed under extremely simplified conditions with very limited applicability. It is only fair to point out that for some purposes of plant and animal breeding, equation (2) is a sufficient predictor because the value of h^2 may evolve very slowly as compared with the mean phenotype, especially if a very large number of genes, each of small effect, are influencing the trait under selection. For a long-term prediction of progress under selection, or for an estimate of the eventual limit to the selection process, however, equation (2) is insufficient. *The sufficient set of state variables for describing an evolutionary process within a population must include some information about the statistical distribution of genotypic frequencies. It is for this reason that the empirical study of population genetics has always begun with and centered around the characterization of the genetic variation in populations.*

PART II / THE FACTS

CHAPTER 2 / **THE STRUGGLE TO MEASURE VARIATION**

It is clear that descriptions of the genetic variation in populations are the fundamental observations on which evolutionary genetics depends. Such observations must be both dynamically and empirically sufficient if they are to provide the basis for evolutionary explanations and predictions. From the schema presented in chapter 1, we see that a sufficient description of variation is necessarily a description of the statistical distribution of genotypes in a population, together with the phenotypic manifestation of those genotypes over the range of environments encountered by the population. The description must be genotypic because the underlying dynamical theory of evolution is based on Mendelian genetics. But the description must also specify the relations between genotype and phenotype, partly because it is the phenotype that determines the breeding system and the action of natural selection, but also because it is the evolution of the phenotype that interests us. Population geneticists, in their enthusiasm to deal with the changes in genotype frequencies that underlie evolutionary changes, have often forgotten that what are ultimately to be explained are the myriad and subtle changes in size, shape, behavior, and interactions with other species that constitute the real stuff of evolution. Thus Dobzhansky's dictum that "evolution is a change in the genetic composition of populations" (1951, p. 16) must not be understood as defining the evolutionary process, but only as describing its dynamical basis. A description and explanation of genetic change in populations is a description and explanation of evolutionary change only insofar as we can link those genetic changes to the manifest diversity of living organisms in space and time. To concentrate only on genetic change, without attempting to relate it to the kinds of physiological, morphogenetic, and behavioral evolution that are manifest in the fossil record and in the diversity of extant organisms and communities, is to forget entirely what it is we are trying to explain in the first place.

AN EPISTEMOLOGICAL PARADOX

If we bear in mind that evolutionary genetics must concern itself with evolutionally significant genetic change, and that the first task of such a science is a sufficient description of the genetic composition of populations, we encounter a paradoxical contradiction between what we wish to measure and what we can measure. In order to provide a description of the statistical distribution of genotypes that will dynamically suffcient, that will fit into a theory of genetic change, it is necessary to *count* genotypes. That is, it must be possible to say in what proportion each genotype is found in a population at each moment in time. With respect to any given locus, it must be possible to state that at some time t, for example, 30 percent of the population were homozygotes A_1A_1, 12 percent heterozygotes A_1A_2, 14 percent heterozygotes A_1A_3, and so on through the catalogue of genotypes. Moreover, in order to predict changes at this locus, it may be necessary to know the genotypic frequencies at many other loci whose gene products interact with the product of the A locus in determining the organism's phenotype. In addition, knowing the frequencies of genotypes separately for each locus may be insufficient to specify the frequencies of their joint genotypes, because the loci may be correlated in their distributions. Thus, the frequency of $A_1A_1B_2B_2$ individuals may not be the simple product of the frequency of A_1A_1 and B_2B_2 genotypes taken separately.

The consequences of such a lack of independence in the distribution of genotypes at separate loci are discussed in chapter 6 and are not of direct concern here. But whether or not loci are correlated, the basic observations about the genotypic composition of the population will be enumerations of genotypes. The necessity of enumerative description arises from the Mendelian nature of inheritance, from its discrete nature, so that the laws of evolutionary transformation are of necessity laws of changing proportions of discrete classes.

The possibility of enumerating discrete genotypic classes depends, as Mendel's original work showed, on whether the substitution of one allele for another at a locus leads to a sufficiently large change in phenotype. Mendel's success depended upon a very careful choice of characters, characters which allowed individuals to be unambiguously classified into a small number of classes. Even Mendel's work was not perfect in this respect since it was impossible to distinguish between the homozygous dominant and the heterozygote, but fortunately Mendel was able to see through this confusion. He would not, however, have been able to see through the confusion of a virtually continuous distribution of phenotypic classes had he chosen to study the inheritance of normal plant height, rather than of a drastic dwarfing gene whose major morphogenetic effect completely obscured the small genetic and environmental differences between individual plants. Not only Mendel, but all the twentieth-century developers of classical and molecular genetics have depended on the availability of gene changes with major and drastic morphogenetic and physiological effects. Imagine the state of molecular genetics if the only difference between various genotypes of bacteriophage were in whether an infected bacterium produced an average of 247.3 or 246.1 progeny phage!

In contradistinction to the discreteness of genotypic classes demanded by Mendelian analysis lies the quasi-continuous nature of the phenotypic differences that are the stuff of evolutionary change. The evolution of species consists in the gradual accumulation of very small changes in physiology, morphogenesis, and behavior. Closely related species may differ on the average in their temperature tolerance or preference, in their size and shape, but it is often impossible to distinguish an individual of one species from another on one or a combination of these characters because of the broad phenotypic overlap between the groups. Even when no closely related species overlaps it, any species will show a range of variation for most of its phenotypic characteristics, variation which has both a genetic and an environmental component. Indeed, for most of the characters whose change the evolutionist wishes to explain, the variation between genetically identical individuals is likely to be of the same or greater magnitude than the variation introduced by a single allelic substitution at one of the loci affecting the character. Even if the heritability of IQ in Caucasians is 80 percent, as is estimated by some studies (Erlenmeyer-Kimling and Jarvik, 1963), the effect of a single gene substitution on IQ is small (except for drastic abnormalities giving rise to mental deficients) as compared with environmental variations. If this were not the case, the Mendelian inheritance of IQ would have been worked out long ago. Evolutionarily significant genetic variation is then, almost by definition, variation that is manifest in subtle differences between individuals, often so subtle as to be completely overwhelmed by effects of other genes or of the environment.

The situation is clear when we consider natural selection itself. Suppose that the difference between a *Drosophila pseudoobscura* genotype A_1A_1, at a certain locus, and a genotype A_1A_2 is that females at the first kind lay an average of 27.3 eggs per day at their peak of egg production at 25 C, whereas females of the second type lay only 26.4 eggs. This 3 percent difference in fecundity is a large one in evolutionary terms and would result, all other things being equal, in a rapid loss of the allele A_2 from the population. But in practice it is absolutely impossible to distinguish females of these two genotypes from each other even in carefully controlled egg-laying trials. The fecundity of individual females is extremely sensitive to both environmental and morphogenetic accidents. A female of either genotype may be totally sterile or may lay 50 eggs per day at her peak egg production. The difference between the averages is simply too small, in view of the huge variance around these averages, to allow individuals to be classified. With respect to such genotypes, then, no enumeration of the population is possible. We cannot say "this female is A_1A_1 and that one is A_1A_2."

When we turn from fecundity to viability, the situation is even worse. Can we distinguish an individual fly whose genotypic probability of survival from egg to newly hatched adult is 95 percent from one whose probability is only, say, 45 percent? Here is a drastic genetic difference indeed. But an *individual* either does or does not survive to a given age. Among adults all have survived and among eggs nothing can be predicted. The probability of survival is an *ensemble* property, not an individual one, so that enumeration of individuals by their phenotype is, in this respect, impossible by definition. Survival and reproduction are the two components of fitness, and genotypic differences with respect to these characters evolve directly under the influence of natural selection. Yet it is impossible to recognize and enumerate the different genotypes in a population with respect to those most basic of all characters.

We see here the fundamental contradiction inherent in the study of the genetics of evolution. On the one hand the Mendelian genetic system dictates the frequencies of genotypes as the appropriate genetic description of a population. The enumeration of these genotypes requires that the effect of an allelic substitution be so large as to make possible the unambiguous assignment of individuals to genotypes. On the other hand, the substance of evolutionary change at the phenotypic level is precisely in those characters for which individual gene substitutions make only slight differences as compared with variation produced by the genetic background and the environment. What we can measure is by definition uninteresting and what we are interested in is by definition unmeasurable. This paradox in one form or another has plagued population genetics since Chetverikov proposed in 1926 that the measurement of hidden genetic variation and its mechanism of maintenance were the central problems of evolutionary genetics. The paradox is not yet resolved, but recent advances in molecular biology and in population genetic theory, which are the substance of this book, offer some hope of a resolution.

THE "CLASSICAL" AND "BALANCE" HYPOTHESES

The problem of the characterization of variation can best be understood if it is formulated in the following manner: If we could examine the diploid genotype of a typical individual chosen from a sexually reproducing population, at what proportion of its loci would it be heterozygous? While this is obviously a more restricted problem than the most general one of "characterizing the variation," a correct answer to this more restricted question would go a long way toward a general solution to the problem of variation. To see that this is so, let us consider two polar predictions about the outcome of such a *gedanken experiment*. Dobzhansky (1955) has named these predictions the "classical" and the "balanced" theory of population structure. No geneticist has with absolute consistency supported the classical or the balanced theory in the extreme forms in which I shall juxtapose them, but most workers who have been concerned with the issue have, at one time or another in their writings, polarized themselves sharply.

The classical theory assumes that at nearly every locus every individual is homozygous for a "wild-type" gene. In addition, each individual is heterozygous for rare deleterious alleles at a handful of loci, on the order of a hundred out of tens of thousands of genes. Some very small proportion of the population will be so unfortunate as to be homozygous for a rare deleterious gene and thus be severely handicapped, and this proportion will be much increased in the offspring of matings between close relatives. This hypothesis is fully described and justified by Muller (1950) in his article descriptively titled "Our load of mutations," in which he estimates that a typical person will be heterozygous for a severely deleterious gene at 8 to 80 loci, depending upon various assumptions about the number of loci in the human genomes, mutation rates, and degrees of dominance. In all cases this corresponds to about 0.1 percent of loci heterozygous per individual.

In the classical view, the genetic description of two randomly sampled individuals from a population would look as follows:

$$\frac{++++m+\ldots+++}{++++++\ldots+++} \qquad \frac{++++++\ldots+m+}{++++++\ldots+++}$$

with + signs indicating wild-type alleles and *m* a deleterious mutant. Each individual is heterozyous for an occasional locus, the particular gene being different in each case. Such a picture accords well with a *a priori* assumption of biochemical genetics that there is one functional or best form of an enzyme and that other forms, specified by alternative alleles at the structural gene locus, would have defective enzyme activity. It has not always been obvious to biochemical geneticists that, like other classical Mendlian geneticists, they were forced to deal with drastic, usually lethal, gene substitutions in order to carry out their experiments and would not recognize slight mutants if they saw them. Indeed biochemical genetics invented the extraordinary class of "leaky" mutants to emphasize that a mutation with partial enzyme activity, as opposed to complete loss of function, was an exceptional and somewhat annoying phenomenon.

The balance hypothesis in its most extreme form postulates, in direct opposition to the classical hypothesis, that individuals from sexually reproducing, cross-breeding populations are heterozygous at nearly every one of their loci, and only rarely will a locus be homozygous except in the offspring of closely related mates. This extreme view is expressed by Wallace (1958b) as follows: "Subject to the limitations imposed by chance elimination of alleles, by mating of close relatives, and by the finite number of alleles at a locus, we feel that the proportion of heterozygosis among gene loci of representative individuals of a population tends toward 100 percent." Such a view has two concomitants. First, there is no allele that can properly be designated wild-type since normal individuals in the population are heterozygotes.

Second, the number of alternative alleles segregating in the population must be large at each locus. Otherwise the ordinary laws of Mendel would cause homozygosity from random segregation and mating, the maximum proportion of heterozygosity at a locus with n alleles being $(n - 1)/n$. On the balance hypothesis, two randomly chosen individuals from a sexually reproducing population would have the following genetic representations:

$$\frac{A_3\ B_2\ C_2\ D\ E_5\ .\ .\ .\ Z_2}{A_1\ B_7\ C_2\ D\ E_2\ .\ .\ .\ Z_3} \qquad \frac{A_2\ B_4\ C_1\ D\ E_2\ .\ .\ .\ Z_1}{A_3\ B_5\ C_2\ D\ E_3\ .\ .\ .\ Z_1}$$

Not only are most loci in heterozygous condition, but different individuals are homozygous for different alleles at different loci when they are indeed homozygous. An occasional locus with only one wild-type form like D is not excluded in the balance view, and an occasional allele, for example B_7, may be very rare and extremely deleterious when homozygous so that, like the classical theory, the balance theory predicts deleterious consequences of close inbreeding.

Manifold consequences flow from the assumption of the classical or balance hypothesis. If the classical hypothesis were correct, the difference between populations would be of far more profound significance than under the balance theory. Since there would be so little genetic variation between individuals within populations, most of the genetic diversity within a polymorphic species would be interpopulational. In man, the manifest genetic differences between geographical races would represent a much greater proportion of total human variation than occurs within races, giving to race a considerable biological importance. A basis for racism may also flow from the concept of wild type, since if there is a genetic type of the species, those who fail to correspond to it must be less than perfect. Platonic notions of type are likely to intrude themselves from one domain into another, and Dobzhansky (1955) was clearly conscious of this problem when he attacked the concept of wild type. "The 'norm' is, thus, neither a single genotype nor a single phenotype. It is not a transcendental constant standing above or beyond the multiform reality. The 'norm' of *Drosophila melanogaster* has as little reality as the 'Type' of *Homo sapiens*." The balance hypothesis, conversely, presumes that a vast amount of hidden genetic diversity exists *within* any population, so that interpopulation differences are less significant.

A second difference between the two views lies in the kind of natural selection that each implies. The classical hypothesis with its picture of virtually complete homozygosity presumes that the chief action of natural selection is to remove deleterious mutations from the population and that the fittest genotypes are the homozygotes for the wild type alleles at all loci. This view does not exclude the occurrence of an occasional favorable mutation, but such a mutation quickly becomes fixed in the population as a new wild type. Moreover, the possibility of mutations without selective effect is also not excluded, but the overwhelming homozygosity of populations demands that the rate of mutation per generation to such neutral alleles must be several orders of magnitude smaller than known rates for drastic mutations. Specifically, the average heterozygosity H per locus for neutral genes in a population of size N, with a neutral mutation rate of μ per generation, is

$$H \sim 1 - \frac{1}{4N\mu + 1}$$

(Kimura and Crow, 1964) so that a heterozygosity of 0.001 per locus, which is generously high under the classical scheme (Muller, 1950) corresponds to

$$\mu = 2.5 \times 10^{-4}\left(\frac{1}{N}\right)$$

For a population of even modest size, for instance 1000, this is a mutation rate of 2.5×10^{-7} as compared with mutation rate to lethal genes in Drosophila of 10^{-5} (Dobzhansky and Wright, 1941; Ives, 1945) or visible mutation, say in maize, of between 5×10^{-4} and 2×10^{-6} (Stadler, 1942). Indeed, unless mutation rates to neutral genes were even smaller than 10^{-7}, there would be a considerable difference in heterozygosity among species with radically different population structures. All in all, then, the classical hypothesis demands what we might call "purifying selection" as the overwhelming rule.

The balance hypothesis, on the other hand, must account for the large standing variation that it supposes in populations by assuming that the diversity of alternative alleles is not selected against. This view of heterozygosity draws its name from Dobzhansky's view (1955) that the alternate alleles are actively maintained in the population be some form of balancing selection, probably selective superiority of heterozygotes, but not excluding other kinds of stabilizing forces. But the hypothesis of high heterozygosity does not, in fact, exclude the possibility that most of the alleles are selectively neutral and that the observed genetic variation is simply the accumulation of neutral mutations in large populations. Again applying Kimura and Crow's formula for heterozygosity, we need only suppose that most species have a population size N that is about equal to the reciprocal of the mutation rate in order to predict 80 to 100 percent heterozygosity. Nor does the balance hypothesis rule out the occurrence of deleterious mutations, since natural selection will keep these in low frequency and they will have very little influence on the observed heterozygosity. Indeed, the *absolute* amount of purifying selection is the same under both the balancing and the classical theories, since it depends only on the frequency of deleterious mutations. What is at issue is the proportion of selective processes of this type. I shall go much more deeply into the contrasting roles of natural selection and random processes in part III of this book. For the moment it is only important to see that different assumptions about the direction and intensity of natural selection do indeed lead to different predictions about heterozygosity in populations.

A third consequence of the different hypotheses about variation is a difference in the view of speciation. If populations are almost entirely homozygous, then speciation must await the occurence of new mutations that may be advantageous in a new environment in which an isolated population is found. Moreover, even advantageous mutations are usually lost in the first few generations because of genetic segregation and random variation of offspring production, so that a mutation whose fitness is $1 + s$ compared with a fitness of 1 for the former wild type has a probability of only $2s$ of eventually becoming fixed (Haldane, 1927). Since it takes more than one gene to make a species, speciation becomes a very difficult event. The classical hypothesis comes close to reintroducing the paradox about speciation that Darwin thought he had solved. Darwin had resolved the problem of passing from one type or mode to another, which is what appears to occur in species formation, by fastening attention on the variation within species and by postulating that this intragroup variation was converted into variation between groups by isolation and natural selection. It was the segregation into a few distinct modes of quasi-continuous variation among individuals that Darwin saw in domesticated animals as the model for what happened in speciation. But if organisms are really very homozygous and therefore genetically identical within species, where is that variation on which Darwin supposed natural selection to operate?

Unlike the hypothesis of homozygosity, the balance hypothesis presumes that the genetic variation for speciation is always present, so that speciation awaits only the appropriate biogeographical and ecological events. An isolated population is nearly certain to find itself in a different biotic environment than is its parental species, since community compositions vary considerably from locality to locality. Selective forces are almost bound to be different as a result, so that steps toward speciation would almost be the rule for any isolated group. If the classical theory seems too conservative in its prediction about speciation, the balance view seems to predict speciation everywhere. Since, however, we cannot make any quantitative predictions of speciation rate from either theory (we have no quantitative genetic theory of speciation at all), there is little to choose on this basis.

I have presented the balance and classical hypotheses of genetic variation as *a priori* predictions about what would be observed if a method could actually be found to describe the distribution of genotypes in a population, rather than first reviewing the evidence and then presenting these hypotheses as alternative interpretations of ambiguous observations. It is a common myth of science that scientists collect evidence about some issue and then by logic and "intuition" form what seems to them the most reasonable interpretation of the facts. As more facts accumulate, the logical and "intuitive" value of different interpretations changes and finally a consensus is reached about the truth of the matter. But this textbook myth has no congruence with reality. Long before there is any direct evidence, scientific workers have brought to the issue deep-seated prejudices; the more important the issue and the more ambiguous the evidence, the more important are the prejudices, and the greater the likelihood that two diametrically opposed and irreconcilable schools will appear. Even when seemingly incontrovertible evidence appears to decide the matter, the conflict is not necessarily resolved, for a slight redefinition of the issues results in a continuation of the struggle. It is part of the dialectic of science that the apparent solution of a problem usually reveals that we have not asked the right question in the first place, or that a much more difficult and intractable problem lies just below the surface that has been so triumphantly cleared away. And in the process of redefinition of the issues, the old parties remain, sometimes under new rubrics, but always with old points of view. This must be the case because schools of thought about unresolved problems do not derive from idiosyncratic intuitions but from deep ideological biases reflecting social and intellectual world views. *A priori* assumptions about the truth of particular unresolved questions are simply special cases of general prejudices.

Attitudes about the kind and amount of genetic variation in populations, like all attitudes about unresolved scientific issues, reflect and are consistent with the intellectual histories of their proponents. People see new problems mirrored in a glass that has been molded by their solutions to old problems. A scientist's present view of difficult questions is chiefly influenced by the history of his intellectual and ideological development up to the present moment, and the resolution of current difficulties will in turn precondition his view of future problems.

The classical school derives from H. J. Muller and consists largely of his students and their students. Muller's experimental work, as is well known, was centered on the induction and manipulation of drastic mutations in Drosophila. He saw the biological world through the medium of the classical genetics and cytogenetics of well-behaved mutants of strong effect, localizable on the genetic or cytological maps. These mutations are nearly always recessive, always lower in fitness, and never exhibit anything that might be thought of as overdominance. Indeed, in the classical school of formal genetics, two mutations are by definition nonallelic if the heterozygous compound between them is closer to wild type than to either mutant. The complementation test for allelism rules out, *a priori*, the possibility that the heterozygote between two alleles at a locus may fall outside the physiological range of the homozygotes. While the techniques that Muller devised for the study of induced and spontaneous mutations were carried over into the study of natural populations, his own point of view was completely colored by the distinction between mutant and wild type that was integral to the program of Morgan's school of genetics, from which he sprang.

The balance school, consisting largely of Dobzhansky and his students in the Americas, and the British school of evolutionists, derives independently in Britain and America from a preoccupation with natural history and wild populations of organisms. Dobzhansky began his career collecting beetles in Central Asia, and the British school, deriving in no little part from E. B. Ford, carries on the genteel upper-middle-class tradition of fascination with snails and butterflies. These workers see the world as Darwin saw it, rich in diversity, and, as confirmed Darwinians, they have assumed that there *must* be immense genetic variation available for adaptation through natural selection.

The balance school is strongly influenced by nineteenth-century optimism about evolution as being essentially progressive (see, for example, Herbert Spencer's *Progress: Its Law and Cause*) and ongoing. For Dobzhansky, natural selection usually "leads to increased harmony between living systems and the conditions of their existence" (1955, p. 12). Muller's view is quite the opposite. It is deeply pessimistic. For the classical school, evolution has already reached its pinnacle, certainly in "those great groups which have long ceased undergoing important evolutionary changes" (Muller, 1949, p. 465). Kimura and Ohta (1971a, p. 166), quoting Muller, believe that "the gene through the long course of evolution has finally found itself in man." In such a world view, genetic change can only be a change for the worse, and the function of natural selection must be to prevent degeneration by maintaining the type.

These points of view about genetic variation are reflected in sociobiological theories as well. Muller believed in a genetic elite and strongly advocated artificial insemination from human sperm banks, the contributors to which would be chosen on the basis of their superior genotypes as manifested in their superior behavioral phenotypes. The balance school is pluralistic, seeing human society as dependent for its functioning on the existence of a variety of genotypes, no one of which is absolutely superior to any other. Both schools are equally "biologistic" in that they believe the nature of human society to be strongly influenced by the distribution of genotypes in the species. For Muller, human progress meant enriching the species for a few superior genotypes while for Dobzhansky it means increasing, or at least maintaining, genetic diversity. Neither view admits the possibility that genetic variation is irrelevant to the present and future structure of human institutions, that the unique feature of man's biological nature is that he is not constrained by it.

It is not the intention of this book to review exhaustively the experimental evidence that the classical and balance schools created and used to support their views before the definitive experiments of the last few years. It is necessary, however, to examine in some detail the kinds of evidence that could be brought to bear on the problem of genetic variation, before those definitive answers were achieved. First, such an examination will reveal the specifically paradoxical features of the problem of measuring variation. Second, the necessary structure of the definitive experiments can only be understood as an attempt to meet the shortcomings of various other approaches. Finally, the "definitive" experiments solve the problem only in a restricted sense and must somehow be integrated with the variety of information obtained from earlier approaches. That is, the experiments that actually characterize the genetic variation in populations turn out to be, in some sense, the right answer to the wrong question and cannot be used alone to solve the problem, How much genetic basis for evolutionary change exists in populations? in a nontrivial way.

VISIBLE MUTATIONS

Direct evidence for the occurrence of genetic variation affecting important morphogenetic processes can be obtained by screening a population for morphogenetic mutants of a classical Mendelian sort. Direct examination of organisms collected in nature does not usually reveal much single-gene variation. In one of the early studies of variation in natural populations, Dubinin and co-workers (1937) examined 129,582 individuals of *Drosophila melanogaster*, collected in three successive years in 11 separate localities. All collections contained a high proportion of flies showing the so-called trident pattern, a dark marking on the dorsal thorax that varies continuously in its expression, is known to be rather sensitive to temperature, and has a complex inheritance including one major gene and many modifiers. If we ignore this unanalyzable and widespread variant, there were 2700 visible variants among the 129,582 flies, or 2.08 percent. In different collections in different years, the proportion varied from 0.49 to 6.84 percent. Three large collections taken in successive years from Gelendzhik are shown in table 1. The change over the three years is much too large to be sampling error, both for the total proportion of variants and for some of the more frequent types, but the causes are unknown. The variations include eye colors, wing variation, bristle changes, and other common variations known from classical mutants of Drosophila. Indeed, the familiarity of the workers with the mutants of Drosophila no doubt influenced their perception and directed their attention. Most of the variants, however, turned out *not* to be genetic, at least in the sense of being the result of single allelic substitution. Of the 28 aberrant phenotypes listed in table 1, 17 were chosen for genetic test and multiple individuals showing these traits were bred. Thus 43 flies with "small bristles," 29 with "small eyes," 11 with "divergent wings," and so on were tested to see whether their aberrant phenotypes were simple Mendelian traits. Of 289 flies tested, only 84 (29.4 percent) were homozygous for a recessive mutant (50 flies) or heterozygous for a dominant mutant (34). The status of the most frequent variant, extra bristle, is uncertain. It is reported as "semi-dominant," but the kinds of genetic tests performed would not distinguish a character that was controlled by many genes, so it must be discounted. The frequency of variant flies owing their phenotype to single-gene substitutions is then between 0.08 percent in 1935 and 1.27 percent in 1933.

The frequency of manifest genetic variation is clearly not great, but since more than half of the simple genetic variants observed in the Gelendzhik study were homozygous for recessive genes, we are observing only a fraction of the genetic variation for such characters

TABLE 1
Morphological variants found among *Drosophila melanogaster* collected at Gelendzhik, U.S.S.R.

Variant	*1933 sample, n = 10,000* %	*1934 sample, n = 14,765* %	*1935 sample, n = 6,960* %
Extra bristle	2.52	0.64	0.21
Small bristle	1.02	0.09	—
Semi-small bristle	1.01	0.22	0.06
Bristle comb	0.19	—	—
Wavy bristle	0.55	—	—
Reduced bristle	0.11	—	—
Small eye	0.37	—	0.07
Rough eye	0.41	0.01	—
Dark eye	0.16	—	0.07
Mottle eye	0.05	0.01	—
Sepia eye	0.02	—	—
Garnet eye	0.01	—	—
Dark body	0.03	0.05	0.01
Yellow body	0.01	—	—
Dachs legs	0.01	—	—
Extra vein	0.05	0.07	—
Analis incomplete	0.16	—	—
X-veinless	0.02	—	0.02
Upturn wing	0.14	0.08	—
Divergent wing	—	0.01	—
Extra x-vein	—	—	0.01
Extra analis-vein	—	0.05	—
Extra media-vein	—	—	0.01
Light eye	—	0.01	—
Notch wing	—	0.01	—
Truncate	—	0.01	—
Comma	—	0.01	—
Tumor	—	0.04	—
Total Changes	6.84	2.07	0.49

Note: Data are of Dubinin et al. (1937), adapted from the summary of Spencer (1947).

unless we detect the heterozygous carriers of these mutations as well. One of the largest attempts to assess the hidden genetic variation for recessive visible mutants was Spencer's (1957) study of *Drosophila mulleri* from Texas. Families were derived from 736 females collected in nature. From these, 263 morphological mutant genes were detected in 224 families. The genes, as expected, covered the full range of morphological abnormalities familiar to the

Drosophila geneticist. Spencer was unable to use the most efficient method of detecting recessive genes because *D. mulleri* does not produce well in single pair culture. From experiments with other species, he estimated that the method used for *D. mulleri* would detect only one-sixth of the recessive visible mutants, so, if we make this correction, there were approximately $6 \times (263/736) = 1.92$ mutants per family. Since each family, in turn, represents four haploid genomes (two from each wild parent), there is approximately half a mutant gene per genome or one per individual in the natural population. This figure agrees well with several smaller studies of a variety of Drosophila species by Alexander (1949, 1952), in which the most efficient method could be used and in which the number of mutants per genome varied from 0.28 (*D. novamexicana*) to 1.19 (*D. hydei*).

As is to be expected, some mutations are represented only once, others occur several times, but none is in a high frequency or accounts for most of the variation. Among the 263 variants, a "scarlet-like" eye color appeared 35 times. Of these occurrences, 21 were tested genetically and found to represent 5 different loci, one locus being represented 9 times. Again assuming that 6 times as many mutants were present as were detected in the sample, the most frequent variant allele can be estimated as having a frequency of $6 \times 9 \times (35/21)$ occurrences per 2944 genomes, or 3.4 percent. No other gene had a variant in nearly this frequency, however. Of 117 different variants, 84 appeared only once and 14 twice.

We can estimate the average frequency per locus of visible mutants by using the fact that 50 years' study of *D. melanogaster* has essentially saturated the genetic map of its X chromosome with mutations of the type detected by Spencer. Lindsley and Grell (1967) list 362 such loci on the X chromosome, which is equivalent to 1800 for the entire genome. Taking 0.5 mutant per genome, we have an average frequency of $0.5/1800 = 0.00028$ as the average frequency of variant alleles per locus in a natural population of *Drosophila*. This value is, of course, completely in agreement with the classical hypothesis, although the high estimated frequency of one of the scarlet genes is not. The detection of an occasional locus with high variant frequencies is a common feature of Drosophila studies; for example, Spencer (1946) found gene frequencies of 10 and 4 percent, respectively, for a bristle variant and an eye color variant in *D. immigrans*. Nevertheless, the average frequency over all loci is of the order of 10^{-3} and only about one in a thousand loci controlling morphological characters of an obvious kind ever has variants in any appreciable frequency.

The advantage of studying visible mutations is the obvious one that they can be enumerated in a population by the simple expedient of inbreeding the offspring of single females from nature. By means of still further allelism tests on apparently identical variants, like those Spencer performed, a complete enumeration, locus by locus, is well within experimental limits. But the objections to using visible variants for assessing overall genetic variation are overwhelming. First, the method is useless if organisms cannot be bred in captivity. Even direct observation of organisms in the wild is useless without tests to distinguish genetic variants from environmentally induced variations, as the work on the Gelendzhik populations showed. This objection applies to most (but not all!) methods of assessing genetic variation, and it is probable that a representative sample of organisms could finally be domesticated if this were important enough.

The far more serious objection to basing any general conclusions on the study of visible variants is that *a priori* they are unlikely to form a representative sample of gene effects. The number of loci at which a genetic change will produce observable and unambiguous morphogenetic effects must represent a small minority of all loci. The number of cistrons in Drosophila must certainly not be less than the approximately 15,162 salivary chromosome bands. Judd, Shen, and Kaufman (1972) have found 16 lethal genes in a small region of the X chromosome with 16 salivary bands. If we take the data of Chovnick (1966) on the *rosy* locus in *D. melanogaster*, we find that one cistron corresponds to 0.0086 centimorgans, giving a figure of about 33,000 genes for the entire genome. In either case the 1800 loci with sharp visible effects make up less than 10 percent of genes.

In addition, most mutations at a locus are unlikely to cause drastic effects, even for the genes at which *some* mutation may cause them. We must presume that the proteins specified by the loci at which morphological mutants are known are, like other enzymes and proteins, capable of a large number of amino acid substitutions that have only small effects on their function. General estimates of the proportion of amino acid sites that would yield "drastic" mutations are not presently possible, especially in view of the unknown relationship between the degree of enzymatic impairment and the degree of morphogenetic disturbance. The one well-documented case, which may be peculiar for some reason, is cytochrome *c*, in which 35 out of 104 amino acids appear to be strictly conserved in evolution. Alterations at the remaining 69 sites have very little effect on reaction kinetics, at least *in vitro*. The sites that have been conserved are not randomly spread along the polypeptide chain, but are clustered and are clearly functional in the three-dimensional structure of the molecule and the attachment of the heme group to the polypeptide (Dickerson, 1971). So, we may provisionally adopt a value of 1/3 for the proportion of amino acid sites that are potentially drastic in a protein that may have severe effects on physiology or morphogenesis.

There is also evidence, which can only be described as impressionistic, that wild-type revertants of morphological mutants are infrequent compared to pseudo-wild-type revertants of single amino acid substitution in known enzymatic proteins. No quantitative comparison exists, however, of relative forward and back mutation rates of the two classes of mutants, and the experimental difficulties would not be trivial in view of the immense heterogeneity in forward and back mutation rates from locus to locus in the same organism. For example, table 2 shows the forward and reverse mutations observed by Schlager and Dickie (1971) in a huge experiment on visible mutations at five coat-color loci in mice. The table illustrates the problem of characterizing these rates for mutations whose molecular nature is unknown. For the one case in which there were back mutations ($d \rightarrow +$) the forward and back rates were not greatly different, and for the other four loci no reversions at all were obtained. This result may simply mean that the particular mutations carried by the stocks at the *a*, *b*, *c*, and *ln* loci were base deletions or even grosser deletions, while the mutant allele at the *d* locus, which did revert, was by chance a base substitution. Reversion rates in excess of mutant rates are also known, as in the case of the *forked* locus in *D. melanogaster* reported by Patterson and Muller (1930), where 4/59,000 mutants from wild type to *forked* were induced by X-rays, but the same dose induced 8/32,588 reversions from *forked* to wild type.

It might be thought that morphological mutants, because they are severe in their effects, must *ipso facto* be rare in populations and therefore necessarily provide no information on the bulk of variation. But this is not certainly the case. It is true that most classical

TABLE 2
Recessive visible mutation rates and reversion rates for five coat-color loci in the house mouse

	Mutations from wild type			*Mutations to wild type*		
Locus	*gametes tested*	*mutations*	*rate* × *10*[6]	*gametes tested*	*mutation*	*rate* × 10[6]
a	67,395	3	44.5	8,167,854	0	0
b	919,699	3	3.3	3,092,806	0	0
c	150,391	5	33.2	3,423,724	0	0
d	839,447	10	11.9	2,286,472	9	3.9
ln	243,444	4	16.4	266,122	0	0
Total	2,220,376	25	11.2	17,236,978	9	0.52

Note: Data are from Schlager and Dickie (1971).

mutants of *Drosophila* lower the viability or fecundity of their homozygous carriers (see, for example, Polivanov, 1964) and many are discriminated against in mating (Merrell, 1949). However, lowered fitness of homozygotes for an alternative allele is not in itself any prediction of its frequency in the population, since the fitness of heterozygotes is critical.

The chief mechanism postulated by the balance school for the maintenance of genic variation is heterosis, so the demonstration of the poor fitness of a homozygote for a drastic allele cannot be used to predict the rarity of that allele. Both balance and classical views are completely compatible with drastic loss of fitness of one homozygote. The issue is the fitness of heterozygotes with respect to the better homozygote. The only well-authenticated case of single-gene balanced polymorphism is that of the sickling gene in man. The homozygote Hb^S/Hb^S is virtually lethal, but the heterozygote Hb^S/Hb^A is more fit than the normal homozygote Hb^A/Hb^A, apparently because heterozygosity affords protection against falciparum malaria (Allison, 1955). In Spencer's survey (1957) of variation in *D. mulleri*, in which the frequency of a variant allele for scarlet eyes was estimated as 3.4 percent, it is significant that two other loci had a variant allele causing a scarlet-like eye color at a frequency of 1.6 percent each.

Thus one cannot dismiss the evidence of morphogenetically drastic variants as having no relevance to our problem simply because they are drastic. Nor can such variants be dismissed *a priori* as belonging to a different class of mutational events than do small variations. Hemoglobin s differs from hemoglobin A only in the substitution of valine for glutamic acid at position 6, and this amino acid substitution is on the outer surface of the molecule, with no relation to tertiary structure or attachment of the heme residue (Ingram, 1963). The substitution happens to lead, however, to crystallization of the hemoglobin at low oxygen tensions, a result that never could have been predicted from structural studies alone.

We are forced to conclude that visible variants, on the face of it, could be more or less polymorphic as a class than are more subtle gene variations, so that even though the visibles are not themselves the "stuff of evolution" their frequencies might be typical of all variation. The very low degree of variation actually observed for this class of alleles requires that supporters of the balance hypothesis explain the lack of balancing selection for such mutations as contrasted with a postulated general balance for more subtle changes. Moreover, this explanation must take into account the fact that visible mutations are not always rare.

LETHAL GENES

I have argued that visible changes must represent a minority (although not necessarily an *atypical* minority) of genetic variants, perhaps as small as 1 percent. Clearly a much more solid foundation for inferences about variation would be built if some quite different class of variants could also be enumerated on a locus-by-locus basis. Lethal alleles offer such an opportunity.

Genes that are lethal when homozygous are obviously a more restricted class in one sense than are genes with visible affects, but they are not restricted to morphogenetic events and their detection is completely mechanical and objective. The general scheme for their detection is shown in figure 2; in essence it is due to Muller (1928). Since this scheme illustrates important properties that arise from the methodological paradox that is fundamental to our problem, and since it will be used to detect other forms of variation, it is worth looking at in some detail.

A single individual from the population to be studied is mated to a stock carrying two different dominant visible marker genes, M_1 and M_2, on homologous chromosomes. These markers are usually chosen to be lethal when homozygous, but this is not necessary if

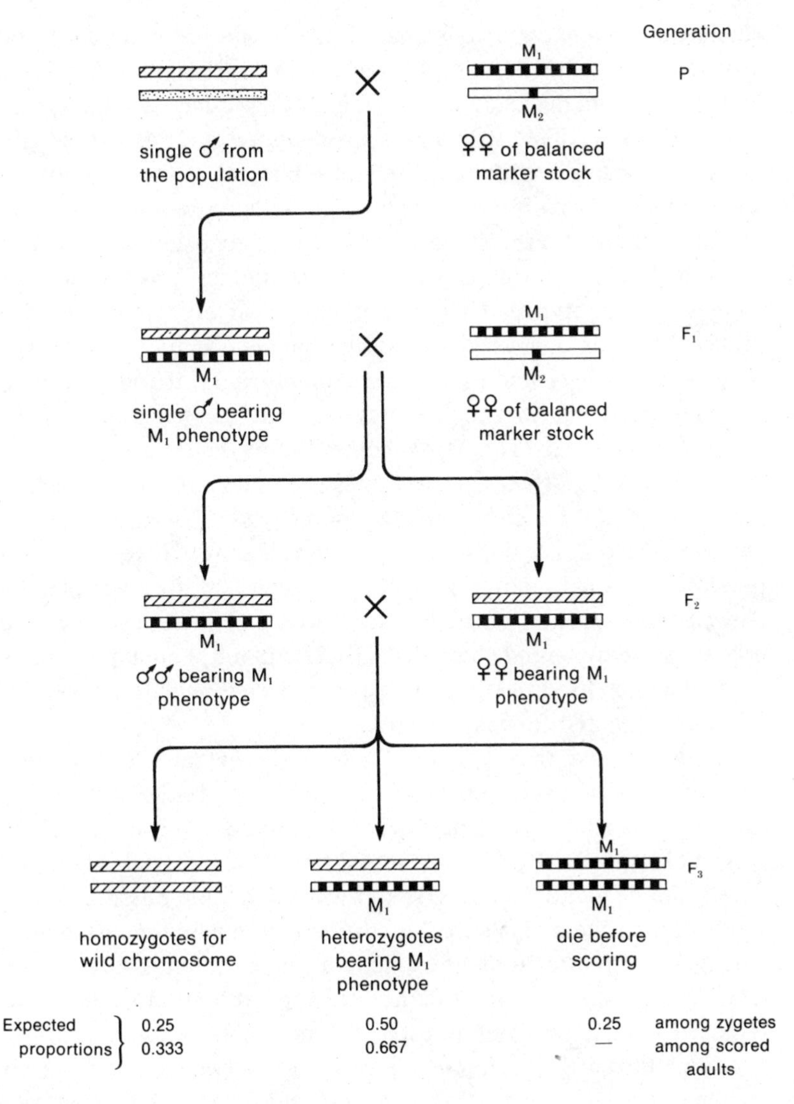

FIGURE 2

Replication scheme for sampling a chromosome from a population and providing a large number of individuals homozygous for that chromosome.

the homozygotes are distinguishable from heterozygotes. We will assume they are homozygous lethal. In addition, either M_1 or M_2 or both should be associated with a complex inversion that will reduce crossing-over to a minimum (ideally to zero). In the F_1 generation a single individual is chosen that has the M_1 phenotype and thus is heterozygous for the M_1 chromosome and one randomly chosen homologue from the gametic pool of his wild parent, for example $+_1$. Since M_1 is associated with a crossover suppressor, this individual's gametic output will consist entirely of replicates of the M_1 chromosome and of the wild chromosome $+_1$, with no recombinants between them. When this heterozygote is crossed back to the M_1/M_2 stock, a large number of males and females in the F_2 generation will have the phenotype M_1 and will all be heterozygous for the replicated chromosome $+_1$. Again the gametic output will consist only of M_1 or $+_1$ gametes with no recombinants; if they are bred together, they will produce the three zygotic classes shown in the last line of figure 2. The M_1M_1 homozygotes die before scoring of the progeny, the $M_1/+_1$ heterozygotes are detectable as M_1 phenotypes, and the $+_1/+_1$ are completely homozygous for all loci on the chromosome that is being controlled and will show whatever phenotype results from homozygosity of those genes. If, for example, the chromosome $+_1$ had carried a recessive visible mutant, all the non-M_1 progeny would show that visible mutant phenotype, so this method can be used to study visible variants although it is unnecessarily complicated for that purpose.

Let us suppose that $+_1$ carried a recessive lethal allele at some locus, a lethal that acted during development before the age of scoring. Then all the $+_1/+_1$ heterozygotes would die and the only observed class in the F_3 progeny would be M_1 phenotypes. We then have a simple and objective system for the detection of lethal recessives. If the scheme in figure 2 is carried out on many separate individuals from nature, a direct estimate of the frequency of recessive lethal-bearing chromosomes in the wild population is obtained. The scheme can be carried out for each element of the haploid set so that a total picture of lethal frequencies over the whole genome can be obtained, provided only that appropriate markers and crossover suppressors can be found for each chromosome in the set.

The scheme has several other advantages. Even though homozygotes $+_1/+_1$ may not survive, the lethal is preserved in the heterozygotes $M_1/+_1$ for further study. It is possible to determine how many different lethal loci are represented in the entire sample, and how many times each locus is represented, by the simple expedient of crossing different lethal heterozygotes. Thus, if the cross

$M_1/+_1 \times M_1/+_2$ yields no wild-type offspring, the chromosomes $+_1$ and $+_2$ must have carried lethals at the same locus. By making all possible $n(n - 1)/2$ crosses among n lethal chromosomes, a complete enumeration can be made. The only information that cannot be obtained is how many different lethals any one chromosome sampled from nature may possess, since the behavior of a lethal-bearing chromosome is the same irrespective of how many lethals it carries. It is usually assumed that lethals are distributed independently at different loci so that the frequency of chromosomes bearing 0, 1, 2, 3, ... lethal genes will be given by the successive terms in a Poisson distribution with mean q, the average number of lethal genes per chromosome. If Q is the proportion of lethal chromosomes in the sample, then $1 - Q$ is the proportion of chromosomes bearing no lethals and

$$1 - Q = e^{-q}$$

so $$q = -\ln (1 - Q)$$

The scheme in figure 2 is designed to circumvent a fundamental problem of methodology. I have discussed at great length the problem of distinguishing genotypes from phenotypes when environmental variation obscures genotypic boundaries. The problem is especially acute when the character is an ensemble property like the probability of survival. If a larva dies before reaching adulthood, this fact does not mean that it is homozygous for a lethal gene. The way around the problem is to create a large number of individuals that have all the same genotype and to characterize the group by its average. The phenotype corresponding to that genotype can be measured with arbitrary accuracy simply by producing a large enough identical group. That is what the mating scheme does. It is a method for replicating a given chromosome and producing an arbitrarily large group that is identically homozygous or heterozygous for the same genes. If all the homozygotes $+_1/+_1$ die before adulthood whereas the $M_1/+_1$ survive, the lethality must be genotypic. The mating scheme is essentially a form of inbreeding because it produces identical homozygotes from a single ancestral genotype. In species that can be clonally propagated, an arbitrarily large group of identical genotype could be produced without recourse to inbreeding, by vegetative reproduction. For a sexually reproducing species, however, some form of inbreeding is the only way in which a genotype can be replicated in order to assess its average performance.

The strength of the inbreeding scheme is also its weakness, for in the process of replicating a genotype, an entire chromosome is replicated and made homozygous. Schemes of inbreeding that depend on mating between close relatives rather than on chromosome manipulation are even worse because the entire genome is being made homozygous. Nor is clonal reproduction any better, since again the whole genome is replicated. In fact, *there is simply no way to make a large number of individuals identically homozygous or heterozygous at one locus while keeping the rest of the genome segregating at random.* Thus we must infer from homozygous *chromosomes* to homozygous *genes.* This problem is not serious for lethal genes because they have such extreme effects that they can be mapped if necessary, and we can translate from lethal chromosomes to lethal genes by assuming the Poisson distribution of lethals. The difficulty will come back to haunt us, however, when we try to apply the inbreeding method to genes of small effect.

Lethal frequencies have been estimated in many populations of many species of *Drosophila*, too numerous to review exhaustively. Table 3 is a representative sample taken from Dobzhansky and Spassky's summary (1954). Because "lethal" chromosomes sometimes on retest produce a few wild type homozygotes, the data are a lumping of lethals and so-called semilethals, the latter being a relatively small class of chromosomes that have a low but non-zero probability of survival as homozygotes.

Table 3 also gives the estimated number of lethal genes per genome for each population, using the Poisson correction and taking into account the proportion of the total genome accounted for by each chromosome. Within a population there are discrepancies because chromosomes do not have lethals proportional to their length. Thus the second and third chromosomes of *D. prosaltans* have lethal frequencies of 32.6 and 9.5 percent, respectively, but their lengths are only in a proportion of 2 : 1. Although there is considerable variation, about one lethal gene per genome is a fairly typical value, and no case predicts more than 2.5 per genome. Lethals are about twice as frequent as visibles, but still rather rare. It is more difficult to know the total number of loci that may mutate to recessive lethals, but minimum and maximum estimates exist. From the frequency of allelism of unrelated lethals in Wallace's (1950)

TABLE 3

Frequencies of lethal and sublethal chromosomes and estimated number of heterozygous lethals and semilethals per genome for several species of Drosophila

Species	*Chromosome*	*Population*	*Number of chromosomes tested*	*% of lethals and semilethals*	*Average numbers of lethals per genome*
D. prosaltans	II	Brazil	304	32.6	0.99
	III	Brazil	284	9.5	0.50
D. willistoni	II	Brazil	2004	41.2	1.33
	III	Brazil	1166	32.1	1.94
	II	Florida	109	31.1	0.93
	III	Florida	122	32.8	1.99
	II	Cuba	25	36.0	1.12
	III	Cuba	39	25.6	1.48
D. melanogaster	II	North Caucasus	795	12.3	0.33
		South Caucasus	2738	18.7	0.52
		Ukraine	2700	24.3	0.61
		Cannonsburg, Pa.	117	28.2	0.83
		Amherst, Mass.	3549	36.3	1.13
		Wooster, Ohio	343	43.1	1.41
		Winter Park, Fla.	468	61.3	2.37
D. pseudoobscura	III	Yosemite, Calif.	109	33.0	2.00
		San Jacinto, Calif.	326	21.3	1.20
D. persimilis	III	Yosemite, Calif.	106	25.5	1.47

Note: Adapted from Dobzhansky and Spassky (1954). Complete primary references given in that paper.

study of second chromosomes of *D. melanogaster*, Lewontin and Prout (1956) estimated there were 340 loci capable of mutating to lethals for this chromosome, which gives 850 for the genome as a whole as a minimum estimate. On the other hand, Hochman (1971) calculated 36 separate lethal loci on the fourth chromosome of *D. melanogaster*, which is only about 2.6 percent of the total euchromatic length of the genome. This gives 1300 as a probable upper limit to such loci, so that the average frequency of lethal alleles per locus at risk is about 0.001, three times as high as visibles. Considering the uncertainties of estimate, there is really no difference between the two values. As for the visibles, there is no compelling reason why recessive lethal genes should be in extremely low frequency, since it is their fitness in heterozygous conditions that will chiefly control their abundance. This point has been clear to everyone and considerable attention has been paid to measuring the heterozygous fitness of lethals, although the rarity of these alleles in general has made it evident that heterosis for lethals could not possibly be the rule.

Many estimates agree that the average dominance of lethals is between 2.5 and 5 percent (see, for example, Cordeiro, 1952; Prout, 1952; Stern et al., 1952; Hiraizumi and Crow, 1960; Temin, 1966), but this value seems to depend very much on the particular lethal chromosome involved, the chromosome against which it is tested in heterozygous condition, and the genetic background on which the comparison is made. Thus Cordeiro (1952) found some lethals to be slightly heterotic, and Dobzhansky and Spassky (1968) found that lethals tested on the genetic background of the populations from which they were sampled showed no deleterious effects in heterozygous condition, whereas if tested on the background of genes from distant localities they were as much as 2 percent less viable than normals. Several studies of *D. pseudoobscura* have failed to show any dominance of lethal heterozygotes at all, or any evidence that heterozygotes for one or two lethals differed in viability from any other heterozygotes (Dobzhansky, Krimbas, and Krimbas, 1960; Dobzhansky and Spassky, 1960, 1963). The totality of the data on this point is ambiguous because of the variety of variables. Nevertheless, in retrospect it is a little difficult to understand why so much attention has been given to the estimation of heterozygous fitness of lethals. Their extreme rarity *in general* is completely incompatible with a usual heterosis, while the possibility that a lethal could be heterotic is proved by the case of sickle-cell anemia.

We can go further. The frequency with which a lethal, heterotic on the average over environments and genetic backgrounds, can arise in nature must be extremely low. As Dobzhansky and Spassky (1968) show, and as must be true from the simplest theory, those lethals with the smallest deleterious effect in heterozygous condition will stay in the population longest and tend to accumulate. If any are heterotic on the average, they will be trapped in stable equilibrium in the population. Then, as time goes on, more and more heterotic lethals will be accumulated until the population is filled with lethal alleles at frequencies of, for instance, 1 percent. Since this has not happened in any population examined, the rate at which such heterotic alleles arise must be extremely low. One evidence of accumulation of such lethals would be a high allelism of lethals in separate collections in various populations. Dobzhansky and Wright (1941) tested this possibility with 123 lethal-bearing third chromosomes of *D. pseudoobscura* collected in a variety of localities in the Death Valley region. They tested allelism of lethals both within and between localities. Most lethals were represented only once, but two were found four times. They summed up their results as follows: "Lethals found repeatedly [i.e., twice] within a population are not significantly commoner in the species at large than are apparently rare lethals while *vice versa*, the lethals recovered from samples from two or three localities show no tendency to accumulate in any particular sample."

The overall picture of variation for lethal genes is then the same as for visibles. One or two loci are heterozygous per individual, but an occasional variant may rise to a few percent in a particular population. These drastic genes appear to confirm the classical picture.

I have discussed lethal chromosomes as if they owed their homozygous lethality to one or more lethal gene substitutions. That is, except for multiple lethals (corrected for by the Poisson calculation), I have equated lethal *chromosomes* with lethal *genes*. But what is the evidence for this widely held assumption? Is it not equally likely that chromosomes are lethal when homozygous because they carry a large number of gene substitutions, each of which is slightly deleterious to development when homozygous? This difference in interpretation of lethals is closely related to the contrast between the classical and balance theories. The former obviously predicts that lethal chromosomes carry lethal genes, rare mutants on the order of one or two per genome. The balance theory, on the other hand, although certainly allowing lethal genes, would also predict a substantial number of chromosomes that are "synthetic lethals," lethal chromosomes that are the sum of many small deleterious gene substitutions. Unfortunately, no quantitative prediction can be made of how many synthetic lethals would constitute a "substantial number." Moreover, classical theory also allows for some synthetic lethals made up of, for example, two semilethal mutations, so the distinction between the predictions is not entirely clear cut.

Not unexpectedly, the issue of synthetic lethals has been hotly debated, since the evidence is equivocal. From allelism tests, some lethal chromosomes are well behaved and clearly carry one or two point lethals. However, the frequency of lethal chromosomes that are allelic to each other is so small, a few percent, that this evidence does not bear on the vast majority of lethal chromosomes. Dobzhansky (1946), on the other hand, has shown that lethals can be created by recombination. From the recombination of a nonlethal chromosome with a temperature-sensitive lethal chromosome (lethal at 25.5 C but not at 16 C), 3 chromosomes out of 100 tested were lethal at both temperatures. In more extensive experiments, 19 out of 450 recombinants between nonlethal chromosomes were homozygous lethal in *D. pseudoobscura* (Dobzhansky, 1955), and 95 out of 4830 in *D. melanogaster* (Wallace et al., 1953). Yet Hildreth (1956) and Spiess and Allen (1961) found no such synthetic lethals in extensive studies. In any event, the demonstration that "synthetic" lethals *can* be produced, or even the frequency with which they can be produced, is not direct evidence on the issue since we do not know the rate at which they are destroyed by recombination. The question is, Of naturally occurring lethal chromosomes, what proportion are carrying point lethal genes or two semilethal genes, as opposed to a large array of slightly deleterious genes that add up to a lethal when the chromosome is homozygous? That question can only be answered in a completely clear-cut way by mapping of lethals on lethal chromosomes from wild populations, a tedious and difficult task that has not been undertaken on a large scale. If the evidence of interaction between different chromosomes of the genome is any indication, synthetic lethals will turn out to be an infrequent class (Spassky, Dobzhansky, and Anderson, 1965; Temin et al., 1969).

There is one piece of evidence which, although not direct, argues rather strongly against synthetic lethals making up a large fraction of naturally occurring lethal chromosomes. This is the remarkable agreement between the number of loci estimated to be capable of mutating to lethals in *D. melanogaster* from the map saturation experiments of Hochman, and the allelism of unrelated second chromosome lethals from the work of Wallace (1950). Synthetic lethals from independent sources will never act as alleles, so that if many lethals are synthetic, the frequency of allelism of independent lethals will be reduced and the apparent number of lethal-producing loci will be exaggerated. More precisely, if p is the proportion of all lethal chromosomes that are synthetic, then the number of loci estimated by the reciprocal of the frequency of allelism will be upwardly biased by a factor of $1/(1-p)^2$. If 50 percent of all lethals were synthetic, the allelism estimate from this source would be four times too high. Wallace's estimate from the reciprocal of allelism is 1070 loci, actually *lower* than Hochman's figure of 1300, so only a very small proportion of lethals can be synthetic in his sample. Other species of Drosophila, whose genome is closely related to *D. melanogaster*, give even lower numbers for the allelism estimate of lethal-producing loci (see, for example, Dobzhansky and Wright, 1941, for natural populations of *D. pseudoobscura*).

The problem of distinguishing lethal loci from lethal chromosomes foreshadows a much greater problem of the same sort when we deal with fitness modifiers.

FITNESS MODIFIERS

Visible variants and lethal alleles have been studied extensively because they satisfy one of the two conflicting requirements for the evaluation of genetic variation in a population. They give unambiguous phenotypes so that individuals from nature can be classified as homozygous or heterozygous for allelic substitution after appropriate genetic tests have been carried out on their progeny. But precisely because visibles and lethals can be used in Mendelian genetics, they are under suspicion. Their drastic gene effects may make them totally unrepresentative of the variation characteristic of much milder and subtler allelic variants. How can we study these more subtle genetic variations?

The technique for replicating a chromosome from nature, discussed in the preceding section, has as its purpose the production of a large ensemble of individuals that are identical with respect to a particular chromosome element. Even though there is considerable environmental and uncontrolled genetic variation influencing a physiological character, for example, development time, the mean development time of such an ensemble will cancel out the uncontrolled variation and allow the effect of the controlled chromosome element to be seen. In this way the difference in development time between homozygotes for two different homologues sampled from nature could be determined, and in fact the entire distribution of chromosomal effects in the natural population could be characterized to any arbitrary degree of accuracy. The chromosomal replication scheme is then a powerful tool for the study of genetic variation of all kinds and has been widely applied for this purpose, with remarkably uniform results.

In the replication scheme of figure 2, applied to *Drosophila*, the three classes of fertilized eggs in the F_3 generation are produced in the ratio of 1 M_1/M_1 : 2 $M_1/+$: 1 $+/+$ according to Mendelian principles. The M_1/M_1 class dies in the egg stage. If the wild homozygotes and the marker heterozygote have equal probabilities of survival from fertilization to the time of scoring (usually of adults that are newly hatched from pupae), then we will observe 2/3 marked flies and 1/3 wild type. If, on the other hand, the wild chromosome carries a gene that when homozygous reduces the probability of survival below that of the $M_1/+$ heterozygote (or increases it, for that matter), there will be a smaller (or larger) proportion of wild type among the scored adults. The detection of lethal genes is obviously a special case of this test.

The viability of wild type homozygotes relative to the marked heterozygotes, however, is not exactly of interest since the marker gene may have a considerable effect on the survival of its carriers. In order to correct for this marker effect, heterozygotes from two independently derived lines are crossed so that wild type *heterozygotes* rather than wild type *homozygotes* are produced. Thus

$$M_1/+_1 \times M_1/+_2$$

will produce two potentially viable classes: $M_1/+$ (a mixture of $M_1/+_1$ and $M_1/+_2$) and $+_1/+_2$ in the expected ratio of 2/3 : 1/3. If the marker gene has any effect on its carriers, this will be observed by an average deviation from the expected ratio when large numbers of different heterozygous combinations are made. In this way the average heterozygote $+_i/+_j$ is taken as the standard of viability against which each homozygote $+_i/+_i$ can be calibrated.

If a large number of chromosomes are sampled from a population and the homozygotes are calibrated against the outcome of the heterozygous control crosses, the distribution of homozygous viabilities that will result is representative of the distribution of genomes in the natural population.

In this way Dobzhansky et al. (1963) extracted 208 second chromosomes from a Bogotá, Colombia, population of C. *pseudoobscura*. The distribution of proportion of wild type appearing in homozygous crosses $M/+_i \times M/+_i$ and heterozygous crosses $M/+_i \times M/+_j$ is shown in table 4 and figure 3. The viability scale is

TABLE 4

Distribution of viabilities of second chromosome homozygotes and heterozygotes of *D. pseudoobscura* from Bogotá, Colombia

		Number of chromosomal combinations	
% of *wild type*	% *viability*	*homozygotes*	*heterozygotes*[a]
0	0	15	0
0–2	0–6	3	0
2–4	6–12	0	0
4–6	12–18	5	0
6–8	18–24	1	0
8–10	24–30	2	0
10–12	30–36	2	0
12–14	36–42	4	0
14–16	42–48	6	0
16–18	48–54	4	0
18–20	54–60	2	0
20–22	60–66	3	2
22–24	66–72	8	0
24–26	72–78	18	9
26–28	78–84	22	8
28–30	84–90	32	20
30–32	90–96	37	36
32–34	96–102	22	52
34–36	102–108	15	37
36–38	108–114	3	36
38–40	114–120	3	4
40–42	120–126	1	5
		208	209

Note: Data are from Dobzhansky et al. (1963).

[a]One semilethal heterozygote has been left out of the tabulation.

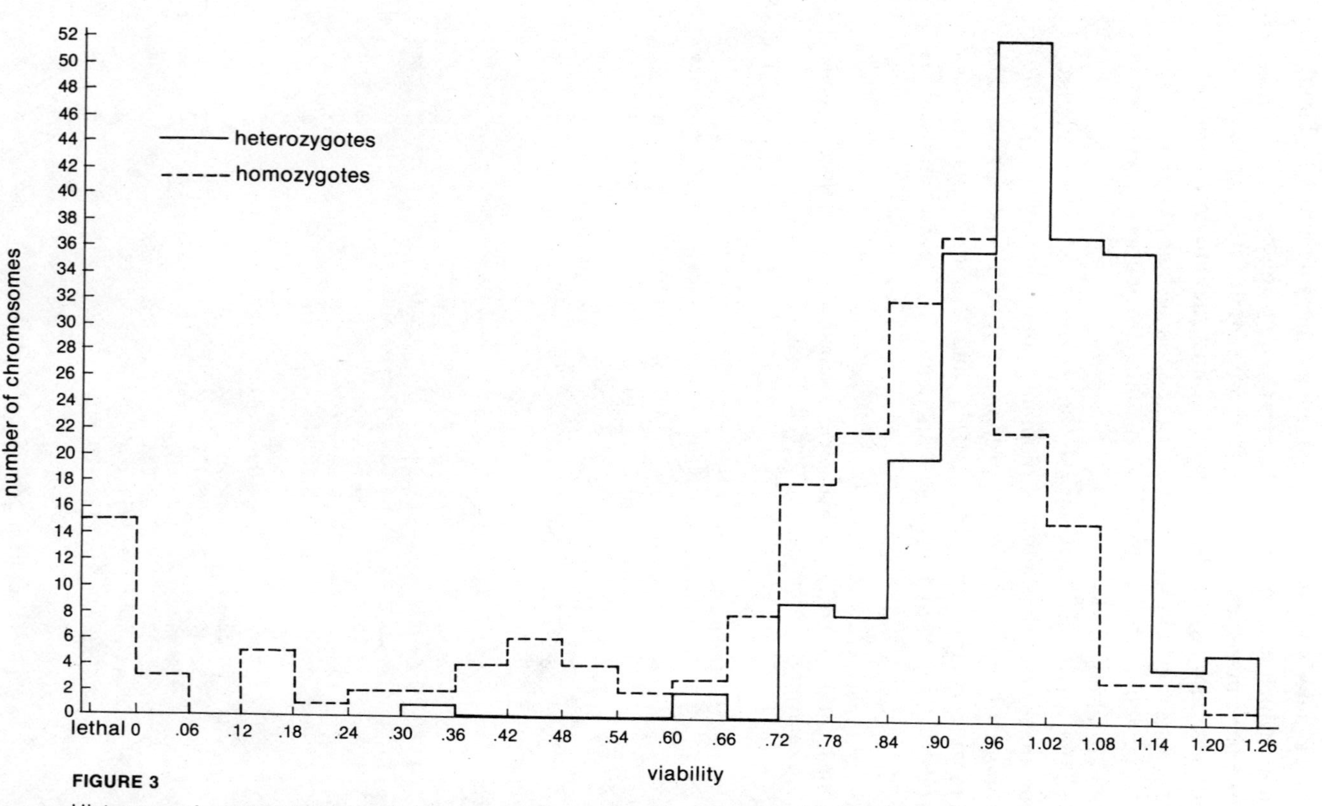

FIGURE 3

Histogram of relative viabilities of 208 second chromosome homozygotes and 209 heterozygotes of *D. pseudoobscura*. Data are from Dobzhansky et al. (1963).

also given. It is calculated by dividing the percentage of wild type by the average figure for heterozygotes, 32.83 percent. Apparently the marker has no deleterious effect in heterozygous condition.

The results shown are typical of the outcome of such experiments for all species of Drosophila ever studied, all populations and all chromosome elements. The distribution of viabilities is bimodal, with one mode at lethality and one somewhat below the mean viability of heterozygotes, and with rather few chromosomes falling in the so-called semilethal class between 10 and 50 percent viability. The mean viability of all chromosome homozygotes in this study is 76.06 percent, while the mean of the "quasi-normal chromosomes," those above 50 percent viability and thus belonging chiefly to the second mode, is 89.91 percent. One must remember that the viability values are relative to the average heterozygote $+_i/+_j$ taken as 100 percent.

The variation in viabilities is not entirely a result of the true differences between chromosomal homozygotes. Some variance arises from the sampling error involved in classifying only a couple of hundred flies in each culture, some from small environmental differences between cultures and segregation of the background genotype. Table 5 shows the partition of the total variance of viability (σ_T^2) of homozygotes and heterozygotes into the environmental-sampling component (σ_{es}^2) and the genetic component ascribable to homozygosity or heterozygosity of the chromosome itself (σ_G^2). Figure 4 gives two normal curves representing the supposed real distributions of chromosomal viability of heterozygotes and quasi-normal homozygotes, after the environmental variance has been removed. The real variance among heterozygotes is not negligible

TABLE 5

Components of variance of viability for second chromosome heterozygotes and homozygotes of *D. pseudoobscura* from Bogotá, Colombia

	σ_T^2	σ_{es}^2	σ_G^2	σ_G
Homozygotes	318.95	194.88	124.07	11.12
Heterozygotes	237.72	197.20	40.52	6.37

Note: The total variance is σ_T^2 ; the variance due to environment, sampling, and residual genotype is σ_{es}^2 ; the net genetic variance and standard deviation among third chromosomes are σ_G^2 and σ_G. Data are recalculated from Dobzhansky et al. (1963).

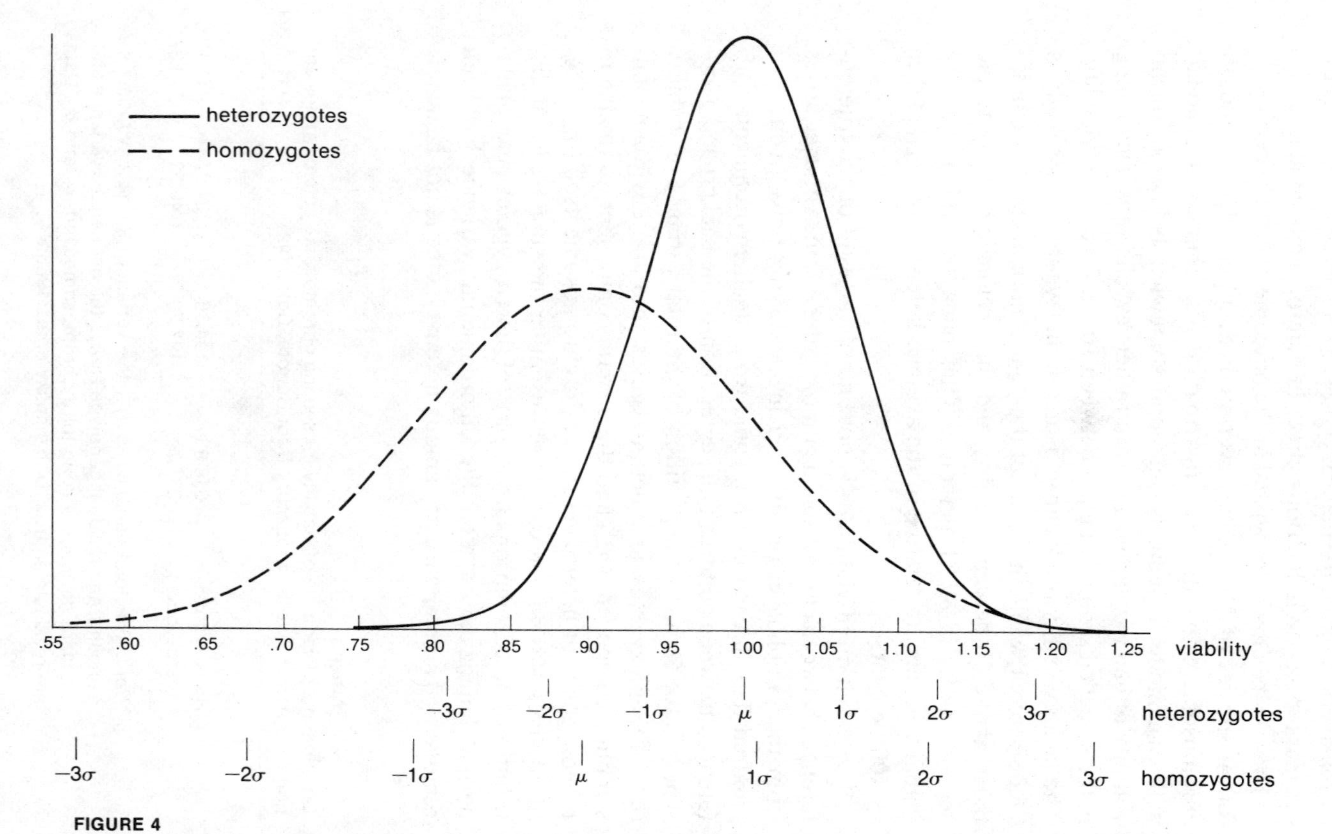

FIGURE 4
Normal curves reconstructing the true distributions of homozygous and heterozygous viabilities for second chromosomes of *D. pseudoobscura* from Bogotá, Columbia. Data are from Dobzhansky et al. (1963).

as compared to homozygotes, which means, presumably, that the viability modifiers are not really recessive.

What do the viability distributions for homozygotes and heterozygotes in figure 4 tell us about our problem? It is essential to bear in mind that the observations are on the viabilities of *chromosomal* homozygotes, not *genic* homozygotes. The replication scheme necessarily makes entire chromosomes homozygous and no method known can make only single loci homozygous while allowing the rest of the genome to segregate. Thus a chromosomal homozygote with a viability of 50 percent and one with a viability of 90 percent may differ at 2 or 2000 loci. Can further analysis of the observations clarify the issue?

A considerable variance in viability among heterozygotes, which presumably represent the typical genotypes of individuals in the natural population, is consonant with both theories of genetic variation. The balance theory necessarily predicts variance among heterozygotes because at each locus heterozygotes are superior in fitness to homozygotes, whereas the classical theory does not demand complete recessivity of deleterious genes but allows that heterozygotes may be intermediate in fitness between homozygous wild type and homozygous mutant. Nor can one build an elaborate quantitative theory of the relative variance of homozygotes and heterozygotes because, as is usual in population genetics, the actual data have an immense range and so are quite ambiguous. Table 6

TABLE 6

Genetic variances (σ_G^2) and standard deviations (σ_G) for a variety of chromosomal homozygotes and heterozygotes in *D. pseudoobscura* and *D. persimilis* compared to Bogotá *D. pseudoobscura*

		Homozygotes		*Heterozygotes*		
Species	*Chromosome*	σ_G^2	σ_G	σ_G^2	σ_G	$\sigma_{hom}/\sigma_{het}$
D. pseudoobscura	II	118.14	10.87	18.41	4.29	2.53
	III	115.30	10.74	52.85	7.27	1.48
	IV	54.74	7.40	27.69	5.26	1.41
D. persimilis	II	61.32	7.83	18.69	4.32	1.81
	III	143.63	11.99	10.85	3.29	3.64
	IV	85.29	9.24	1.25	1.12	8.25
D. pseudoobscura (Bogotá)	II	124.07	11.12	40.52	6.37	1.76

Note: Data are from Dobzhansky and Spassky (1953).

gives the relationship between the standard deviations for homozygotes and heterozygotes for a set of experiments carried out simultaneously and therefore strictly comparable. The ratio for the Bogotá population is not atypical, but the range is so great that almost any value is "reasonable."

If the distribution of heterozygous viabilities in figure 4 is taken as a standard, the mean of homozygotes lies about 1.6σ below the standard viability, and about 42 percent of quasi-normal homozygotes are more than 2σ below the average heterozygote. These latter are generally referred to as "subvital" combinations, but the designation is purely arbitrary. There are, in addition, chromosomes that appear to be "supervital," that is, significantly higher than heterozygotes in viability. Such homozygotes are rare. In the Bogotá sample, only four homozygotes appeared to be more than 2σ above the heterozygous mean. The normal curves in figure 4 predict about 2 percent of homozygotes to be in this class. Even though rare, their existence would appear to be an anomaly, especially under the balance hypothesis, since advantageous genes should sweep through the population and become fixed. It has been shown, however, that such supervital chromosomes invariably lose their supervitality when tested in other environments (Dobzhansky et al., 1955).

Supervitality in natural conditions under a fluctuating environmental regime is probably an extremely rare event, and its appearance in controlled conditions in the laboratory is not informative. The classical hypothesis predicts such supervitals, since they are presumably the chromosomes that are free of deleterious mutations and are therefore the best, wild-type genotype. Nor is their rarity at all unexpected under the classical view, since the frequency of homozygous chromosomes with a viability 2σ greater than the average heterozygote will be very low if there are even a few deleterious genes per chromosome on the average.

The bimodality of viabilities in experiments on homozygous chromosomes might have either of two underlying causes, related to the problem of the genetic nature of the variation. The lethal mode and the quasi-normal mode might arise from quite different sources. Lethal chromosomes are mostly single-gene effects, although we can be less certain about the semilethals, which appear in rather larger numbers in the synthetic-lethal experiments than do lethals. The quasi-normal mode, however, might be a result of a large number of small gene effects summed up on each chromosome, giving a roughly normal distribution of homozygous viabilities. In this view, the two modes are the result of two quite different phenomena—the lethal mode reflecting rare deleterious "classical mutants," the quasi-normal mode reflecting an underlying distribution of common allelic variation at the bulk of loci. This is the view of the balance theory.

The classical view would be to assume that the entire distribution results from rare single-gene substitution with a range of effects from slight to severe, and that the apparent bimodality is an artifact of the way in which gene effects are detected, i.e., viability. If one assumed, for example, that there was a single skew distribution of deleterious recessive gene effects with a very long left tail, measured as percentage of normal enzyme activity, any activity below a certain value would be lethal, so that there would be an apparent pile-up of chromosomes at the lethal point because of this threshold effect. If the lethals and semilethals are simply a pile-up of the distribution at a threshold, there ought to be a negative correlation between the average viability of quasi-normal chromosomes and the proportion of lethals and semilethals, since the pile-up at the lethal threshold will become greater as the curve of gene effects is slid leftward along the abscissa in figure 3. The proportion of lethals and semilethals and the average viability of quasi-normal chromosomes from several populations of several species of Drosophila are shown in table 7, and a scatter diagram of the values is given in figure 5. The correlation between viability of quasi-normals and the proportion of lethals is indeed negative ($r = -0.2$) but only weakly so and far from statistical significance. Note that although the lowest frequency of lethals (9.5 percent) is associated with a very high mean viability of quasi-normals (96.6 percent) in the second chromosome of *D. prosaltans*, the other chromosome of that species shows a high frequency of lethals and a high viability of quasi-normals. Thus the evidence is ambiguous on the relation between lethals and quasi-normals.

The data of table 7 are a fairly good sample of the results of chromosomal inbreeding experiments for major autosomes in *Drosophila*. The viability of X chromosomes when homozygous is much higher and X chromosome lethals are rare, as is to be expected from the fact that the X chromosome is exposed to selection in hemizygous condition in males in each generation. Thus "recessive" genes are on the average, over both sexes, semidominant. An extensive table given by Dobzhansky (1951) for the X chromosome

TABLE 7

Proportion of lethal and semilethal chromosomes and mean viability of "quasi-normal" homozygotes of Drosophila

Species	Chromosome	Locality	% of lethals + semilethals	Mean viability of "normals"	Reference
D. pseudoobscura	II	Colombia	18.3	89.9	Dobzhansky et al. (1963)
	II	Yosemite, Calif.	33.0	75.0	Dobzhansky and Spassky (1953)
	II	San Jacinto, Calif.	21.3	99.2	Dobzhansky, Holz, and Spassky (1942)
	III	Yosemite, Calif.	25.0	77.1	Dobzhansky and Spassky (1953)
	III	Death Valley region	15.0	91.7	Dobzhansky and Queal (1938)
	III	Central America	30.0	89.2	Dobzhansky (1939)
	IV	Yosemite, Calif.	25.9	85.6	Dobzhansky and Spassky (1953)
	IV	San Jacinto, Calif.	25.5	87.6	Dobzhansky, Holz, and Spassky (1942)
D. persimilis	II	Yosemite, Calif.	25.5	87.9	Dobzhansky and Spassky (1953)
	III	Yosemite, Calif.	22.7	83.4	
	IV	Yosemite, Calif.	28.1	77.9	
D. prosaltans	II	Bahia, Brazil	32.6	91.7	Dobzhansky and Spassky (1954)
	III	Bahia, Brazil	9.5	96.6	
D. willistoni	II	Brazil	41.2	86.7	
	III	Brazil	32.1	88.8	
D. melanogaster	II	Wisconsin	25.1	79.0	Greenberg and Crow (1960)
	II	Long Island	20.5	83.1	

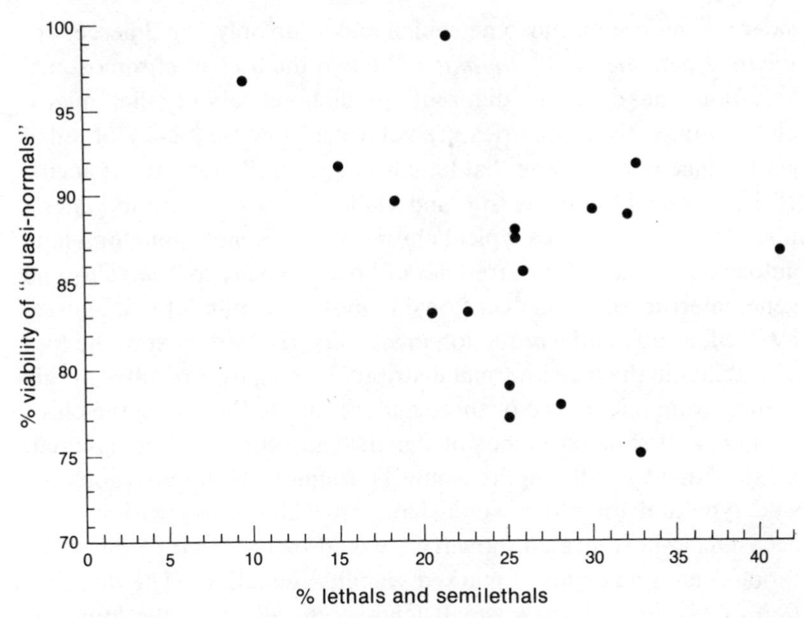

FIGURE 5

Scatter diagram of the relationship between the frequency of lethal and semilethal chromosome homozygotes and the average viability of "quasi-normal" homozygotes in a number of studies of *Drosophila* populations. Data are from table 7.

of *D. melanogaster* from 27 populations over the world shows lethal chromosome frequencies ranging between 0.035 and 1.18 percent but with the values strongly clustered between 0.1 and 0.3 percent, and with a median of 0.19 percent. Most of these are newly acquired lethals, of course.

One peculiarity of the data in table 7 is the weak correspondence between chromosome size and the viability distribution. All the species in the table have five long euchromatic chromosome arms, each comprising about 1/5 of the total genome. Chromosome II of *D. willistoni, D. prosaltans,* and *D. melanogaster* is meta-centric and consists of two of these arms, so it represents 40 percent of the genome in each species. The lethal frequencies for the second chromosome of *D. prosaltans* and *D. willistoni* are indeed the highest in the table, but their quasi-normal viabilities are close to the average. The values for *D. melanogaster*, however, are unexceptional.

A more striking and informative size effect is observed if we consider the micro-chromosome, which makes up only 2 to 3 percent of the total genome in *Drosophila.* The two theories of chromosomal variation make quite different predictions about the micro-chromosome. Both theories predict a very low frequency of lethal genes since both assume that lethals are generally rare at any locus. If the classical theory is true and viability modifiers are in general rare at any locus, then a typical chromosomal homozygote for a long autosome will have long stretches of homozygosity for the wild-type gene, interrupted by an occasional homozygous mutant that causes, by itself, a substantial reduction in viability. Indeed most of the loss of viability in the quasi-normal distribution in figure 3 results, on this theory, from one or two or three mutants at random along the chromosome. If small stretches of the chromosome could be assayed, most of these small lengths would be found to be homozygous for wild type and therefore would have no viability depression. An occasional short stretch, however, would include one of the deleterious genes and show a marked viability reduction. The distribution of viability of these small lengths chosen from the long autosome ought, then, to be bimodal, with one mode indistinguishable from the average heterozygous fitness and the other mode with lower mean viability.* That is, small random pieces of a long autosome ought to be heterogeneous in their viabilities. If the balance theory is correct, however, all the small lengths will carry viability modifiers to about the same extent, since this theory assumes variation at nearly all loci. Then the viability distribution would be unimodal, with a mean very slightly lower than the average heterozygote.

Using the micro-chromosome as a model for a random short stretch of autosomal genomes requires extremely careful work if the real genetic variance is to be observed among such short chromosomal segments. Such a careful and highly replicated experiment has been done by Kenyon (1967), who compared second with fourth (micro) chromosomes in a population of *D. melanogaster.* Appropriate data, transformed to the scale used in tables 5 and 6, are given in table 8. The particular population of second chromosomes used

*The relative proportion of pieces in the two modes would depend upon the size of the pieces examined and the number of deleterious mutants per chromosome. If the pieces are $1/n$ of the chromosome length and the mutants per chromosome are m, then the proportion in the less viable mode is $1 - e^{-m/n}$.

TABLE 8

Means (σ_G^2) and variances (σ_G) of viability for 126 second and 125 fourth chromosome homozygotes and heterozygotes in a *D. melanogaster* population

	% of lethals	*Quasi-normal, % viability*	σ_G^2	σ_G
II homozygote	12.7	91.6	201.67	14.2
II heterozygote	—	100.0	7.64	2.76
IV homozygote	0	102.7	60.95	7.81
IV heterozygote	—	100.0	36.09	6.01

Note: Calculated from Kenyon (1967).

as a base of comparison has only 12.7 percent lethals, and the quasi-normals are only 8.4 percent below the heterozygotes in fitness. Since the fourth chromosome is only 7 percent of the length of the second, the absence of lethals is reasonable. Moreover, the observed slight increase in fitness of fourth chromosome homozygotes is neither statistically significant nor significantly different from the expected decrease of 0.6 percent. A histogram of fourth-chromosome homozygous viabilities is given in figure 6, and there is no trace of a second mode at a viability less than 100 percent or even of a long tail of viabilities toward the lower values. If anything, the distribution is skewed to the right. Apparently, quasi-normal fourth chromosomes are homogeneous in their distribution but are not all alike, since the genetic variance among fourth chromosomes given in table 8 is highly significantly different from zero.

Although these results make it highly unlikely that quasi-normal chromosomes of normal length owe their viability variation to one or two subvital mutations on each chromosome, the test has no power against the alternative that there are, say, 10 such substitutions per chromosome. For if there were, pieces of the size of chromosome IV either would carry none, in which case they would form a mode at 100 percent, or would carry one such mutant, in which case they would fall in a mode at 99.2 percent since their viability loss of such a small homozygous piece would be 1/10 that of the average second chromosome. But these two modes would be completely confounded, especially in view of the genetic variance of viability. Thus we have ruled out a model in which viability variation results from one or two point mutations per chromosome, but we cannot distinguish between the hypotheses of 10 substitutions

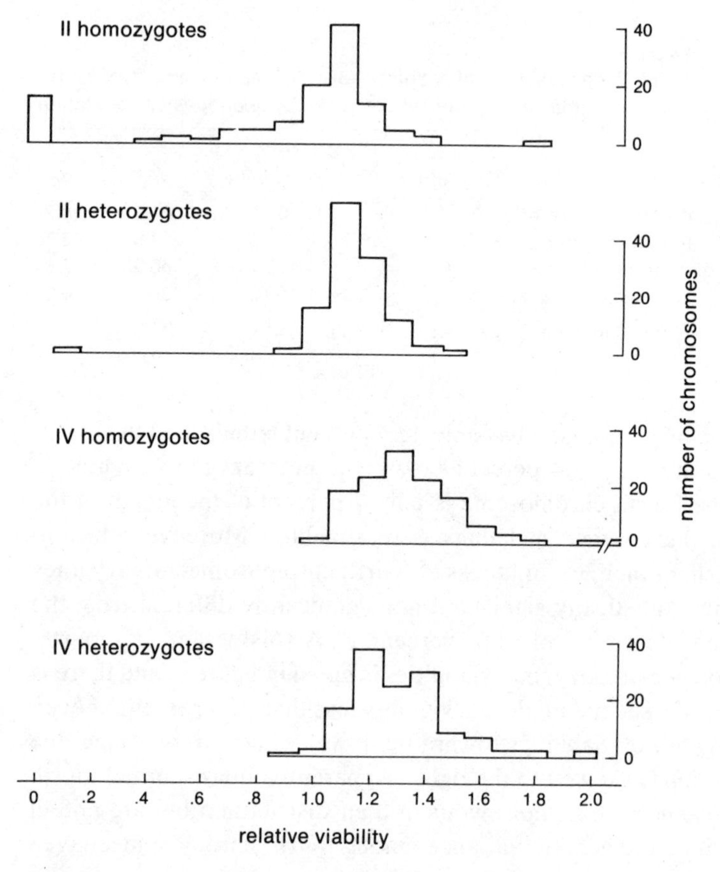

FIGURE 6

Observed distribution of viability of second and fourth chromosome homozygotes and heterozygotes in a population of *Drosophila melanogaster.* From Kenyon (1967). Reprinted with permission of *Genetics.*

per chromosome or 1000. Once again actual experimental practice does not have the power to distinguish between classical and balance hypotheses.

Viability is only one component of fitness. In Drosophila another obvious component is fertility, but this is much more difficult to characterize. There is tremendous environmental and developmental variance in fertility between females carrying the same homozygous chromosome. Since each female must be characterized by several weeks of egg production, the characterization of the genetic variation of the quasi-normal range of fertility is difficult but not impossible. Table 9 shows the result of a study by Marinkovic (1967) of fecundity in 211 *D. pseudoobscura* chromosomal homozygotes and heterozygotes. The distribution of fecundities is obviously a good deal wider than for viability and at least part of that breadth is the result of the greater environmental variance of fecundity. As for viabilities, the mean of homozygotes is distinctly less than for heterozygotes, in this case only 81 percent of the heterozygous mean. Curiously enough, no completely sterile homozygotes were found, although four had extremely low fecundity. This lack of complete sterility may be partly the result of the optimal conditions under which the tested females were raised. A more usual result, when females develop under crowded conditions, is that about 10 percent of homozygotes are completely sterile. Table 10 shows the sterility frequencies from the experiment of Dobzhansky and Spassky (1953). Since male and female sterility were

TABLE 9

Fecundity of females homozygous and heterozygous for second chromosomes in *D. pseudoobscura*

Eggs per female per day	*Number of heterozygous combinations*	*Number of homozygous chromosomes*
0–49	4	0
50–99	8	0
100–149	13	5
150–199	37	3
200–249	36	22
250–299	38	19
300–349	36	19
350–399	24	30
400–449	10	14
450–499	5	4
500–549	0	0
550–600	0	2
Mean fecundity	260.4	322.2
Relative fecundity	0.807	1.000

Note: From Marinkovic (1967).

TABLE 10
Homozygous chromosomes and sterility in *D. pseudoobscura* and *D. persimilis* from Yosemite, Calif.

Species	*Chromosome*	*% of females sterile*	*% of males sterile*
D. pseudoobscura	II	10.6	8.3
	III	10.6	10.5
	IV	4.3	11.8
D. persimilis	II	18.3	13.2
	III	14.3	15.7
	IV	18.3	8.4

Note: From Dobzhansky and Spassky (1953).

generally independent in their study, the total sterility is about equal to the lethality (see table 7). All in all, the loss in fitness from variation in fertility appears to be about the same as for viability.

A third physiological trait that has complex relations to fitness is developmental rate. In a species with a fixed small number of generations per year and a synchronization of generations because of a diapause or other cessation in breeding activity, development time is not likely to be a critical variable; but in a continuously breeding species, or one in which all stages of the life cycle are present at all times, or in which development must be finished before a critical temperature or rainfall is reached, developmental rate is positively correlated with fitness. Marinkovic (1967) also studied the rate of development from egg to adult in his population sample, and the data are given in table 11 for homozygotes and heterozygotes. The data are expressed as development time from egg to adult, relative to the $M/+$ class in the segregating F_3 generation, and are arranged in decreasing order, corresponding to increasing fitness. As compared with the mean, the variance is rather small, a characteristic of genetic variation in developmental rates (Lewontin, 1965), while the mean for homozygotes is 2 percent slower than for heterozygotes.

A clever and straightforward scheme has been used by Sved and Ayala (1970) to look at the distribution of total net fitnesses among chromosomal homozygotes. If F_2 individuals, $M/+_i$, from the chromosome replication schemes are allowed to form a continuously breeding population over a number of generations, two things may happen. If the chromosomal homozygote $+_i/+_i$ is less fit than the het-

TABLE 11
Development time from egg to adult in chromosomal homozygotes and heterozygotes of *D. pseudoobscura*

% of standard development time	*Number of homozygotes*	*Number of heterozygotes*
112.0–114	2	0
110.0–111.9	5	0
108.0–109.9	4	1
106.0–107.9	9	0
104.0–105.9	9	3
102.0–103.9	25	11
100.0–101.9	30	20
98.0– 99.9	18	18
96.0– 97.9	8	6
94.0– 95.9	8	3
92.0– 93.9	1	0
Mean development time	102.3	100.4

Note: Adapted from Marinkovic (1967). Values are expressed as percentage of the mean development time of marked heterozygotes.

erozygote $M/+_i$, a stable equilibrium frequency of marker and wild chromosome will result, since the homozygote M/M is lethal. Moreover, the equilibrium frequency, $\hat{q}$, of the wild type provides an estimate of net fitness of the homozygote relative to the marker heterozygote, although some ambiguities arise since genotype frequencies cannot be estimated among fertilized eggs. If, on the other hand, the homozygote $+_i/+_i$ is more fit than $M/+_i$, the marker will be lost and the rate of loss can be used to estimate fitnesses by a slightly more complex procedure. As a standard of comparison, F_2 flies $M/+_i$ from many different lines can be used to set up a population so that nearly all wild-type individuals are $+_i/+_j$. The average $+_i/+_j$ fitness relative to $M/+_i$ will then be estimated from the rate of loss of the marker and will serve as a standard against which to compare the chromosomal homozygotes.

This technique was applied by Sved (1971) to 24 second chromosomes from a population of *D. melanogaster*. The populations with heterozygotes $+_i/+_j$ gave an estimate of relative fitness of 0.5 for $M/+$, which, it should be noted, is much less than the viability estimates for such markers, usually close to 1.00. Table 12 shows the estimates of net fitness of each of the chromosomal homozygotes

TABLE 12
Viabilities and net fitnesses relative to marker heterozygotes from population cage experiments

	Viability	*Net fitness*	*Viability*	*Net fitness*
	.14	(−.92)	.79	.59
	.40	.34	.80	(−.25)
	.41	(−.16)	.81	.73
	.52	.32	.87	.56
	.53	(−.03)	.87	.49
	.53	.32	.88	.49
	.55	.47	.91	.56
	.67	.08	.92	.42
	.70	.65	.93	.61
	.73	.54	.96	(−.02)
	.75	.10	1.03	.15
	.79	.23	1.03	.47
Mean			.73	.28(.34)

Note: Data are from Sved (1971).

compared with the viability values for the chromosome. They have been arranged in increasing order of viability to emphasize the lack of correlation between viability and total fitness. The negative fitness values are artifacts of the estimation technique and the mean net fitness has been calculated as 0.28 including them and 0.34 if the negative values are set to zero. Since these net fitnesses are relative to the marker heterozygote, they must be multiplied by 0.5 to standardize them against wild-type heterozygotes. Although the mean viability of chromosomal heterozygotes is 0.73, the mean net fitness is only 0.14 (or 0.17 if the negative values are set to zero). Thus the loss of fitness from viability alone is only about one-third of the total effect of homozgosity, and fertility variation is the more important component of total fitness. This "fertility" variation must be thought of in the broad sense as including male mating activity and the age distribution of female fecundity. The kind of data given by Marinkovic only provides for the total fecundities of live females. It does not take into account either female longevity or the distribution of fecundity through the female lifetime. In a continuously breeding population like the populations used by Sved and Ayala, earliness of egg production counts for more than total egg production in determining fitness. At the total fecundities given by Marinkovic, a one-day decrease in the age at which egg laying commences would be worth the same in fitness as a two-thirds increase in total egg production (Lewontin, 1965).

From the data on fecundity and net fitness it would appear that a truly "supervital" homozygote must be an extraordinary rarity indeed, at least in Drosophila.

The revelation that a large fraction of fitness variation in *Drosophila* is a result of variations in female fecundity and male fertility seems to destroy any hope of dealing with fitness modifiers as mappable and countable Mendelian entities. By recovering recombinants between wild and marked chromosomes, it is possible to map, crudely, the distribution of viability effects along a chromosome, much as was done by Breese and Mather (1960). To do so requires that from each wild chromosome a number of recombinant chromosomes be derived, and each of those recombinants must then have its viability in homozygous condition determined. Unless the recombination breaks up the chromosome into a dozen or so segments, the mapping will be extremely crude. In the study of Breese and Mather, one parental chromosome had a viability of 55 percent and the other a viability of about 80 percent (no direct estimate of the latter figure is available), yet the experiments could not distinguish between the hypothesis that one parental chromosome contained two deleterious genes and the other a single mutant, and the hypothesis that there was a more or less continuous distribution of viability modifiers over the length of the chromosome, the hypothesis Breese and Mather themselves favored. Attempts on their part to further subdivide the chromosome into six regions gave results that were uninterpretable because of inconsistencies and differences that were too small for reliable analysis. To try to carry out such an analysis on net fitness by the technique of Sved and Ayala for many parental chromosomes on a sufficient scale to get unambiguous information on the distribution of fitness modifiers along the chromosome would require many thousands of times more effort than expended by Breese and Mather. And even then, the sensitivity of fitness to environmental variables, the statistical complexities of estimating net fitnesses when observations cannot be made on the phenotypes of newly formed zygotes (Prout 1965), and the large standard errors of fitness estimates compared to the average differences between genotypes (Wilson, 1970; Prout, 1971a) make it fairly certain that this immense effort would produce only ambiguous results.

THE GENERATION OF VARIATION

One remarkable feature of variation for fitness characters, and indeed for other quasi-continuous characters, is the high rate at which new variation is produced by mutation and recombination. Mukai (1964) tested the viability of chromosomal homozygotes, derived from a strain of *D. melanogaster* that itself was initiated as a chromosomal homozygote, at various numbers of generations after the original establishment of the strain. Table 13 gives the average homozygous viability of the rederived chromosomes and the estimated genetic variance among the homozygotes after generations of accumulation of spontaneous mutations.

The loss in mean viability, although rather erratic, is sufficient to produce a mean viability of quasi-normals close to that observed for natural populations of *D. melanogaster* (see table 7), after only 25 generations. The variance accumulates somewhat more slowly, however. The genetic variances given in table 13 are in the same units as those in tables 5, 6, and 8, for natural populations. Although the variances vary widely, they are on the order of 100 units in natural populations as compared to the 40 units accumulated in 25 generations of Mukai's experiment. Yet only about 60 generations would be required, at this rate, to approach the variance of an equilibrium population. The discrepancy between the rate of loss of mean fitness and the rate of accumulation of variance is not easily

TABLE 13
Mean viability of chromosomal homozygotes and estimated genetic variance in viability after various numbers of generations of accumulation of mutations

Generation	*Viability as % of control*	*Genetic variance among homozygotes*
0	100.0	0
10	95.9	6.55
20	95.9	17.34
25	86.1	39.82

Note: From Mukai (1964).

TABLE 14
Mean viability of homozygotes of several *Drosophila* species for recombinant chromosomes (expressed as percentage of the viability of the parental chromosomes when homozygous) and the proportion of the genetic variance among homozygotes derived from natural populations that is generated by one generation of recombination between "normal" chromosomes

	% Viability		*% Variance (recombinant/natural)*		
Species	*recombinant/parental*	*natural/parental*	*all chromosomes*	*quasi-normals*	*Reference*
D. pseudoobscura	81	70	43	75	Spassky et al. (1958)
D. persimilis	89	72	24	27	Spiess (1959)
D. prosaltans	88	67	25	28	Dobzhansky et al. (1959)
D. willistoni	97	52	35	(100)	Krimbas (1961)

explained, although it is no doubt partly due to the more rapid loss, through natural selection, of the more deleterious mutations.

A strikingly similar estimate of the generation of new variance comes from a consideration of two morphological characters, abdominal and sternopleural bristle number, in *D. melanogaster*. From previously published data on the response to artificial selection in initially inbred lines, Clayton and Robertson (1955) estimated that between 0.1 and 0.2 percent of the equilibrium variance for bristle number is produced in each generation by spontaneous mutation, so that between 500 and 1000 generations would suffice to bring a homozygous population to equilibrium.

The generation of genetic variance by recombination is more rapid by two orders of magnitude than generation from mutation alone, as shown by a series of coordinated experiments on four species of Drosophila (table 14). Chromosomes that were essentially "normal" in viability when homozygous were allowed to recombine, and the homozygous viability of a sample of recombinant chromosomes was tested. There were two consistent effects, shown in table 14. First, the recombinant chromosomes had a lower mean viability than the parental genomes from which they arose. These parental genomes were a biased sample of high-viability homozygotes; we see from a comparison of the first two columns of table 14 that the recombinant chromosomes regressed considerably toward the mean viability of a random sample of homozygotes taken from nature. Except in *D. willistoni,* more than a third of the excessive viability of the parental homozygote was lost among the recombinants. Such a loss implies that a considerable fraction of the viability depression of homozygotes arises from epistatic effects between loci. Second, a considerable fraction of the variance among homozygotes sampled from a natural population can be produced by a single generation of recombination between two chromosomes, both of which are "normal" in their viability. Moreover, the variance produced by the recombination between particular chromosomes had no significant correlation with the viability difference between those chromosomes when they were tested in homozygous condition.

Unfortunately, the fact that around one-third of the total genetic variance between homozygous chromosomes can be generated by a single generation of recombination does not tell us very much about the average amount of heterozygosity in a population. Whereas these results make it fairly certain that chromosomes do not differ by one or two mutations only, there is nothing in the outcome of the experiments that is inconsistent with, for example, a dozen or a score of mutations on each chromosome. It is an unhappy fact of genetics, one that appears over and over in any attempt to interpret the outcome of experiments, that the segregation of a handful of loci on a chromosome will produce a spectrum of genetic variation among *chromosomes* that is indistinguishable from the spectrum produced by the segregation of thousands of genes.

STANDING THE PROBLEM ON ITS HEAD

I have tried to emphasize that there is a close relation between theories of the amount of genetic variation in natural populations and theories of the way in which natural selection operates. Because of the close connection between the action of natural selection and the amount of allelic variation, the classical and balance theories are theories about both variation and selection, dual aspects of a single question. If selection operates almost entirely to "purify" the genome, if there is a wild-type state for a locus and every mutation is deleterious in homozygous condition and very slightly deleterious or even neutral in heterozygous condition, then genetic variation will be rare at any locus. What genetic variation there is will be the as yet uneliminated deleterious mutations that have arisen relatively recently. Even under the most favorable conditions for variation, complete recessivity of mutant genes, the heterozygosity at a locus will be only

$$H = 2\sqrt{\frac{u}{1-w}}$$

where u is the total mutation rate to deleterious alleles and w is the fitness of the homozygous mutants. With mutation rates of the order of 10^{-5} and with fitnesses of homozygotes between 90 and 95 percent of wild type, the heterozygosity at a locus will be between 1 and 3 percent. Although this is a long way from the nearly 100 percent heterozygosity expected under the balance hypothesis, it is rather larger than the classical hypothesis predicts. If, however, we assume that deleterious mutants have a slightly deleterious effect in *heterozygous* condition, that is, if deleterious mutants show a very slight dominance, the picture changes radically. If the fitnesses of the three genotypes are

$+/+$	$+/m$	m/m
1	$1-hs$	$1-s$

then the heterozygosity at equilibrium is very close to

$$H = \frac{2u}{hs}$$

For example, if a deleterious mutant had a homozygous fitness of 90 percent ($s = 0.1$), then even if the dominance of the deleterious gene was as little as 10 percent so that the fitness of the heterozygote was 99 percent of normal, the average heterozygosity would be only 0.2 percent, an order of magnitude lower than for the completely recessive gene and more in line with Muller's estimates. Thus, even if it is not possible to enumerate heterozygotes directly in a population, it necessarily follows that they are rare and that most individuals are homozygotes for a wild-type allele at nearly all loci, if it can be shown that genes which are deleterious when homozygous are either recessive or, better, slightly dominant in heterozygous condition.

Alternatively, suppose that a heterozygote between two alleles has a *higher* fitness than either homozygote. This is equivalent to letting h be negative. Then at equilibrium both alleles are maintained in intermediate frequencies and the heterozygosity will be

$$H = \frac{2h(h-1)}{(2h-1)^2}$$

For example, if $s = 0.1$ and $h = -0.1$, so that the heterozygotes are 1 percent *more* fit than the homozygote $+/+$ rather than 1 percent less fit as in the previous example, the heterozygosity is 16.3 percent at the locus and the balance theory is correct. In such a case it becomes incorrect to label one allele the "wild-type," and since the heterozygote is the most fit, a more convenient parameterization of the fitnesses is

A_1A_1	A_1A_2	A_2A_2
$1 - t$	1	$1 - s$

At equilibrium, the frequency of the allele A_1 is

$$p(A_1) = \frac{s}{s+t}$$

and the heterozygosity is

$$H = \frac{2st}{(s+t)^2}$$

Translating the example just given, $t = 0.01$, and $s = 0.11$, so that

$$p(A_1) = 0.875$$

and

$$H = 0.163$$

as before.

An important consequence of higher fitness of the heterozygote is that it leads to a stable equilibrium of intermediate allele frequencies with an accompanying high heterozygosity. If mutations should arise that have this property of *overdominance* or *single-gene heterosis*, they will accumulate in the population at intermediate frequencies, while the "classical" mutants will be eliminated by selection. Even if only a small minority of mutations were of this overdominant sort, the population genotype would come to be dominated by them because of their preferential accumulation, and the population would become highly heterozygous at every locus. Even a low frequency of overdominant mutations is incompatible with classical theory because a low *a priori* frequency of overdominant mutations becomes at equilibrium a high *a posteriori* frequency of overdominant segregating alleles. If it could be shown that overdominant mutations arose with any frequency, then even though it was not possible to enumerate heterozygosity in a population, it would necessarily be the case that heterozygosity was high and that allele frequencies were intermediate.

To recapitulate, if virtually all heterozygotes between alleles lie between the homozygotes in their fitness, or even if many are equal to the better homozygote, heterozygosity in populations will be rare and all individuals will be virtually identical genetically. If, on the other hand, overdominant combinations of alleles occur repeatedly, even though not often, these overdominant alleles will accumulate in stable equilibrium at intermediate frequencies and heterozygosity will be high. I have deliberately left vague, for the moment, the concepts of "virtually all" and "repeatedly, even though not often." There are clearly ranges of parameter values for which both theories of selection would predict the same amounts of heterozygosity.

The possibility of distinguishing between the classical and balance theories of population structure by looking at the fitness of heterozygotes rather than by counting genes has led to several ingenious approaches to the analysis of chromosome replication experiments.

Correlations Between Homozygotes and Heterozygotes If the dominance h of a deleterious gene is the same for all genes, or if it varies in a way unrelated to the intensity of selection s against the homozygote, then from the scale of fitnesses

$+/+$	$+/m$	m/m
1	$1 - hs$	$1 - s$

we can predict a positive correlation between the fitness of homozygotes and heterozygotes. Since h is constant or unrelated to s in its value, the greater the loss of fitness of the homozygote, the greater the loss of the heterozygote. If we carry this line of argument to overdominant cases, however, then we would predict a negative correlation between homozygous and heterozygous fitnesses since, on this hypothesis, h is negative so the greater the loss of fitness of m/m, the greater the increase of fitness of $+/m$ relative to $+/+$. On this scheme, the slope of the line relating homozygous to heterozygous fitness is, in fact, a direct estimate of h, the negative of the average dominance of deleterious genes.

This kind of argument has led several investigators to look carefully at the relation between the fitness (usually the viability) of pairs of chromosomal homozygotes and the heterozygote between them. The results are completely equivocal. I have already reviewed briefly on page 44 the question of the average dominance of lethals. Part of the ambiguity of the data on the dominance of lethals arises from the class of "normal" heterozygotes to which they are compared. The notion of the dominance of lethals assumed a comparison with homozygotes for a perfect wild type, but in fact some heterozygotes have to be used and since there is a real variance in viability among heterozygotes, it may matter very much which chromosomes are involved in the standard. When we turn to general estimates of the fitness of heterozygotes between different classes of chromosomes, there is considerable variation from study to study. Dobzhansky, Krimbas, and Krimbas (1960), for example, in a study of nearly 600 chromosomes of *D. pseudoobscura* involving two different chromosome elements, two temperatures, and two populations, found an overall positive, small correlation between the homozygous fitness of a chromosome and its fitness when heterozygous with a marker chromosome. The relationship between homozygote and heterozygote was curvilinear, however, with heterozygous viabilities being minimum for chromosomes with intermediate homozygous viabilities. It is their claim that all the apparent regression is due to the supervital chromosomes, which are much more viable in heterozygous condition with the marker than is the average chromosome. A much more sophisticated treatment of these data and other results by Wallace and Dobzhansky (1962) confirmed this claim. It should be noted, however, that these results concern the performance of chromosomes when heterozygous with a dominant marker, rather than random heterozygotes between two wild chromosomes.

Dobzhansky and Spassky (1963) found no significant regression of heterozygotes on homozygotes when neither chromosome carried a marker and when the heterozygote was compared with the average of the two homozygotes. But Kenyon (1967) found a strong positive correlation in her studies of the fourth chromosome of *D. melanogaster*. Wills (1966) found lethals to be completely recessive in *D. pseudoobscura*, but subvitals to be partly dominant, and Marinkovic's study of fecundity (1967) showed a positive correlation between heterozygotes and homozygotes. To complete the confusion, the study of Tobari (1966) showed that there was no dominance for viability of *D. melanogaster* at 25 C, but at the stress temperature of 29 C the correlation between homozygous and heterozygous viability was negative, apparently showing overdominance!

The unsatisfactory state of the experimental evidence arises from the fact that the effect being sought, even if it exists, must be of a magnitude too small to observe consistently. Suppose we take the value of 2.5 percent dominance of lethals as a reflection of reality and assume this degree of dominance to be general. Then heterozygotes for lethals will have, on average, a viability of 97.5 percent of the best "heterozygote" (which will in fact turn out to be a wild-type homozygote) so the entire range of mean heterozygous viabilities from lethal to normal is only 2.5 percent. But we have already seen that the genetic standard deviation for heterozygous viabilities averages about 5 percent (table 6). Even if all environmental variance could be eliminated, the regression could only account for 10 percent of the variance of heterozygous viabilities, which means a correlation of about 0.3. Such a correlation is much too small to be significant, in view of the small number of viability classes (ten at the most) into which the homozygotes can be accurately sorted.

Putting aside the statistical problems, the theoretical relation between regression and dominance is very weak. It assumes that h is the same, on the average, for all classes of homozygous viabilities. If this assumption falls, the entire analysis falls. Suppose we assume, as does Muller (1950), that the average degree of dominance increases with homozygous fitness so that mildly deleterious genes are much more dominant than lethals. If h should be inversely proportional to s, then the fitness of a heterozygote would be

$$1 - hs = 1 - \left(\frac{k}{s}\right)s = 1 - k$$

which is a constant, and there would be no regression at all. If the dominance should decrease at any rate faster than proportional, say as $1/s^2$, then the fitness of a heterozygote would be

$$1 - hs = 1 - \left(\frac{k}{s^2}\right)s = 1 - \frac{k}{s}$$

and would be *negatively* correlated with the homozygote if there were partial dominance ($h > 0$) and *positively* correlated if there were overdominance. It appears from this consideration that the correlation between homozygous and heterozygous viabilities is not only unlikely to be consistent, but is uninterpretable anyway.

The B/A Ratio An interesting suggestion of Morton, Crow, and Muller (1956) for the detection of heterosis depends upon the great difference in inbreeding depression between a locus with several alleles in intermediate frequency in the population and a locus with only rare mutant alleles. To appreciate this point we need the concept of the *genetic load* in a population, by which is meant the loss of fitness in the population as the result of segregation of unfavorable genotypes. In particular, we will define the genetic load as the difference in fitness between the population and a hypothetical population composed solely of the fittest genotype.

Suppose we consider a deleterious allele with a small deleterious effect in heterozygous state. The fitnesses of the genotype can be scaled as

$+/+$	$+/m$	m/m
1	$1 - hs$	$1 - s$

and if the population is random mating, the genotypes at zygote formation will be in the proportion

$+/+$	$+/m$	m/m
p^2	$2pq$	q^2

where $p = 1 - q$ is the frequency of the + allele. The genetic load in a population as a result of this locus will be

$$L = hs(2pq) + q^2 s$$

But p will be very close to one and q close to zero since a deleterious, semidominant mutant will be in very low frequency. To a very close order of approximation the genetic load in the random-mating population L_R will come entirely from heterozygotes and will be

$$L_R = 2hsq$$

a very small number.

If we should inbreed this population completely without changing the frequency q, the result will be homozygotes $+/+$ and m/m in the ratio $p : q$. The genetic load L_I in this inbred population will derive entirely from the homozygotes m/m and will be

$$L_I = sq$$

L_I is also a small number, but it is a good deal larger than L_R, and the ratio

$$\frac{L_I}{L_R} = \frac{sq}{2hsq} = \frac{1}{2h}$$

If the dominance of a deleterious recessive were, for example, 2.5 percent, as estimated by some experiments for lethals, the ratio of loads would be $1/.05 = 20$. An even larger ratio would result if the deleterious gene were completely recessive. By our previous reasoning, for a completely recessive gene

$$\frac{L_I}{L_R} = \frac{sq}{sq^2} = \frac{1}{q}$$

For a large population, the frequency of a deleterious recessive gene is unlikely to be greater than about 1 percent, so the ratio would be of the order 100 or greater. In theory, a deleterious recessive gene at equilibrium has a frequency $\sqrt{(u/s)}$ where u is the rate of mutation to the deleterious allele. Then the load ratio will be $\sqrt{(s/u)}$ which, for $u \sim 10^{-5}$ and $s = 0.1$, will be 100.

A very different situation pertains for a heterotic gene. Again assuming random mating, the fitnesses and frequencies of the genotypes can be represented as

A_1A_1	A_1A_2	A_2A_2
$1 - t$	1	$1 - s$
p^2	$2pq$	q^2

Now the genetic load arises from both homozygotes, which are both reasonably common unless s and t are extremely different in size so that one homozygote is very bad and the other almost as fit as the heterozygote. In general

$$L_R = p^2t + q^2s$$

a reasonably large number, of the order of s and t.

If we inbreed this population to complete homozygosity, the two homozygotes will be in frequency p and q and the inbred load will be

$$L_I = pt + qs$$

also a large number, but not a great deal larger than L_R. The genetic load in the random-mating population is already large because of the segregation of two common alleles, and it is not made much larger by getting rid of the heterozygotes. More precisely, if the alleles are at equilibrium under selection,

$$p = \frac{s}{s+t}$$

and

$$q = \frac{t}{s+t}$$

If these values are substituted into the expressions for L_R and L_I, we get the surprising result that

$$\frac{L_I}{L_R} = \frac{(st+ts)/(s+t)}{(s^2t+t^2s)/(s+t)^2} = 2$$

The load ratio for a heterotic gene is precisely 2, at equilibrium, irrespective of the values of the parameters.

Morton, Crow, and Muller (1956) suggested that the genetic load be estimated in a random-mating population and in a completely inbred set of progenies from that population, the ratio between them examined, and a judgment about partial dominance of deleterious genes as opposed to heterosis be made from the size of that ratio. They applied the method by comparing the pre-adult mortality in the population at large for several human samples with the pre-adult mortality in offspring of cousin marriages. By linear extrapolation from the nonrelated marriages ($F = 0$) and the cousin marriages ($F = 1/16$) it was possible to estimate the mortality rate that would obtain in a completely inbred group ($F = 1$). If B is the mortality in the completely inbred group, and A the mortality in the outbred population,* then

$$\frac{B}{A} = \frac{L_I + E}{L_0 + E}$$

where E is the "environmental" mortality, the mortality that occurs irrespective of genotype. The ratio B/A is an underestimate of the load ratio L_I/L_R, because the added constant mortality in both numerator and denominator causes the ratio to be spuriously close to 1. If E were very large compared with the genetic loads, the ratio B/A would be nearly 1, whatever the values of L_I and L_0. Morton, Crow, and Muller found B/A ratios ranging from 7.9 to 24.4 and concluded that since these are underestimates of the true load ratio, the evidence is strongly in favor of partial dominance of deleterious genes, of the order of 10 percent and strongly against overdominance, which predicts a ratio of 2. Neel and Schull (1962), including both mortality and nonlethal disease in their estimates of genetic load, found B/A values of 4.48 and 4.63 for Nagasaki and Horoshima. Dobzhansky, Spassky, and Tidwell (1963) used unrelated ($F = 0$), half-sib ($F = 0.125$), and full-sib ($F = 0.250$) matings in four populations of *D. pseudoobscura* to estimate A and B by least squares regression of the viabilities of the three classes. Their estimates of B/A were 3.57, 3.63, 4.60, and 5.82 for the four populations.

The values estimated by Neel and Schull and by Dobzhansky and co-workers are disconcerting, because, like so many estimates in population genetics, they do not fall clearly into the range of one prediction or the other. They are too high for the simple heterotic prediction and too low for the simple dominance prediction, falling precisely in the region where ambiguity is complete. The explanation of the ambiguity arises from a closer look at the assumptions of the B/A argument.

First, the predicted ratio of 2 for the heterotic case is restricted to a locus with two alleles. If there are k alleles in heterotic equilibrium, the predicted ratio is $L_I/L_R = k$. An observed ratio of 4 is com-

*As given by Morton, Crow, and Muller, B is the *added* mortality in the outbreds, but the usage of these symbols in the literature is confused and I use a form that simplifies the presentation.

patible with an average dominance of 12.5 percent but it is equally compatible with the heterotic maintenance of four alleles. Whether this latter possibility seems reasonable depends upon one's initial position on the balance and classical theories.

Second, there is a confusion between single loci and whole genomes that may bias B/A ratios in either direction because of gene interaction. The estimation of B by extrapolation from low levels of inbreeding is correct for a single locus since it is a simple algebraic result (with no assumptions hidden in it) of the load being proportional to the inbreeding coefficient. The assumption that whole genomes will have a load proportional to inbreeding depends upon the additivity of loads at different loci. If loads were more than additive, the extrapolated load at $F = 1$ would overestimate the deleterious effect of making a single locus homozygous, while if epistatic interactions between genes caused a canceling out of deleterious effects, the extrapolated load would underestimate the effect of homozygosity for each locus separately. As usual, the data on this point are ambiguous.

Special experiments to test interactions in viability were performed independently by Spassky, Dobzhansky, and Anderson (1965) and Temin and co-workers (1969) with different results. The Spassky group, working with simultaneous homozygosity of chromosomes II and III in *D. pseudoobscura*, found the average interaction of the two chromosomes to be about as large as the average viability loss for each chromosome separately. Letting the viability of a doubly homozygous line be $(1 - s)(1 - t)(1 - i)$, where $1 - s$ and $1 - t$ are the viabilities of the second and third chromosomes separately, the average values in *D. pseudoobscura* were

$$\bar{s} = 0.103$$
$$\bar{t} = 0.146$$
$$\bar{i} = 0.094$$

The Temin group, performing the same experiment in *D. melanogaster*, found no significant interaction, their values being

$$\bar{s} = 0.09$$
$$\bar{t} = 0.11$$
$$\bar{i} = 0.03$$

In both cases there is a great deal of variation around the mean from chromosome combination to chromosome combination, but the error variance of these values is very great and not much can be said about the real distribution of interactions.

Third, there is the problem of averages. If we take the B/A ratio at face value, the numerator and denominator will each be the sum of many different effects. Each heterotic locus will make a large contribution to both numerator and denominator, while each semidominant locus makes a small one, so that the ratio will be dominated by a relatively small number of heterotic genes. For example, if 90 percent of the loci have a deleterious recessive with 2.5 percent dominance, maintained by a recurrent mutation rate of 10^{-5}, whereas the remaining 10 percent of loci are heterotic with the homozygotes 95 percent as fit as the heterozygote, the load ratio will be only 2.15, essentially indistinguishable from a completely balanced load.

Fourth, the B/A ratio based on viabilities is an overestimate rather than an underestimate of L_I/L_R. The theory of load ratios depends upon total fitness, of which viability is only a part. In Drosophila it seems to be only a minor part, if we accept the results of Sved (1971) who, it will be recalled, found net fitness of homozygotes to be 0.14 while their viability was 0.73. On this basis alone the total genetic load is 3.2 times as great as the viability load. The problem is deeper than this. Genetic load is defined as the deviation of fitness from the fitness of the optimum genotype in the population. Whereas an absolute viability of 100 percent may be taken as the obvious maximum for the optimum genotype, no similar limit can be placed on the fecundity of the optimum genotype. The fecundity of the random-bred populations and of the inbreds derived from them lie at an unknown and unknowable distance below the fecundity of the "best" genotype. The fitness estimates of Sved are *relative* fitnesses, but the theory of load ratios demands a measure of *absolute* fitness, because it requires an estimate of the deviation of a random load population from that absolute. Thus, B/A ratios based on viabilities really have the form

$$\frac{B}{A} = \frac{L_I + E - L_{IF}}{L_R + E - L_{RF}}$$

where L_{IF} and L_{RF} are the true absolute losses of fecundity in the random and inbred groups, relative to the "optimum" genotype.

How large are L_{IF} and L_{RF}? We do not know in man, mouse, or Drosophila, except to say that they are larger than the viability component in Drosophila and in man. It is at least conceivable that a more realistic load ratio could be estimated in species of birds that have a rigorously determined clutch size and more or less fixed number of breeding years, so that the variance in fecundity would be extremely low. In such cases load could be estimated entirely from the probabilities of reaching adulthood and forming a stable nesting pair that is fertile.

This last objection goes to the heart of methods that involve the use of "genetic loads." Let us contrast a completely recessive deleterious gene whose fitness, $1 - s$, is 0.8 and whose mutation rate is 5×10^{-5}, with a slightly heterotic gene for which one homozygote has a fitness of 0.8 and the other a fitness of 0.99. The equilibrium frequency of the recessive deleterious gene will be

$$q = \sqrt{u/s} = 0.016$$

and the equilibrium frequency of the rarer of the two alleles for the heterotic case will be

$$q = \frac{s}{s+t} = 0.048$$

which is not very different. However, the load ratio will be 66 for the recessive gene but exactly 2 for the heterotic gene. We have the peculiarity that the load ratio jumps discontinuously from 2 to 66 with only a minute change in the fitness of the heterozygote from completely recessive to very slightly overdominant. This apparently absurd behavior of the load ratio results entirely from the peculiarity of definition of genetic load as the loss of fitness relative to the optimum genotype. In changing from a completely recessive to a slightly dominant gene, nothing of any biological significance has happened. The gene frequency has slightly increased and the mean reproductive rate of the population has changed very little, but the optimum genotype has changed *by definition* from the abundant homozygote to the heterozygote which is rare. As a result, the load ratio *by definition* has changed drastically, *but nothing that is measurable has changed.* We must be specially wary in population genetics of idealized quantities that owe their existence to definitions or scaling properties but that cannot be measured or have no biological significance to the population.

D/L Ratios Another suggestion for the use of genetic loads to distinguish heterotic from partly dominant genes comes out of the differing expectations for the distribution of homozygous fitnesses among newly arisen mutations and among mutations that have been screened by natural selection. The distribution of homozygous chromosomal viabilities that results from the usual chromosomal replication experiment can be divided, as we have seen, into two modes. Let us designate by L the total loss of fitness (again from some optimum genotype) that results from the lethal and semilethal homozygote in the array, and by D the loss of fitness of the "quasinormal" or "detrimental" homozygotes. Then the ratio D/L has different expectations for different degrees of dominance of the genes when the chromosomes are taken from an equilibrium population, since the equilibrium frequencies of genes with different degrees of debility will be different. Moreover, D/L should be still different for homozygotes of newly produced mutations, since selection has not yet acted to change their frequencies, and the ratio should be independent of dominance or overdominance, reflecting only the total occurrence of the different types of mutations. Table 15 shows the

TABLE 15
Expected D/L ratios among homozygotes for newly arisen mutations and for genes at equilibrium

	D/L	$\dfrac{D/L \text{ in equilibrium}}{D/L \text{ new mutants}}$
New mutants	$n\bar{s}/N\bar{S}$	
Equilibrium populations	$n/\tilde{h}/N/\tilde{H}$	
I. $h > 0$		
h constant	n/N	$\bar{S}/\bar{s}$
hs constant	$n\bar{s}/N\bar{S}$	1
h/s constant	$n\bar{S}/N\bar{s}$	$(\bar{S}/\bar{s})^2$
II. $h = 0$	$n\overline{\sqrt{s}}/N\overline{\sqrt{S}}$	$(\bar{S}/\bar{s})^{1/2}$
III. $h < 0$		
k alleles	$n\bar{\tilde{s}}/N\bar{\tilde{S}}$	1
2 alleles with $S_1 >> S_2$	n/N	$\bar{S}/\bar{s}$

Note: From Greenberg and Crow (1960).
s = loss of fitness for a homozygous detrimental.
S = loss of fitness for a homozygous lethal or semilethal ≅ 1.
n = number of detrimental mutations per genome.
N = number of lethal and semilethal mutations per genome.
s = harmonic mean of s.
h = dominance of detrimentals.
H = dominance of lethals.

expected D/L ratios for different cases of dominance and for new mutations. According to these results, it would be possible to distinguish a constant partial dominance from a partial dominance that grew greater as the selection grew less intense, as suggested by Muller. In the former case the D/L ratio for equilibrium mutations would be many times greater than for new mutations, the more so the less deterimental was the average quasi-normal, while in the latter case the D/L ratio would be the same in new and equilibrium mutations. Unfortunately, one cannot distinguish the latter case from the case of heterosis with k alleles, nor the constant dominance case from the situations where there are two heterotic alleles with one much more frequent than the other and only slight heterosis.

Greenberg and Crow (1960) calculate D/L ratios for a variety of data on Drosophila and also provide fresh experimental evidence themselves. The range of D/L in equilibrium populations is from 0.587 to 3.204, with a median value of 1.54, whereas for newly arisen mutations it varies from 0.103 to 1.053, with a median of about 0.30. The ratio of these ratios is about 5, which would favor either constant but very high dominance or two heterotic alleles with the poorer homozygote having a fitness of about 0.8. Greenberg and Crow give reasons for preferring one of the higher D/L ratios for newly arisen mutations and one of the lower ones for equilibrium mutations so that the ratio comes out closer to 1, favoring the alternative hypotheses with hs a constant. This formulation does not deal with the possibility that h is negative for deleterious genes (heterosis) but slightly positive for lethals, perhaps 2–3 percent. In the case of k heterotic alleles, this would produce a ratio somewhere between the two alternatives.

The method suffers from all the difficulties attendent on ratios of genetic loads discussed in the preceding section. Since the optimum genotype is not known, the genetic load is not estimable for total fitness. There is, moreover, a great heterogeneity in D/L ratios so that the ratio of these ratios, the diagnostic statistic, may vary from 30 down to 0.5, depending upon the choice of data. It does not seem that the D/L ratio offers any unambiguous evidence on the problem at hand.

The Heterozygous Effect of New Mutations The most daring and original suggestion for solving our problem by characterizing the intensity and degree of natural selection at the typical locus was made by Wallace (1957). The difficulty, as I have pointed out, is to distinguish the effect of single-locus substitution, since no method exists for following segregation at a single locus with small effects. Wallace proposed reversing the direction of the experimental comparison by starting with a completely homozygous stock and inducing a few random mutations in one of the two homologous genomes. One could then compare the original homozygote, the homozygote for the newly mutated genome, and the heterozygote between them. The daring of this proposal lies in two features. First, the number of mutations induced must be very small so that interaction between loci does not become important. Therefore the effect to be expected must be small. Second, the technique is a one-sided test for heterosis, and negative results would be meaningless. If the *average* heterozygote for a newly induced mutation were superior to the unmutated homozygote as well as the mutated homozygote, then heterosis would have to be very general in natural equilibrium populations. Natural selection, it must be remembered, will act as a screen and enrich the population for heterotic mutants if they occur. If the average heterozygote for newly induced mutations were not heterotic, however, nothing would be proved because heterotic mutation might make up a minority of new mutations and so be lost in the average, but in *equilibrium* natural populations heterozygotes might predominate because of enrichment by natural selection. Wallace had hoped, in his original proposal, that even if the average of new mutations were not superior in heterozygous condition, that he could, by suitable statistical analysis, estimate the proportion of heterotic mutations induced. In view of the multiple assumptions in such analyses and their evident lack of power, such an estimation seems very unlikely to lead to clear-cut results.

The essence of the experimental technique is straightforward. In the F_2 generation of the standard chromosome-replication scheme, some males from a given line are irradiated with X rays at a dose calculated to induce a few mutations per chromosome. These irradiated males are then crossed to their unirradiated sisters to produce the usual F_3 segregation, except that the wild-type offspring are not completely homozygous for the chromosome under test but are heterozygous at a few randomly mutated loci. At the same time unirradiated males in the F_2 generation are mated to their sisters to produce the usual F_3 segregation of homozygous wild types. These wild types, the controls, are *completely* homozygous since there was no irradiation of their fathers. A comparison of irradiated "homozygotes" with control homozygotes will show the effect of newly induced mutations in heterozygous condition.

Wallace's actual experimental design was much more complex since he arranged the crosses to avoid irradiating the marker chromosome, he made sure that the wild-type chromosome in the test was a quasi-normal before irradiation, he transferred the chromosome used onto a common genetic background with the marker stock, and he had two markers segregating in the F_3. The final generation, in which counts were made, segregated for four genotypes

Control: $M_1/M_2 : M_1/+_i : M_2/+_i : +_i/+_i$

Experimental: $M_1/M_2 : M_1/+_i^* : M_2/+_i : +_i/+_i^*$

where $+_i$ is a particular wild chromosome and $+_i^*$ is that chromosome with a few random mutations induced in it. The results of the first large-scale experiment (Wallace, 1958a), using 500 R of X rays, gave the following weighted average viabilities relative to the doubly marked class:

Control:	M_1/M_2	$M_1/+$	$M_2/+$	$+/+$
	1.000	1.094	1.146	1.008
X ray:	M_1/M_2	$M_1/+^*$	$M_2/+$	$+/+^*$
	1.000	1.115	1.137	1.033

The 2.5 percent *increase* in viability of $+/+^*$ relative to $+/+$ is remarkable. The amount of radiation (500 R) was enough to induce detectable homozygous viability changes in only about 20 percent of the irradiated chromosomes, so the heterosis would appear to be very strong per mutation.

Although these results were striking, and statistically significant, subsequent results were less satisfactory. Muller and Falk (1961) and Falk (1961), in an experiment that differed in some technical details from Wallace's and in which much higher doses of X rays were given, failed to find any increase in the mean fitness of the fractional heterozygotes, but they failed to find any significant decrease either. These results do not differ significantly from Wallace's, but neither do they give any support to average heterosis. The failure to find a significant result with a much higher dose was in part matched by Wallace's own later results (1963), which showed a marked increase in viability with 750 R irradiation, but no further increase with 2250 R. However, the failure of Muller and Falk to find a significant *decrease* in fitness after a very large dose of irradiation is at variance with the usual claim of the classical school that most mutations are semidominant and that most chromosomal heterozygotes in nature are homozygous for only a few mutations, the rest of the loci being wild-type homozygotes.

In 1963 Wallace summarized a very large series of experiments involving 8189 cultures and 2.5 million flies. The result was a small (0.9 percent), nonsignificant increase in the viability of fractional heterozygotes over their unirradiated controls. Clearly the original report of a 2 percent advantage was too strong, and at the present time the issue is in doubt. The lack of any significant depression of the viability of fractional heterozygotes would be as meaningful as an actual positive result if we knew whether a significant number of mutations had been induced in Wallace's experiments, but we do not.

An approach similar to Wallace's, but using spontaneous mutations, was taken by Mukai, Chigusa, and Yoshikawa (1964, 1965). They compared homozygotes for chromosomes that had accumulated spontaneous mutations for 32 generations with heterozygotes between these chromosomes and a chromosome in which homozygous viability had not changed during the 32 generations. When this comparison was carried out on a homozygous genetic background, the heterozygotes had a relative viability of 103.02 as compared with 99.92 for heterozygotes between the five *best* homozygotes, 84.12 for the average of mutated homozygotes, and 98.12 for the "unmutated" homozygote. Spontaneous mutations that produced bad homozygous effects were apparently beneficial in heterozygous condition. When the same comparison was made on a heterozygous genetic background, however, the apparent heterosis disappeared. The phenomenon of heterosis of new mutations on a homozygous background, but not on a heterozygous background, appeared again when Mukai, Yoshikawa, and Sano (1966) looked at radiation-induced mutations. There was a heterosis of 3 percent after treatment with 500 R, and effect larger than that found by Wallace, provided the genetic background was homozygous for chromosomes derived from the natural population. The heterosis disappeared, indeed there was no significant effect of any kind, when the background was heterozygous or when it was homozygous for an arbitrary inbred line from the laboratory.

Mukai argues from his results that there is an "optimum" level of heterozygosity, a hypothesis that Wallace (1958b), too, introduced for the interpretation of his results. It would also appear that a considerable epistasis of yet another kind must be operating if Mukai's results are generalizable, since it is not only the *level* of homozygosity that determines the overdominance of new mutations, but its *identity* as well.

What are we to make of these experiments on new mutation? The results of Falk, Wallace, and Mukai are at variance with each other in spirit, if not statistically, with the largest and most exhaustively analyzed experiment (Wallace's) leaving the whole matter in doubt. If there is an optimum level of heterozygosity, what is it? Do the few mutations per chromosome induced by Wallace and Mukai bring the genome close to that level? If so, the classical school is vindicated, since even in Mukai's spontaneous-mutation experiment the average number of mutations accumulated per chromosome is between 4 and 5, using Mukai's (1964) estimate of 0.1411 as the gross mutation rate per chromosome per generation for viability modifiers. If there are specific epistatic interactions between particular background homozygotes and the overdominance of new mutations, is there any chance that the optimum heterozygosity level could be estimated? In view of the immense effort that these experiments required, and the uncertain result they produced, such work is not likely to be repeated and must remain suggestive but ambiguous in its meaning for the problem of heterozygosity.

EVIDENCE FROM SELECTION

We wish to assess the genetic variation among organisms for both retrospective and prospective reasons. To reconstruct the historical processes that have led to the present differentiation among species and populations, we need to know how different they are genetically and how much variation exists from individual to individual within populations. We need the same information if we are to make any sensible prediction about the biological future of populations and species, most especially if we are interested in their controlled breeding in plant and animal improvement. It is the first rule (and perhaps the only rule) of plant and animal breeding that some kind of assessment of genetic variation ought to be made before a tedious and expensive program of selection is undertaken. If the heritability of a character in the population to be selected is very low, for instance 5 percent or less, then special forms of selection, say family selection, must be used, or else some environmental regime that will cause a major increase in heritability must be sought. Additive genetic variation is, after all, the key to success in selection, at least to success in selection on an individual basis, and some kind of genetic variance in appreciable amounts is needed for selection on any scheme. If all the loci relevant to a trait are genetically invariant, or nearly so, then no form of selection, no matter how sophisticated, can succeed.

The argument about selection and variation can be turned around. Suppose artificial selection is practiced in a population and succeeds in changing, in a heritable way, the phenotypic distribution in the population. Then it follows that there must have been nontrivial amounts of genetic variation for that character in the population to begin with. It might even be argued that success in selection is all we really care about anyway, and that assessment of genetic variation is unimportant once a character has been demonstrated to be selectable. Such an argument ignores the fact that we cannot predict the eventual limits to selection or the way in which the heritability will change over the course of selection, so that any long-term view of plant and animal improvement necessarily requires information about allele frequencies and gene numbers.

Even though success in selection does not tell us everything we need to know about genetic variation, it does prove that genetic variation was present to be selected. What is remarkable about the history of experiments and of real practice of artificial selection is the high frequency of success. It is well known, indeed part of the lore of genetics, that selective breeding has been responsible for immense changes in the qualitative and quantitative characters of domesticated plants and animals. What is not so strongly stressed is that a good deal of the major increases in yield of agricultural organisms has been the result of startling changes in the technology of husbandry, including the availability of fertilizers and the mechanization and capitalization of agriculture. But even putting these major environmental changes aside, organisms can be selected for a great diversity of characters. Table 16, taken from two tabulations, shows a variety of characters in domesticated species in which there is sufficient heritability and variance for successful selection. It is not meant to be exhaustive, but only exemplary.

What does such a tabulation prove? Where are all the characters for which selection has been practiced unsuccessfully? As Gordon Dickerson points out, "Published reports of sluggish response to selection in animals are rare indeed, possibly due to editorial frowning or author reluctance concerning publication of negative results!" (1955). Dickerson was less reluctant than most, and his editor's brow remained unwrinkled as he reported negative results for selection on viability and egg production, equivocal progress in speeding

TABLE 16
Heritabilities of a number of traits in animals and plants

	Initial h^2
Cattle	
white spotting in Frisians	.95
butterfat percentage	.6
milk yield	.3
conception rate at first service	.01
Swine	
back fat thickness	.55
body length	.5
180-day weight	.3
litter size	.15
Merino sheep	
wool length	.55
fleece weight	.4
body weight	.35
Poultry (White Leghorn)	
egg weight	.6
age at first egg	.5
egg production of surviving birds	.3
body weight	.2
viability	.1
Corn	
plant height	.70
yield	.25
ear length	.17

Note: Adapted from Falconer (1964) and Brewbaker (1964).

up sexual maturity, and unexciting progress in egg weight over a period of 20 years in a closed flock of chickens. This lack of progress was not the result of the lack of additive genetic variance for the characters, however, since estimates of heritability for different characters ranged from 8 to 59 percent, based on correlations between relatives. It is probable that natural selection and negative correlations between characters, as well as major environmental changes (some years were poor for all characters) were responsible for the overall lack of selection progress. We will never know how many cases of failure in selection because of a lack of genetic variation have remained unreported. It is certain from Dickerson's article that he would not have reported his results if the cause of the failure had been so straightforward.

We are on more certain ground if we turn to controlled model selection experiments in Drosophila. It is a commonplace of Drosophila population genetics that "anything can be selected for" in a non-inbred population. Indeed, the commonness of this experience, the restricted size of the group of people working in Drosophila population genetics, and the mechanisms of informal but efficient dissemination of nonpublishable information (Drosophila Information Service and Drosophila Research Conferences) make it likely that a failure to select for a character would receive some notice. Of course, some cases must go unreported, but what is most important about the results in Drosophila is the enormous *variety* of traits that can be selected for rather than the frequency of success for any particular trait. The following list, which is far from exhaustive, is meant to provide an impression of the extraordinary variety of physiological, behavioral, and morphogenetic processes for which there is genetic variance in substantial quantities in outbreeding Drosophila populations.

Size. Both body size and wing size can be increased and decreased by selection, with the divergent lines differing by about 25 percent of the initial size after 15 generations of selection (Robertson and Reeve, 1952). Moreover, the divergence in wing length may be due to changes in either cell size or cell number (Robertson, 1959).

Bristle number. Both abdominal bristles (Mather and Harrison, 1949) and sternopleural bristles (Thoday and Boam, 1961) can be increased and decreased. The divergence between high and low lines reached two-thirds of the original character value in 25 generations for abdominals, and eventually the lines diverged by 100 percent of the original value.

Developmental rate. A divergence of 1.2 days was created between fast and slow lines of development from egg to adult as compared with the control development period of 17.4 days, after 11 generations of selection (Clarke, Smith, and Sondhi, 1961).

Fecundity and egg size. Bell, Moore, and Warren (1955) succeeded in selecting both increased and decreased fecundity, producing in 11 generations a divergence of 26 eggs per female per day as compared with the original average of 88 eggs. They also produced a 13 percent difference in egg size.

Behavior. Whereas *D. pseudoobscura* is normally somewhat negatively geotactic and negatively phototactic at low levels of nervous activity, strongly positive and negative geo- and phototactic lines have been produced by selection (Dobzhansky and Spassky, 1967). Selection for mating preference can be carried out by allowing free mating in a mixture of two mutant stocks and then destroying all hybrid progeny each generation. In this way, for example, Knight, Robertson, and Waddington (1956) changed the mating pattern of *ebony* and *vestigial* mutants from a random one to one in which the ratio of homogametic to heterogametic mating was 1.6 : 1.

Pattern. The normal pattern for Drosophila is one anterior and two posterior ocelli, accompanied by two anterior and four posterior ocellar bristles, but lines can be selected with only the two posterior ocelli and four posterior bristles (Maynard Smith and Sondhi, 1960). Wing venation pattern can be altered in a variety of ways by selection; for example, removal of the posterior cross vein (Milkman, 1964) occurred in up to 96 percent of individuals after 22 generations of selection.

"Invariant" patterns. Many characters show no phenotypic variation at all under normal genetic and environmental conditions. If normal morphogenesis is severely disrupted by an environmental shock (Waddington, 1953) or by a mutant gene (Rendel, 1959), phenotypic variation is induced which can then be selected. In this way Waddington produced populations with 80 percent of the halteres transformed into rudimentary wings (the so-called bithorax phenotype) and Rendel selected lines with six or more scutellar bristles in place of the normally constant four bristles.

Variance. Not only can the mean of a character be selected, but the variance can be directly selected without significantly altering the mean. In a remarkable series of selection experiments, Rendel (1967) showed that the entire distribution of phenotypic classes could be tailored to any arbitrary shape by selection. For example, Rendel and Sheldon (1960) canalized development of scutellar bristles around the mean of 2, doubling the probit distance spanned by that phenotype. Moreover, both the genetic and the environmental components of variation may be altered. Scharloo (1964) selected the extremes of the distribution (*disruptive* selection) or the mean of the distribution (*stabilizing* selection) of the length of the fourth wing vein, which had been shortened by the introduction of a mutant gene. The results after only six generations of selection were as shown in the accompanying table.

	Variance			
	Total	*Additive genetic*	*Fly*	*Residual*
Control	157.4	91.3	28.2	37.9
Stabilizing	91.2	45.6	29.7	15.9
Disruptive	510.9	449.6	26.8	34.5

After another seven generations of selection, the total variance of the disruptive line was 1100 and of the stabilizing line only 57.

Mutant expression. The experiments of Maynard Smith and Sondhi, of Rendel, and of Scharloo, described as being selection on pattern and on variance, also show that a major morphogenetic mutant can have both its average effect and its dominance altered by the selection of modifier genes. It is the general experience that morphogenetic mutants come more and more to overlap wild type in Drosophila in successive generations of laboratory culture, but that full expression can be restored by outcrossing the mutant stock to an unrelated line and then re-extracting the mutant.

Developmental sensitivity to environment. The number of eye facets in Drosophila is a decreasing function of temperature. This effect is more pronounced in strains with the mutant Bar. Waddington (1960) selected lines in which the difference in eye-facet number at high and low temperatures was considerably less (11.0) than the difference in controls (100.8).

Resistance. Resistance to DDT has been produced several times in Drosophila. In twice-replicated experiments, Bennett (1960) produced, after 15 generations of selection, resistant lines that had a median lethal dose 30 times that of the controls, and sensitive lines that were 5 times as sensitive as controls.

The genetic system. Even the method of reproduction or the pattern of recombination can be radically altered by selection in *Drosophila*. Although *Drosophila* is typically a sexually reproducing genus, Carson (1967) succeeded in selecting a self-sustaining parthenogenetic line of *D. mercatorum* in which 6.4 percent of the eggs developed successfully without fertilization, producing fertile females. The frequency of parthenogenesis in unselected lines was on the order of 10^{-3}.

Perhaps more remarkable is the success in selecting to increase and decrease the total amount of recombination in an arbitrarily selected region of the third chromosome of Drosophila (Chinnici, 1971 a, b). Whereas the normal map length of the segment is 15.4 units, selection increased it to 22.1 and decreased it to 8.5 yet had no effect on recombination in regions outside the interval directly selected for. Moreover, the genetic control of the recombination appeared to be spread throughout the genome. A much more striking increase in the same chromosomes was attained by Kidwell (1972), who doubled the recombination rate in the centromere region after only 12 generations.

There appears to be no character—morphogenetic, behavioral, physiological, or cytological—that cannot be selected in Drosophila. The only known failure is the attempt of Maynard Smith and Sondhi in their pattern experiment to select for left-handed flies. They did succeed, however, in increasing the asymmetry, although it was a fluctuating asymmetry not biased toward the right or left.

The suggestion is very strong, from the extraordinary variety of possible selection responses, that genetic variation relevant to all aspects of the organism's development and physiology exists in natural populations. Although it is not proven that genetic variation for all characters is present in any single population, there has been no special search for suitable populations in which to carry out each selection experiment. On the contrary, investigators have used whichever material was convenient, so there is good reason to suppose that any outbred population or cross between unrelated lines will contain enough variation with respect to almost any character to allow effective selection.

TABLE 17

The realized heritability in four schemes of selection for abdominal and sternopleural bristles in *D. melanogaster*

		Abdominals	*Sternopleurals*
Tandem		.37	.45
Index	rep. A	.25	.32
	rep. B	.32	.30
Independent	rep. A	.34	.40
	rep. B	.26	.29
Separate		.21	.32

Note: From Sen and Robertson (1964).

Moreover, the genetic variation with respect to different characters is probably at different sets of loci. That this is so is suggested by an experiment on selection for two different bristle characters in the same population (Sen and Robertson, 1964). Sternopleural and abdominal bristles were selected in the same line by tandem selection (characters selected in alternating generations), independent selection (characters selected simultaneously in every generation), and index selection (the two character values were added to give one index), and these were compared with the effect of separate selection on the two bristle characters in separate lines. Table 17 shows the realized heritability for each character under the different schemes of selection. Apparently it makes no difference whether the characters are selected separately or together. More experiments of this sort would be most illuminating, as would experiments in which the effect of a long history of selection of one character on the response to selection for some other character was examined.

In general, the results of artificial selection experiments remain the strongest evidence we have of widespread genetic variation for genes that are relevant to characters of adaptive significance. These results do *not* prove, however, that large numbers of genes are segregating relevant to any particular character. It is sometimes mistakenly assumed that a slow and steady selection response for a quantitative character is evidence of the segregation of many genes of small effect. But this is certainly not the case. It is impossible to make any statement about the number of loci segregating from the speed of selection response. Even one locus will provide a slow and steady response if heritability is low or if alternative alleles at the locus are near fixation. Nevertheless, if nearly any character can be selected for rather easily in nearly any population, then it is certain that a large number of genes of different function are segregating in natural populations, even if these genes collectively do not represent a large fraction of the total genome. Certainly the most extreme form of the classical hypothesis, which allows only a handful of rare mutations to be heterozygous in each individual, is contradicted by the selection results. Some substantial number of loci contributing to adaptive morphological and physiological characters must be segregating at intermediate allelic frequencies.

CHAPTER 3 / **GENIC VARIATION IN NATURAL POPULATIONS**

A METHODOLOGICAL PROGRAM

By characterizing the struggle to measure variation as a struggle between two opposing views, the classical and balance theories of population genetics, I have in part obscured the underlying problem of making a functional whole out of theory and observation. Suppose that the experiments of Wallace and of Mukai had proved beyond any reasonable doubt that new mutations were, on the average, heterotic, or that we found the evidence from the great variety of selection experiments absolutely compelling. Then the balance hypothesis would be admitted and we would know that there was a tremendous amount of genetic heterozygosity at most loci. Yet even then we would be no closer to an empirically and dynamically sufficient genetic description of populations.

Knowing that populations in general must be highly heterozygous will not tell us how fast or how far natural selection can go in changing a particular character, nor will it allow us to make any inferences about the past history of populations and geographical races, or about the genetic processes in speciation. The answers to these latter questions, which comprise the task that evolutionary geneticists have set themselves, require the estimation of the quantities that appear in population genetic theory; they require the characterization of *gene frequencies*. The whole of population genetic theory remains an abstract exercise unless the frequencies of alternative alleles at various loci can be determined in different populations and at different times in the history of a given population. We cannot escape from the fact that the theory of evolutionary genetics is a theory of the historical changes in the frequencies of genotypes. The observations of evolutionary genetics must, then, also be observations on the frequencies of genotypes and their variations in time and space.

I have already discussed at length (p. 21) the deep structural difficulties that face us in any attempt to estimate allelic frequencies for evolutionarily significant variation. For phenotypes of evolutionary interest, like size, shape, metabolic rates, and probabilities of survival and reproduction, the average effects of gene substitutions are small compared with the variation from environmental fluctuation. The counting of genotypes in a population, however, requires that the differences in phenotype produced by allelic substitutions be large enough to allow an unambiguous classification of individuals into genetic classes. Any technique that is to enumerate genotypes in populations must satisfy the following program:

1. Phenotypic differences caused by the substitution of one allele for another at a single locus must be detectable as an unambiguous difference between *individuals*.

2. Allelic substitutions at one locus must be distinguishable in their effects from allelic substitution at other loci.

3. All, or a very large fraction of, allelic substitutions at a locus must be detectable and distinguishable from each other, irrespective of the intensity or range of their physiological effects.

4. The loci that are amenable to attack must be a random sample of genes with respect to physiological effects and with respect to the amount of genetic variation that exists at the locus.

Requirements 1 and 2 really amount to requiring that there be a one-to-one correspondence between phenotype and genotype so that ordinary Mendelian genetic analysis can be carried out on the phenotypes and so that allelic frequencies can be estimated in populations by counting individuals. It is important that gene substitutions be detectable even in heterozygous condition, that there be imperfect dominance. Although Mendelian analysis obviously can be carried out on alleles showing complete dominance (after all, Mendel did it), dominance makes population sampling much more difficult and introduces a bias. If, for example, a dominant allele were in frequency 0.8 in a population and ten different recessive alleles were in frequency 0.02 each, the dominant phenotype would make up 96 percent of the population and no one of the recessive homozygotes or combinations would be as frequent as 0.1 percent. In any reasonable sample the variation at the locus would be greatly underestimated, and in a sample of 100 individuals, for example, there would be a 30 percent chance that all would have the dominant phenotype.

Requirement 3 is our demand that the detectable variation be the "stuff of evolution," the genetic basis of the subtle changes in development and physiology that make up the bulk of evolutionary change. This demand is, at first sight, in contradiction to requirements 1 and 2 and lies at the root of our methodological problem.

Requirement 4 arises because we can never hope to characterize the tens or hundreds of thousands of genes in the genome of a higher organism, so we must be able to sample a relatively small number of loci as being representative of the kind and amount of genetic variation in the genome as a whole. At first sight, requirement 4 is paradoxical, for it demands that we sample loci irrespective of their variation in a population. An unbiased estimate of the amount of heterozygosity in a population must include a proportionate sample of those loci that are invariant. But the science of genetics is built upon *differences*. If all organisms were identical and showed no heritable variation, there would be no science of genetics; indeed, there would be no problem of genetics. While requirements 1 and 2 are in conflict with requirement 3, requirement 4 is in contradiction to Mendelian genetics as a whole. It is little wonder that the struggle to measure variation has been such an unhappy one.

The sorts of variation reviewed in chapter 2 fail in one way of another to satisfy the demands of a correct method. Visible mutations and lethals are in reasonable accord with requirements with 1 and 2 although the problems of dominance and the necessity for allelism tests of lethals, and even of visibles, make gene counting tedious. However, lethals and visibles are neither a random sample of allelic substitution nor a random sample of loci since they are of such drastic effect. The same objection applies to the classic visible polymorphisms such as banding in snails (Lamotte, 1951), pattern polymorphism in Lepidoptera (Ford, 1953) and ladybirds (Timofeeff-Ressovsky, 1940), or to strongly selected biochemical polymorphisms such as sickle-cell anemia or thalassemia in man (Allison, 1955). The issue of whether it is possible and reasonable to extrapolate from visibles and lethals to substitutions of small effect lies at the basis of the disagreement between the classical and balance schools. On the other hand, the distribution of chromosomal viabilities and fertilities certainly includes a broad range of both drastic and subtle effects, but the method of chromosomal replication makes it impossible to detect single allelic substitution at single loci (requirement 1) and was specially designed to cope with the fact that the genotype of single individuals could not be characterized (requirement 2).

A hint of the kind of observations that would satisfy our methodological program is given by human blood group polymorphisms. The technique for the detection of these polymorphisms, reaction against specific antisera, resolves the methodological contradiction between the demand for Mendelizing distinguishable loci and the demand for a random sample of allelic effects. The classification of individuals according to their blood group genotypes is unambiguous, except for certain problems of dominance, and there is no confusion between loci. The physiological and developmental functions of human blood groups are unknown, and the method for detecting them does not depend on their physiological effect but only on the molecular configuration of glycolipids bound to the red-cell membranes. These differences are presumably the direct result of small differences in the glycosyl transfer enzymes coded by the blood group genes, so that primary gene effects are being detected, irrespective of the subsequent biological effects of the blood group substance. Although the immune reaction undoubtedly detects even small changes in molecular configuration, we are not certain that it detects all of them; at any particular stage of the development of technique, some genetic variations go unnoticed. Thus, the difference between the A_1 and A_2 subgroups was not worked out until 30 years after the discovery of the ABO blood groups.

Unfortunately, the antigen-antibody method of detecting blood groups guarantees that criterion 4 of the methodological program will not be fulfilled, because the detection of a locus for a blood group substance depends absolutely on the existence of a difference. A total of 33 genes have been detected that code for human blood groups, because at least one variant individual has been discovered for each gene. But how many genes are there for which no variants have been discovered? Since the detection of a blood group depends upon the existence of antigenic variation within it, we cannot know. Once again, the methodological program is too demanding.

THE METHOD

The solution to our dilemma lies in the development of molecular genetics. It appears certain that the sequence of nucleotides that make up a structural gene is translated with a high degree of accuracy into a sequence of amino acids making up a polypeptide chain.* Therefore, putting aside redundant nucleotide substitutions, any change in the base sequence of a DNA molecule will be reflected in a substitution, deletion, or addition of an amino acid in the polypeptide chain coded by the gene in which the alteration has occurred. Since the polypeptide chains from a single gene or occasionally from two genes make up a species of enzyme or structural protein, any change in the amino acid sequence of an enzyme or structural protein can be directly ascribed to an allelic substitution at the locus coding for the polypeptide in which the change has occurred. Moreover, if an individual is a heterozygote, both forms of the enzyme or protein will appear since both forms of the gene will be transcribed and translated into protein. The exceptions to this rule of no dominance are the chain-terminating mutations or other DNA changes that interrupt or suppress transcription or translation (which represent only 3 percent of all single-step base substitutions) and the rare severe deletions.

The amino acid sequence of proteins is a phenotype that satisfies all the requirements of the methodological program. A single allelic substitution is detectable unambiguously since it results in a discrete change in phenotype—the substitution, deletion, or addition of an amino acid. Except for redundant code changes, which are irrelevant for our problem, every substitution is detectably different, and the gene effects of different loci cannot be confused with each other

*But not necessarily with *perfect* accuracy. Thus, suppressor genes are known to introduce a high noise level into the translation of all other genes, by coding for unusual transfer RNAs.

since they code different proteins. The physiological or morphogenetic effect of allelic substitution, on the other hand, may be arbitrarily small and environment may interact in an arbitrary way with the gene products to determine the total phenotype of the organism, yet this interaction will be irrelevant to the detection of the genetic difference. The conflict between the discrete phenotypic effects demanded by Mendelism and the subtle phenotypic differences relevant to evolution is resolved by looking directly at the gene products and not at their physiological and morphogenetic effects.

Finally, the apparent paradox of trying to detect invariant genes is also resolved. Most enzymes and proteins are the products of single genes, although some, like hemoglobin and lactate dehydrogenase in vertebrates, may be composed of polypeptide chains from two different genes, joined by covalent or weaker bonds. If a particular protein should be invariant in its structure in a population, this can be equated to an invariant gene as a first approximation, although it might, in fact, indicate two invariant genes. Molecular genetics, by equating one gene to one polypeptide and thus, in most cases, to one protein, makes it possible to detect genes without variation. Genetics, which began by considering the hereditary transmission of differences between organisms, has reached the stage where it no longer depends for its inferences upon those differences. The conditions for progress in genetics in the past have been eliminated by the very progress that those conditions made possible.

How can we turn this theoretical knowledge about the amino acid sequence of proteins into a practical program for measuring variation? At the present time it is not feasible to use the primary amino acid sequence of proteins directly as a phenotype. To determine the phenotype of an individual with respect to a single locus would require a high stage of purification of the protein in question, followed by a sequence determination, or at least a peptide "fingerprint" study. Current technology is simply not far enough advanced to allow such a procedure to be carried out on hundreds of individuals for scores of proteins, with a reasonable expenditure of labor. We are forced to consider some characterization of proteins that is sensitive to single amino acid substitutions but allows reasonably rapid examination of large numbers of individuals and many proteins.

One possibility is immunological techniques, by which many impure samples can be tested against a standard antiserum induced against a highly purified and homogeneous protein. There are two difficulties with this procedure. First, although it might be possible to detect differences from the standard, it is much more difficult to characterize differences among variants, and impossible to tell heterozygotes from homozygotes. Second, antigen-antibody reactions against single amino acid substitutions differ quantitatively rather than qualitatively, so the clear-cut and simple amino acid sequence difference is converted into a continuously varying character. For example, Salthe (1969) attempted to use micro-complement fixation to characterize genetic variation for lactate dehydrogenase in frogs, and although he certainly demonstrated heterogeneity within and between populations, the resolution of the technique was inadequate for classifying genotypes.

The alternative to this biological detection scheme is to use the physicochemical properties of the proteins, and it is to this alternative that geneticists have turned.

When condensed into a polypeptide chain, amino acids are in the general form

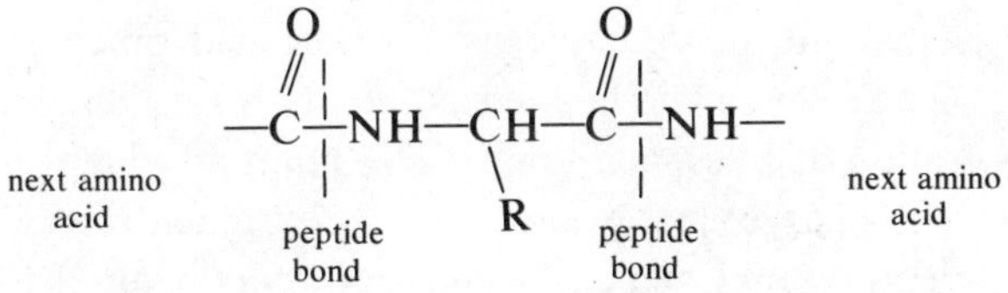

```
                    O                 O
                   //|               //|
                 —C—|—NH—CH—C—|—NH—
   next amino        |        \        |        next amino
      acid           |         R       |           acid
                  peptide          peptide
                   bond              bond
```

where R is specific to the amino acid. The 20 common amino acids in proteins fall into three groups with respect to this R side chain. Sixteen of them, listed at the left in table 18, have non-ionizable (nonpolar) R chains. They are electrostatically neutral. Two,

TABLE 18
Classification of amino acids by side groups, with standard abbreviations

Neutral	*Neutral*	*Basic (positive)*
Alanine (Ala)	Methionine (Met)	Arginine (Arg)
Asparagine (Asn)	Phenylalanine (Phe)	Lysine (Lys)
Cysteine (Cys)	Proline (Pro)	
Glutamine (Gln)	Serine (Ser)	
Glycine (Gly)	Threonine (Thr)	*Acidic (Negative)*
Histidine (His)	Tryptophan (Try)	Glutamic acid (Glu)
Isoleucine (Ile)	Tyrosine (Tyr)	Aspartic acid (Asp)
Leucine (Leu)	Valine (Val)	

arginine and lysine, have an ammonia group that is in a dynamic equilibrium between a neutral and a positively charged form, depending on the concentration of hydrogen ions in its immediate milieu

$$H^+ + \text{—C—NH}_2 \rightleftharpoons \text{—C—NH}_3^+$$

The third group, consisting of aspartic and glutamic acids, has a carboxylic acid R group and so is in a dynamic equilibrium between neutral and negatively charged forms

$$H^+ + \text{—C—C(=O)—O}^- \rightleftharpoons \text{—C—C(=O)—OH}$$

A polypeptide made up of a mixture of the three types of amino acid will have a net negative or positive charge, depending on the balance of charges and the folding of the molecule. As the pH is lowered (hydrogen ion concentration is increased), more and more of the NH_2 groups will become positively charged NH_3^+ ions, while the acidic COO^- ions will be saturated and become neutral. The result is that the polypeptide as a whole will take on a positive charge. The reverse will happen as the pH is raised (decreasing the concentration of hydrogen ions). The point at which the negative and positive charges just balance out to give a neutral polypeptide is the iso-electric point, which for most proteins in animals is slightly alkaline, around pH 8.

If an allelic change at a locus results in the replacement of an amino acid in one group in table 18 with an amino acid from another, the iso-electric point of the protein will be altered, as will the net charge on the protein at any given pH. For example, a single-step change in the DNA codon AAC to the codon AAA results in the substitution of the positively charged lysine for the neutral asparagine. An even more drastic single-step change is from AAG to GAG, with resulting substitution of a negatively charged glutamic acid for the positively charged lysine. Such changes in net charge can be used to separate proteins and thus to identify the products of allelic forms of the same gene. The technique by which this separation is achieved is *gel electrophoresis*.

Figure 7 is a diagram of a typical gel electrophoresis apparatus during the course of an experiment. It consists essentially of a slab

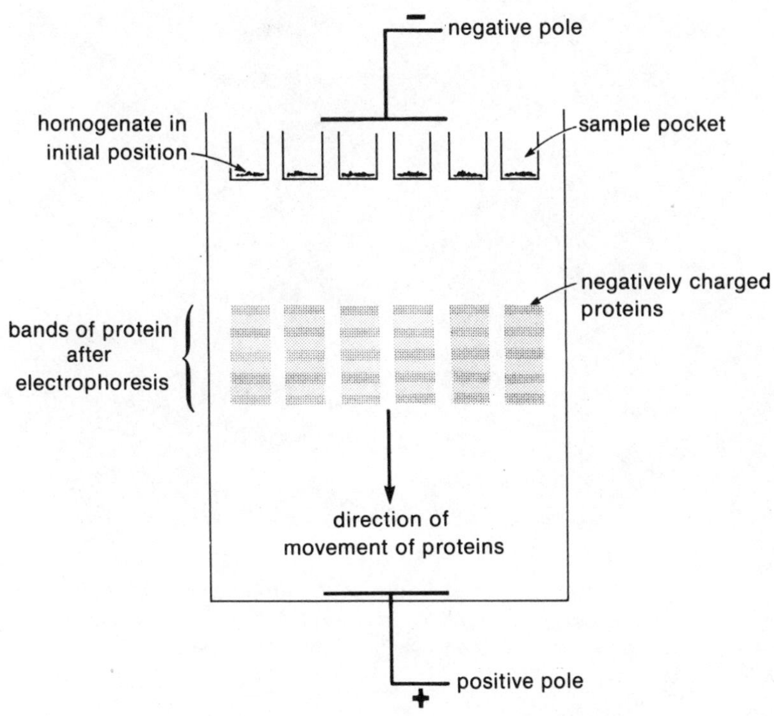

FIGURE 7

Diagram of a vertical slab gel-electrophoresis aparatus.

of some jellylike material (starch, agar, or a synthetic polymer) whose two ends are in contact with the opposite poles of an electric potential. Material for electrophoresis is introduced into wells at one end of the gel and any charged molecules will move down along the gel under the force of the electric field applied. The gel is enclosed in a cooling jacket to prevent its overheating and the consequent denaturation of the proteins moving through it. In practice, a bit of tissue, or, in the case of Drosophila, a whole individual, is ground up to break down the cells, the solid material is centrifuged or otherwise removed, and the liquid phase containing a mixture of the soluble proteins of the organism is placed in one of the slots in the gel. The pH of the gel and the grinding solution is adjusted to be on the alkaline side of the iso-electric point of the proteins being studied, so that the proteins will be negatively charged and migrate toward the positive pole of the apparatus. The speed of migration of any particular protein will depend upon its molecular size and the

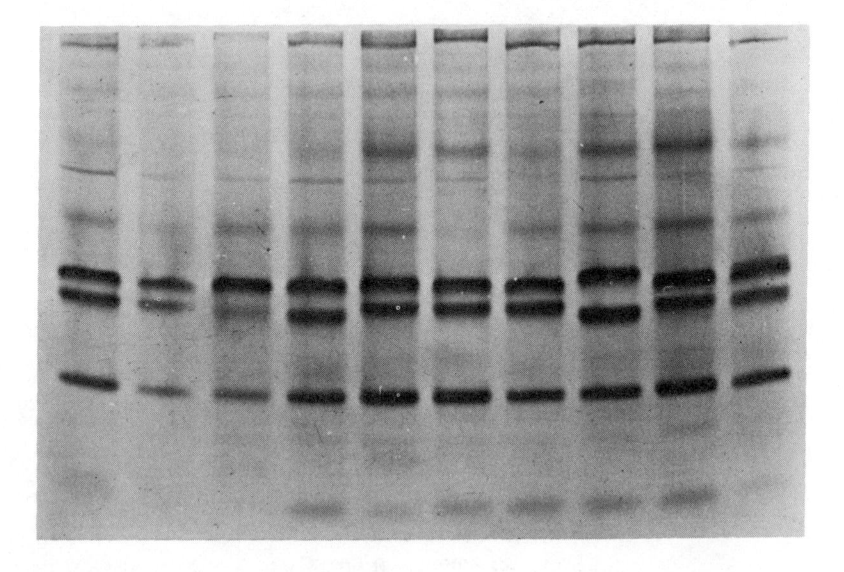

FIGURE 8
Larval hemolymph proteins of *D. pseudoobscura* after electrophoresis, stained with coomassie blue. Each vertical column contains the proteins of a single larva, each band being a different protein.

net charge it carries. After a high voltage has been applied to the gel for one to two hours, the various proteins that migrated at different rates will be concentrated at different points along the gel, as indicated in the diagram, although of course they are invisible. The problem then is to visualize them.

Proteins in high concentration can be seen simply by staining the gel with a dye that is a general protein stain. Figure 8 shows such a gel made from *Drosophila pseudoobscura* larvae. There are a large number of bands, and the high repeatability of the method is shown by the identity of the pattern of bands from larva to larva.

Enzymes, however, generally exist in such low concentrations that there is insufficient material in a single fly or tissue sample to be visualized with a general protein stain. Each enzyme can be found on the gel by immersing the slab in a bath containing a substrate for that enzyme and a dye that will be bound or colorized at the place where the substrate is split. For example, the oxidized form of a dye may be colorless, but may become colored if it is reduced by electrons, which are passed to the dye when a dehydrogenase enzyme

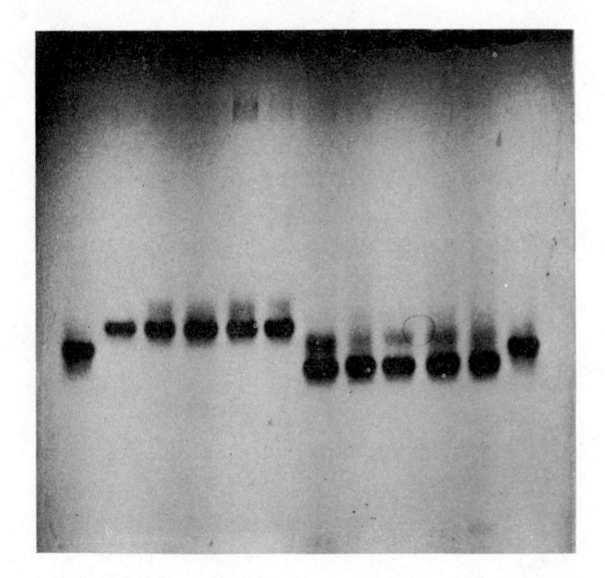

FIGURE 9

Adult esterases from *D. pseudoobscura. Sample 1,* standard strain; *samples 2–6,* five individuals from a second strain; *samples 7–11,* five individuals from a third strain; *sample 12,* standard strain. From Hubby and Lewontin (1966). Reprinted with permission of *Genetics.*

and its cofactor split. Although many reactions are more complicated and require various intermediates, the principle is the same for all. The result is a colored band in the gel at the place where the particular enzyme is concentrated. Several different enzymes can be visualized on the same gel, provided the conditions for the reactions are not mutually exclusive or inhibitory. Figure 9 depicts a gel of *D. pseudoobscura* adults stained for an esterase enzyme. Each sample is an individual fly. Samples 1 and 12 are a standard strain, while samples 2 to 6 and 7 to 11 are from two different homozygous strains. Again we see the high repeatability within strains and a clear-cut difference between strains, which turned out, on later analysis, to be the result of allelic differences at a single locus. Thus the electrophoretic mobility of an enzymatic protein is a repeatable phenotype that gives discrete, unambiguous differences between genotypes.

The method is capable of distinguishing different homozygotes not only from each other but also from heterozygotes. Figure 10 is the result of comparing individuals of a strain segregating at the

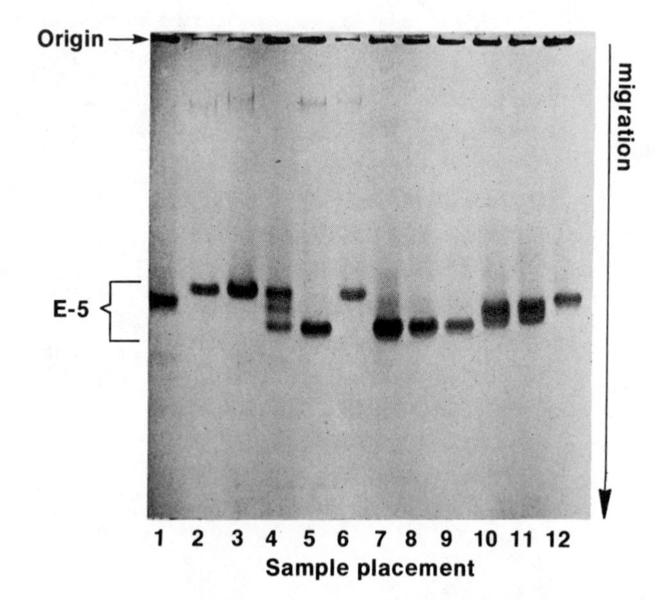

FIGURE 10

Adult esterases from *D. pseudoobscura*. Sample 1, standard strain *est-5*$^{1.00}$/*est-5*$^{1.00}$; samples 2 and 3, *est-5*$^{0.95}$/*est-5*$^{0.95}$; sample 4, *est-5*$^{0.95}$/*est-5*$^{1.12}$; sample 5, *est-5*$^{1.12}$/*est-5*$^{1.12}$; sample 6, *est-5*$^{0.95}$/*est-5*$^{0.95}$. From Hubby and Lewontin (1966). Reprinted with permission of *Genetics*.

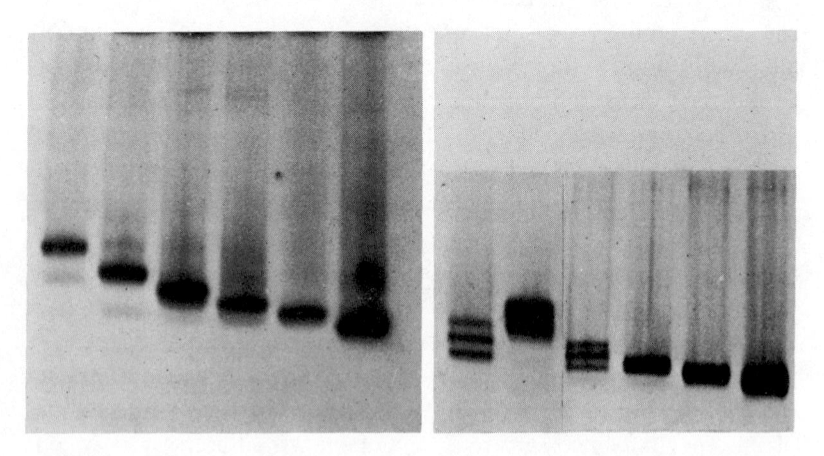

FIGURE 11

Allozyme phenotypes. Left, homozygotes for six different alleles at the *esterase-5* locus in *Drosophila pseudoobscura*; right, several different heterozygotes between alleles. From Hubby and Lewontin (1966).

esterase-5 locus in *D. pseudoobscura*. Sample 1 is again a homozygote standard strain, and the allele it carries has been designated $est\text{-}5^{1.00}$, the superscript standing for the relative electrophoretic mobility of the protein specified by that allele. Samples 2, 3, and 6 have a more slowly moving protein, while sample 5 moves faster than the standard. Sample 4 shows three bands, two identical with the fast- and slow-moving forms and one intermediate between them. Further genetic analysis of individuals from the strain from which this last sample comes shows that it is segregating for two alleles, $est\text{-}5^{0.95}$ and $est\text{-}5^{1.12}$, and that sample 4 is the heterozygote, $est\text{-}5^{0.95}/est\text{-}5^{1.12}$. The presence of the intermediate band in the hybrid indicates that esterase-5 is a dimeric enzyme, so that in heterozygotes three sorts of dimers are produced: homo dimer *0.95—0.95*, with the same mobility as the dimer made by the $est\text{-}5^{0.95}/est\text{-}5^{0.95}$ homozygote; homo dimer *1.12–1.12*, with the same mobility as the dimer made by the $est\text{-}5^{1.12}/est\text{-}5^{1.12}$ homozygote; and a hetero dimer *0.95–1.12*, with a mobility halfway between. Even if the enzyme were monomeric so that "hybrid" molecules were not produced, a heterozygote would still produce two different forms of the enzyme, each corresponding to one of the homozygotes. There is no dominance at the level of production of the polypeptides, and each allele in the heterozygote functions to make its product. The different enzyme forms produced by different alleles at the same locus have been called *allozymes* by Prakash, Lewontin, and Hubby (1969) to distinguish them from the more general phenomenon of *isozymes* that are the different molecular forms of an enzyme arising from any cause (Markert and Møller, 1959). Figure 11a shows the allozymes of homozygotes from six different alleles recovered from a population of *D. pseudoobscura*, and figure 11b gives some of the heterozygous combinations between them. When the mobility of two allozymes is very close, the three bands in the heterozygote cannot be resolved but appear as a single thick band. Nevertheless, heterozygotes are distinguishable from homozygotes.

THE FIRST APPLICATIONS OF THE METHOD

The fourth point in our methodological program requires that we score loci at random with respect to the amount of variation present in populations. All that is required in order to apply gel electrophoresis of proteins and enzymes to the problem of the amount of heterozygosity in natural populations is a large repertoire of dye-coupled enzyme-substrate reactions that have been accumulated for reasons other than the study of polymorphisms. By about 1964, an appropriately large repertoire existed, the methods having been developed to study the ontogeny of enzymes, their tissue localization, their sensitivity to inhibitors, their molecular weight, their purification, but not, in general, their genetic variation. Yet, in the course of accumulating these methods, a large amount of genetic variation turned up incidentally. A survey of the literature by Shaw (1965) recovered 16 different enzymes in some 20 species of organisms, including flagellates, insects, amphibia, birds, and mammals, that had revealed some variation in electrophoretic mobility. Of 58 cases, 35 were reported as "polymorphic," but the meaning of this term was vague in most cases since genetic studies had not been done, or population samples had not been taken. Nevertheless, the stage had clearly been reached at which an appropriate tool, gel electrophoresis, existed to fulfill the exacting methodological program required for an answer to the central problem of evolutionary genetics.

Two groups of workers simultaneously, and completely independently, set about estimating the heterozygosity and allelic distributions in natural populations by carrying out large-scale surveys of electrophoretic variation in randomly chosen enzymes and proteins. One group, in London, studied ten randomly chosen human enzymes in the English population (Harris, 1966). They found three to be polymorphic. Red-cell acid phosphatase had three alleles with frequencies 0.36, 0.60, and 0.04; phosphoglucomutase had two alleles with frequencies 0.74 and 0.26; and adenylate kinase (not yet completely analyzed in 1966) had two alleles at frequencies 0.05 and 0.95. The proportion of individuals heterozygous for these three loci was then 0.509, 0.385, and 0.095, respectively; averaging these with the seven monomorphic loci gives an estimate of heterozygosity in the English population of 0.099.

The second group, in Chicago, made a more extensive study in *Drosophila pseudoobscura*, including eight enzymes and ten larval hemolymph proteins (Hubby and Lewontin, 1966; Lewontin and Hubby, 1966). Their initial survey was not made on natural populations directly, but on 43 strains derived from single fertilized females taken from 5 natural populations. Of these strains, 33 had been in the laboratory for approximately five years (75 generations) and the remaining 10 were the F_2 and F_3 generations from newly trapped wild females.

Table 19 shows the results of this study in detail. The most striking feature that is immediately apparent is the great genetic variation present in these strains. Of the 18 different gene loci represented, 7 have electrophoretic variants and, of these, 5 loci have three or more alleles represented. This variation is all the more striking because of the small number of strains examined and the fact that, with the exception of *alkaline phosphatase-4*, all the variable loci have more than one variant strain, from more than one population. Clearly, the variation is widespread in natural populations. An interesting feature of the variation is the persistence of some segregation in strains that have been maintained in the laboratory for 75 generations with repeated episodes of breeding from very small numbers of parents. Among the variable loci, the sample from Strawberry Canyon, California, fresh from the wild, is segregating in 43 percent of the strains although among the strains with long laboratory histories only 9 percent are still segregating. Yet this persistent segregation is not at random over loci, most of it being at the *pt-7* and *pt-8* loci.

An extremely important aspect of the observations in table 19 is that, on genetic analysis, the variation between strains turned out in every case to be genetic and unifactorial. That is, *for every variable protein and enzyme, the variation was the result of the segregation of alleles at single loci. This fact is the cornerstone of the method.* It allows us to calculate allelic frequencies at each locus and therefore heterozygosities in the population, but most important, it allows us to equate each of the invariant proteins, as a close approximation, to an invariant locus. Unless we can make such an equation, the entire procedure is useless for estimating the proportion of loci that is polymorphic and the amount of heterozygosity per locus that exists in a population, since the number of monomorphic loci would be unknown. Genetic analysis is therefore essential, especially in view of the fact that minor variations in experimental conditions can in some cases cause enough variation in electrophoretic mobility to obscure or mimic genetic variation (G.B. Johnson, 1971).

From the data of table 19, allelic frequencies in each population

TABLE 19

Number of strains from each population of *Drosophilla pseudoobscura* that are either homozygous or segregating for various alleles at different loci

Locus	Allele	Strawberry Canyon[a]	Wild Rose[a]	Cimarron[b]	Mather[a]	Flagstaff[c]	Bogotá[d]
esterase-5	.85	0	0	0	1	0	0
	.95	0	1	0	1	1	0
	1.00	0	3	3	0	4	1
	1.03	0	1	0	2	0	0
	1.07	0	0	2	1	4	0
	1.12	0	1	0	2	0	0
	.95/1.00	1	0	0	0	0	0
	.95/1.07	1	0	0	0	0	0
	.95/1.12	0	0	1	0	0	0
	1.00/1.07	4	1	0	0	0	0
	1.00/1.12	3	1	0	0	0	0
	1.03/1.07	1	1	0	0	0	0
	1.03/1.12	0	1	0	0	0	0
	1.07/1.12	1	0	0	0	0	0
malic dehydrogenase	.90	0	0	0	1	0	0
	1.00	6	10	6	4	8	1
	1.11	2	0	0	0	0	0
	1.22	0	0	0	0	1	0
	.90/1.00	0	0	0	2	0	0
	1.00/1.11	2	0	0	0	0	0
glucose-6-phospate dehydrogenase	1.00	9	10	4	6	9	1
alkaline phosphatase-4	.93	0	0	0	0	1	—
	1.00	9	11	6	7	8	—
alkaline phosphatase-6	+	9	10	5	7	9	—

alkaline phosphatase-7	+	9	9	5	7	9	—
	−/+	0	1	1	0	0	—
α-glycerophosphate dehydrogenase	*1.00*	*10*	*10*	*6*	*6*	*8*	*1*
leucine aminopepidase	*.95*						
	.97	*2*[f] *alleles*	*3*[g] *alleles*	*2 alleles*	*2*[h] *alleles*	*3 alleles*	*1 allele*
	1.00						
	1.02						
pt-4	*.45*	*10*	*10*	*6*	*6*	*8*	*1*
pt-5	*.55*	*1*	*4*	*4*	*6*	*2*	*1*
pt-6	*.62*	*10*	*10*	*6*	*6*	*8*	*1*
pt-7	*.73*	*0*	*0*	*0*	*0*	*1*	*0*
	.75	*9*	*10*	*5*	*5*	*6*	*1*
	.77	*0*	*0*	*0*	*0*	*0*	*0*
	.73/.75	*0*	*0*	*0*	*0*	*1*	*0*
	.75/.77	*1*	*0*	*1*	*1*	*0*	*0*
pt-8	*.80*	*0*	*0*	*0*	*0*	*0*	*1*
	.81	*2*	*2*	*3*	*2*	*1*	*0*
	.83	*1*	*4*	*1*	*1*	*5*	*0*
	.81/.83	*7*	*4*	*2*	*3*	*2*	*0*
pt-9	*.90*	*3*	*8*	*4*	*1*	*0*	*0*
pt-10	*1.02*	*0*	*0*	*0*	*0*	*0*	*0*
	1.04	*4*	*9*	*6*	*4*	*8*	*0*
	1.06	*0*	*0*	*0*	*0*	*0*	*1*
	1.02/1.04	*0*	*1*	*0*	*0*	*0*	*0*
	1.04/1.06	*6*	*0*	*0*	*2*	*0*	*0*
pt-11	*1.12*	*4*	*10*	*6*	*6*	*8*	—
pt-12	*1.18*	*5*	*10*	*6*	*6*	*8*	*1*
pt-13	*1.30*	*7*	*10*	*6*	*6*	*8*	*1*

Note: From Lewontin and Hubby (1966).
[a]California. [b]Colorado. [c]Arizona. [d]Colombia. [e]Both loci segregating in the same strain. [f]Three strains segregating. [g]One strain segregating. [h]Two strains segregating.

can be estimated by assuming that the strains were roughly representative of the populations from which they were sampled. The frequencies are very rough estimates indeed, since so few strains were taken from each locality and since, except for Strawberry Canyon, there was a long history of laboratory culture. These frequency estimates, which are not worthy of serious consideration in themselves, can, however, in turn be used to estimate the proportion of the genome that is heterozygous in each population, a figure that is a great deal more reliable than the individual allele frequencies.

Table 20 gives both the proportion of loci polymorphic and the average heterozygosity per individual. A locus was counted as polymorphic only if a variant allele appeared in more than one strain out of all 43 examined. The variation is impressive. A third of all loci are polymorphic, and the average individual is a heterozygote at one out of eight of his loci, estimates that are remarkably similar to Harris's preliminary results for man. Even at those loci for which an individual is homozygous he is likely to differ from another randomly chosen individual in the population, since about one-third of all loci are segregating.

The estimates of variation become more impressive, indeed staggering, when one realizes that a majority of amino acid substitutions do *not* involve charge changes. From table 18 it is easy to show that 136 out of the possible 380 amino acid substitutions, or 36

TABLE 20

Proportion of loci, out of 18, that are polymorphic and proportion of genome that is estimated to be heterozygous in an average individual, for 5 populations of *D. pseudoobscura*

Population	*Proportion of loci polymorphic*	*Proportion of genome heterozygous per individual*
Strawberry Canyon, Calif.	.33	.148
Wild Rose, Calif.	.28	.106
Cimarron, Colo.	.28	.099
Mather, Calif.	.33	.143
Flagstaff, Ariz.	.28	.081
Average	.30	.115

Note: From Lewontin and Hubby (1966).

percent, involve a charge change. If we consider only single-base substitutions in DNA, then of the 399 nonredundant single-step missense mutations possible, 128, or 32 percent, involve a change in charge. These values take no account, however, of the inequalities in amino acid content of proteins. Although the empirical formulas differ considerably from protein to protein, a correction for amino acid content gives values of 26–28 percent for the proportion of substitutions that would be charge changes. But these calculations are of uncertain worth because in actual practice many neutral amino acid substitutions may make sufficient conformational or resonance changes to make the substitution detectable electrophoretically. For example, all of the substitutions in cytochrome *c* over the whole range of organisms from yeast to man, involving multiple substitutions at 70 percent of all amino acid sites, have not caused any change in the arithmetical sum of charges, all cytochromes having compensatory substitutions. Yet, with sufficient diligence, it is possible to distinguish these "neutral" substitutions electrophoretically. Given that Hubby and Lewontin did not undertake extensive investigations of various buffer systems and other technical alterations that might have further resolved apparent identities, we must assume that they detected only about a third of the total sequence variation actually present. On this basis, the average heterozygosity per locus for natural populations of *D. pseudoobscura* is about 35 percent, and essentially every gene is polymorphic. At least on the face of it, the classical hypothesis of population structure is firmly and directly refuted, and the balance theory is revealed as correct.

Even if we take Lewontin and Hubby's and Harris's values directly, without trying to correct for the undetected variation, natural populations of flies and men are immensely variable genetically, members of the same population and even the same family differing from each other at thousands of loci. The first direct measurements of genetic variation in natural populations have proved, in Dobzhansky's words, that the norm "is not a transcendental constant, standing above or beyond the multiform reality" (1955). As we shall see later, they may have proved too much.

SURVEYS OF VARIATION

The first estimates of heterozygosity in natural populations were followed by much more extensive and accurate surveys for both man and *D. pseudoobscura*, as well as by a large number of studies in other organisms. There is considerable variation in the kind of information that can be obtained from the different surveys, depending on their experimental methods and sampling techniques. One of the earliest, a preliminary report on what was eventually an extensive study of *D. ananassae*, showed 9 polymorphic loci out of 20, representing six different enzyme functions (Johnson, et al., 1966), but no gene frequencies were given.

A study of ten egg white proteins in domestic chicken flocks and breeds (Baker, 1968) showed that four loci were polymorphic in at least one breed and that 24 out of 37 breeds were segregating for at least one locus of the four. Because only a few, often only two or three, birds could be tested from most breeds, no quantitative estimate of heterozygosity can be made from the bulk of the sample. A few flocks were more extensively sampled, however, and the results are given in table 21. While allele frequencies may occasionally differ widely between two flocks of the same breed or between two breeds that are closely related, the average heterozygosity per individual is a characteristic of breeds rather than individual flocks. It is not at all clear what is meant by a "popula-

TABLE 21
Frequency of one of the two alleles segregating at each of the four polymorphic egg-white protein loci among ten loci tested in several flocks of domestic chickens

Flock	*n*	*Ovalbumin*^{S}	G_1^A	G_2^D	Tf^S	*Heterozygosity per individual*
Australorp A	19	1.00	0.63	1.00	1.00	0.047
B	18	1.00	0.17	0.94	1.00	0.040
Lt. Sussex A	51	1.00	0.49	0.61	0.71	0.139
B[a]	24	1.00	0.52	0.79	0.98	0.087
⎡Maran[b] A[a]	24	0.81	0.83	0.65	1.00	0.105
⎢B	30	0.77	0.87	0.42	0.95	0.116
⎣N. Holland Blue[a]	25	0.86	0.24	0.62	1.00	0.108
⎡Barnevelder[a]	22	1.00	0.86	0.52	1.00	0.074
⎣Welsummer[a]	15	1.00	0.63	0.93	1.00	0.060
Average						0.086

Note: From Baker (1968).
[a]Flocks kept by the same breeder.
[b]Bracketed breeds were derived from a common ancestral type in recent times.

tion" of domesticated chickens, and so it is difficult to say what universe is being sampled.

One difficulty with the interpretation of surveys is illustrated by the very large studies of Ruddle et al. (1969) and Roderick et al. (1972) on inbred and feral mouse populations. These workers studied 16 enzyme loci in 35 inbred strains collected from a variety of laboratories, and 17 loci in 6 collections of mice from the wild. Among the inbred lines, not a single locus was constant over all strains, 4 loci showed a single variant strain, and the remaining 12 loci varied among many strains. But the loci studied were chosen because they had already been shown in published reports to have some genetic variation. Thus it is impossible to assess the relevance of this variation for our problem. In the samples from natural populations, 5 out of 17 loci were variable, and the average heterozygosity per population varied from 7.6 to 14.7 percent. But again the loci were previously known to have some genetic variation in laboratory strains.

No survey of protein variation is completely free of difficulties and biases, but reasonably reliable estimates of genetic variation in natural populations require:

1. Adequate sample sizes, large enough to provide about 50 wild genomes for each locus.

2. Genomes tested either directly from natural populations or only a few generations removed from the original samples.

3. A *large* sample of loci, more important even than a large sample of individuals, because of the very considerable differences in heterozygosities at different loci. For example, Harris's original estimate of 0.099 for heterozygosity in man, from the first ten loci looked at, turned out to be 50 percent too large (Harris and Hopkinson 1972).

4. A *diverse* sample of loci, not weighted heavily by one or two enzymatic functions. It may be very misleading, for example, if in a sample of 14 loci, 6 code for enzymes that are nonspecific esterases, attacking the same range of nonbiological substrates (Ayala et al., 1970) or if 10 out of 18 genes code for high concentration globular proteins of unknown function (Lewontin and Hubby, 1966). Of course, it may not matter, but with our present rudimentary understanding of the distribution of variation over different kinds of loci, a survey that is spread over the widest possible range of enzymatic functions is preferable.

5. An *unbiased* sample of loci, not chosen consciously or unconsciously because of known variability. In the very first surveys this was not a problem, but as time goes on and the literature comes to be filled with reports on studies of enzyme polymorphism, it becomes progressively more difficult to choose a sample of loci in a new organism or population that is *not* tainted by the workers' knowledge about what has gone before. Geneticists like variation and find genetic uniformity rather dull. The excitement of seeing a new genetic segregation in a new organism is real and seductive. It is surely no accident that, in virtually every organism examined in the last six years, even when only a few enzymes are chosen the nonspecific esterases turn up and are nearly always polymorphic. Often the bias is unconscious and only appears as trends in data. If, for example, the estimated heterozygosity goes down as more and more loci are looked at in an organism, or if many populations have been examined (but only some loci in each) and a negative correlation between the heterozygosity and the number of loci tested is found, then there is strong reason to suspect an unconscious bias toward variable loci. A. D. Hershey is reported to have described heaven as "finding an experiment that works and doing it over and over and over." Population geneticists too have found paradise.

I have summarized in table 22 the data known to me from those allozyme surveys that seem to approach most closely the requirements of adequacy and unbiasedness. None is perfect, and others, not included, are nearly as good. I have chosen arbitrarily to include only surveys of at least 18 loci. The table shows the proportion of loci that are polymorphic, defined as the proportion of loci in which the most common allele does not exceed a frequency of 0.99, and the estimated heterozygosity per locus per individual in the population. Since the proportion of "polymorphic" loci is somewhat arbitrary and has a high variance (because a relatively small number of loci are tested), the heterozygosity per individual is the more informative figure and no survey has been included from which this value cannot be estimated.

The original results of Lewontin and Hubby and of Harris are completely confirmed. In a dozen species shown in table 22, between 20 and 86 percent of loci are segregating within a population, and the heterozygosity per individual falls in a relatively narrow range over all species, between 5.6 and 18.4 percent. The median

TABLE 22
Surveys of genic heterozygosity in a number of organisms

Species	Number of populations	Number of loci	Proportion of loci polymorphic per population	Hetero-zygosity per locus	Standard error of hetero-zygosity	Reference
Homo sapiens	1	71	.28	.067	.018	Harris and Hopkinson (1972)
Mus musculus musculus	4	41	.29	.091	.023	Selander, Hunt, and Yang (1969)
M. m. brevirostris	1	40	.30	.110		Selander and Yang (1969)
M. m. domesticus	2	41	.20	.056	.022	Selander, Hunt and Yang (1969)
Peromyscus polionotus	7 (regions)	32	.23	.057	.014	Selander et al. (1971)
Drosophila pseudoobscura	10	24	.43	.128	.041	Prakash, Lewontin, and Hubby (1969); Prakash, Lewontin, and Crumpacker (1973)
D. persimilis	1	24	.25	.106	.040	Prakash (1969)
D. obscura	3 (regions)	30	.53	.108	.030	Lakovaara and Saura (1971a)
D. subobscura	6	31	.47	.076	.024	Lakovarra and Saura (1971b)
D. willistoni	2–21	28	.86	.184	.032	Ayala et al. (1972)
	10	20	.81	.175	.039	Ayala, Powell, and Dobzhansky (1971)
D. melanogaster	1	19	.42	.119	.037	Kojima, Gillespie, and Tobari (1970)
D. simulans	1	18	.61	.160	.052	Kojima, Gillespie, and Tobari (1970)
Limulus polyphemus	4	25	.25	.061	.024	Selander et al. (1970)

proportion of polymorphic loci is 30 percent and the median heterozygosity per individual is 10.6 percent, with man somewhat on the low side and *Drosophila pseudoobscura* somewhat on the high side of the median variation. Thus we will not be far off if we characterize sexually reproducing species of animals as being polymorphic for a third of their genes, and individuals within the species as being heterozygous for about 10 percent of their loci. Again we must bear in mind that, for the genes sampled, these are *minimum* estimates since they are based on only those gene substitutions that are detectable electrophoretically.

The standard errors of average heterozygosity given in table 22 are meant to emphasize the importance of examining a large number of loci. These standard errors are calculated from the variance among loci, after the heterozygosities are averaged over populations within a species, and they are between one-fifth and two-fifths as large as the mean heterozygosities. Such large standard errors arise because different loci differ markedly in their genetic variation, with the majority being monomorphic, and the polymorphic minority being very heterozygous.

In table 23 the average heterozygosities for different loci in man and *Drosophila pseudoobscura* have been listed in ascending order. Although the range of heterozygotes for *Drosophila* is greater, there is a strong suggestion of bimodality among the 20 polymorphic genes in man. Because of such heterogeneity, it is impossible to interpret differences in heterozygosity between species until large numbers of loci have been tested. For example, two earlier studies of *D. melanogaster* and *D. simulans*, one based on 10 loci (O'Brien and MacIntyre, 1969) and one on 6 loci (Berger, 1970), gave heterozygosities for *D. melanogaster* of 0.23 in both cases, but only 0.07 and 0.00 for *D. simulans*. Ingenious explanations can be made for the apparent lack of heterozygosity in *D. simulans*, based upon the differences in population structure and economy of the two species. There is even a temptation to speculate about genetic heterogeneity as an adaptive strategy that is optimal for *D. melanogaster*, whereas homozygosity is a better strategy for *D. simulans*. But all such speculations are idle in view of the larger study of Kojima, Gillespie, and Tobari (1970), cited in table 22, which completely reverses the apparent facts, showing *D. simulans* to be more heterozygous than *D. melanogaster*. Nor is the difference in het-

TABLE 23
Average heterozygosity in populations of *Drosophila pseudoobscura* and man, for different loci

D. pseudoobscura		*Man*	
Locus	*Heterozygosity*	*Locus*	*Heterozygosity*
12 monomorphic loci	.000	51 monomorphic loci	.00
		Peptidase C	.02
Acetaldehyde oxidase	.012	Peptidase D	.02
		Glutamate-oxaloacetate transaminase	.03
Protein-7	.063		
Protein-13	.070	Leucocyte hexokinase	.05
Malic dehydrogenase	.102	6-Phosphogluconate dehydrogenase	.05
Octanol dehydrogenase	.109	Alcohol dehydrogenase-2	.07
		Adenylate kinase	.09
Leucine aminopeptidase	.155	Pancreatic amylase	.09
		Adenosine deaminase	.11
Protein-10	.229	Galactose-1-phosphate uridyl transferase	.11
Protein-12	.234		
α-Amylase-1	.353	Acetyl cholinesterase	.23
Xanthine dehydrogenase	.492	Mitochondrial malic enzyme	.30
		Phosphoglucomutase-1	.36
Protein-8	.513	Peptidase A	.37
Esterase-5	.741	Phosphoglucomutase-3	.38
		Pepsinogen	.47
		Alcohol dehydrogenase-3	.48
		Glutamate-pyruvate transaminase	.50
		RBC acid phosphatase	.52
		Placental alkaline phosphatase	.53

Note: Data from Prakash, Lewontin, and Crumpacker, (1972) and Harris and Hopkinson (1972).

erozygosity between the two species significant (0.160 as against 0.119), being less than one standard error.

Surveys of polymorphism have not yet reached the stage at which comparisons of total heterozygosity between species can be meaningful; if such comparisons are to be made for comparative evolutionary studies, something on the order of 100 randomly chosen loci will be required. Only in *Homo sapiens* has anything like a sufficient number of genes been studied, and unfortunately only Europeans have been extensively surveyed.

The variety of organisms in table 22 (man, mouse, Drosophila, and horseshoe crab) gives us some confidence in the generality of the result, and a number of less extensive surveys widen the taxonomic range even more, without contradicting the general picture. *Drosophila athabasca* and *D. affinis* (Kojima, Gillespie, and Tobari, 1970), *D. ananassae* and *D. nasuta* (Johnson et al., 1966; Gillespie and Kojima, 1968; Stone et al., 1968), and *D. mimica* (Rockwood, 1969), the harvester ant, *Pogonomyrmex barbatus* (Johnson et al., 1969), the snails, *Cepaea nemoralis* and *C. hortensis* (Manwell and Baker, 1968), the salamander, *Ambystoma maculatum* (T. Uzzell, unpublished data), the chicken (Baker, 1968), the pheasant, *Phasianus calchicus* (Baker et al., 1966), and the quail, *Coturnix coturnix* (Baker and Manwell, 1967) have all been shown to be polymorphic for a number of protein-specifying genes in natural populations, although the data do not allow accurate quantitative values to be calculated (comparable to those in table 22), usually because too few loci have been studied.

Thus far there has been no extensive survey of allozyme variation in plants, although the few studies that have been made leave no doubt that plants, too, are highly polymorphic. Marshall and Allard (1970) and Clegg and Allard (1972) have found clear polymorphism for several genes in two species of wild oats, *Avena barbata* and *A. fatua*, despite the fact that these are strongly inbreeding species. A survey of a population of *Oenothera biennis* revealed 5 out of 19 enzymes to be polymorphic, but every individual in the population was a heterozygote for these five genes, presumably because *O. biennis* is a permanent, multiple-translocation heterozygote and the polymorphic genes are markers of the translocated segments (Levin, Howland, and Steiner, 1972). *Taraxacum officinale*, a completely apomictic dandelion, also shows a good deal of allozyme variation between clones growing in different habitats (Solbrig, 1971), but no genetic analysis is possible.

With their immense variety of breeding systems, plants will be extremely important for comparative studies and for sorting out the forces influencing allozyme variation. They are also far more easily manipulated for population experiments, both in nature and in artificial cultivation, than are elusive and willful animals. In proportion to their potential, plants have been greatly neglected as material for studies of genetic variation.

The very consistency of the data in table 22 and the support they receive from the other, less extensive, studies is a little worrisome. If the polymorphism so widely observed is truly related to the evolutionary processes that have molded and will mold the history of various species, there ought to be some variation among species in the degree of their genetic variation. In particular, there ought to be some species which, for reasons of their unique population structure or recent evolutionary history, are rather less variable then most. There must be species that have recently expanded from a very small genetic base, or have undergone severe selection for certain specialized genotypes, or whose breeding structure constantly gives rise to high local inbreeding, so that a substantial fraction of the genome should be monomorphic.

There are several reports of electrophoretic surveys in which abnormally low heterozygosity has been observed. Most, unfortunately, are not extensive or intensive enough for one to be certain that the difference is not the result of a small sample of genes or a difficulty in technique. The most extreme case is the report by Serov (1972) that 138 individuals of *Rattus rattus* from a wild population in Novosibirsk were all completely homogeneous for 21 loci. The sample of both loci and individuals seems large enough to make this observation reliable, but the lack of technical details, and the startling difference from the house mouse, despite the great similarity in natural economy between the two rodents, make this report tantalizing but uncertain.

Of 7 loci studied in three Southeast Asian populations of a macaque, *Macacus fascicularis* (Weiss and Goodman, 1972), 5 showed detectable genetic variation in one or more populations, but in most cases one allele was in overwhelming preponderance. In two localities, Thailand and the Philippines, only 2 out of 7 loci were polymorphic and the heterozygosity per individual was 0.019 and 0.034, respectively. In Malaysia, however, 3 loci were polymorphic and heterozygosity was 0.077, values that are well within the usual range.

Finally, a species of Pacific salmon, *Oncorhyncus keta*, is reported to be polymorphic for 11 to 14 percent of its loci, with heterozygosity lying between 0.0195 and 0.032 (Altukhov et al., 1972). Even these limits are uncertain, however, because a large number of monomorphic hemoglobin, lens protein, and muscle protein bands are included.

Three reports of low heterozygosity are convincing. A wide geographical sample of the cricket frog, *Acris crepitans*, in the United States showed this species to be polytypic but not very heterozygous (Dessauer and Nevo, 1969). Thus 12 out of 20 enzymes and proteins were variable in the species as a whole, but most of the 27 populations were fixed or nearly fixed for one allele or another. The average proportion of polymorphic loci per population was only 12 percent and 4 loci were completely homozygous within populations yet differed between populations. Three distinct geographical races could be discerned on the basis of the preponderant electrophoretic alleles, an Appalachian race, a Gulf race, and a Midwestern race; considering the large number of populations sampled and the reasonable number of loci, these divisions are probably real. *Acris crepitans* is probably a truly polytypic species with low heterozygosity within populations.

A second case concerns species of the kangaroo rat, *Dipodomys*. In a survey of 17 loci in several forms of this desert-dwelling rodent, Johnson and Selander (1971) studied large samples of two species, *D. merriami* and *D. ordii*, in a number of populations. For *D. merriami*, the proportion of loci polymorphic varied between populations from 12 to 29 percent and the heterozygosity was between 0.020 and 0.071, with a mean of 0.051. *Dipodomys ordii* was even less genetically varying, with polymorphism between 6 and 24 percent of loci and heterozygosity between 0.004 and 0.017 per individual, the mean being 0.0081.

A fossorial rodent, the mole rat, *Spalax ehrenbergi*, was surveyed for genetic variation at 17 loci by Nevo and Shaw (1972), with similar results. Polymorphism varied from 6 to 29 percent, with an average of 19 percent, and heterozygosity was from 0.018 to 0.056 in different geographic races, with a mean of 0.037.

With the possible exception of *Acris crepitans*, the data are at present insufficient to be sure that any of these species is significantly more homozygous than those in table 22, and we must bear in mind that, based on the earlier, less extensive studies, *Drosophila simulans* would have been regarded as a monomorphic species. I must emphasize again that we cannot yet compare the levels of heterozygosity in different species and be all all confident that the observed differences are real. Species as diverse physiologically, evolutionarily, and ecologically as man, mouse, and the phylogenetic "relic," *Limulus*, the horseshoe crab, are virtually identical in both the proportion of loci that are polymorphic and the average heterozygosity per individual.

HOW REPRESENTATIVE ARE THE GENES?

The methodological program set out at the beginning of this chapter demanded that a satisfactory method for measuring genic variation be capable of detecting most or all of the allelic substitution at a locus, and that loci be chosen "at random" with respect to physiological function and degree of variation. Clearly, the method of electrophoretic separation of proteins does not detect all, or perhaps even most, of the allelic substitution at a locus. Perhaps it detects as little as one-quarter of all substitutions. Thus our estimates of heterozygosity are downwardly biased, perhaps by as much as four times, and this makes even more startling our estimate that a third of all loci are polymorphic and 10 percent of all genes are heterozygous in each individual. Are there any possible biases in the other direction, biases that may inflate our estimate of the total variation in the genome as a whole? We must consider the requirement that genes be sampled "at random" from the genome.

The constitution of the genome, in the sense of the functional array of gene products, is still a mystery in higher organisms. It is certain that many genes code for enzymes, and one by-product of the investigation of natural populations of Drosophila has been the localization of many new genes, coding specific enzymes, on the genetic map of different species. But the sperm of *D. melanogaster* contains enough DNA for 10^8 base pairs. With three base pairs per codon and an average of 150 amino acids per polypeptide, there is enough DNA to code about 2×10^5 polypeptides. For man, with 16 times as much DNA, there is enough to code 3×10^6 polypeptides. We can hardly believe that higher organisms are capable of manufacturing between a quarter million and 3 million different enzymes! Even to suppose that 10 percent of *Drosophila*'s DNA codes for enzymes (20,000 enzymes) stretches our credulity in the light of present knowledge of biochemistry. Of the remaining 90 percent of the DNA, some may not be informational at all, and we need not concern ourselves about this non-genic fraction. Yet a further fraction, perhaps as much as 50 percent, making up the "medium" and "fast" reannealing DNA, consists of a relatively small number of different genes that are reduplicated scores, hundreds, or thousands of times. The function of this DNA is unknown. It might be assumed to code for proteins that are required in very large amounts, like hemoglobin or other respiratory pigments. Yet the hemoglobin genes are certainly not present in multiple copies in the basic haploid genome of vertebrates, since hemoglobin variants behave as simple all-or-none characters in inheritance. Some unknown fraction of the genome codes for so-called structural protein, the protein of membranes, cell organelles, muscle, and lens. There is no compelling reason to suppose that the majority of genes might not code for such molecules. Finally, there is no information on the genes in higher organisms that may be important in control of protein synthesis, in a manner analogous to the operon of bacteria. Some large fraction of the DNA of higher organisms may be of this type.

Thus far the "random sample" of genes tested for electrophoretic variation has been restricted to the genes coding enzyme proteins and a few nonenzymatic molecules such as egg white, serum albumins, and larval hemolymph proteins of Drosophila. We know nothing of structural protein or of controlling genes. Moreover, most but not all of the enzymes tested have been soluble enzymes easily extractable from the liquid phase of cells. A few particle-bound enzymes have been studied, and they do not appear to differ in degree of polymorphism from soluble enzymes. Thus, in man, two out of three mitochondrial enzymes are polymorphic (average heterozygosity 0.11), in *Peromyscus polionotus* none of three mitochondrial enzymes is polymorphic, while in *Limulus* two out of two are (average heterozygosity 0.14). It is likely, then, that soluble enzymes are representative of enzymes in general

Even among enzymes there may be differences between groups of functionally related molecules in their degree of heterozygosity. Gillespie and Kojima (1968) postulated that enzymes directly involved in energy metabolism would be less variable because their functions were more "essential" and needed to be more coordinated than would other, peripheral enzymes. An extensive examination of this thesis was made by Kojima, Gillespie, and Tobari (1970) for several species of *Drosophila* and for Selander and Yang's data on the house mouse. I have made the same comparison for man, using Harris and Hopkinson's data, and the results are shown in table 24. Group I enzymes include all those that catalyze steps in glycolysis

TABLE 24
Polymorphism and heterozygosity for enzymes of energy metabolism (group I) and other enzymes (group II) in *Drosophila*, mouse, and man

Species	*Number of loci*	*Proportion polymorphic*	*Heterozygosity*
Drosophila melanogaster[a]			
group I	11	.36	.094
group II	8	.50	.156
D. similans[a]			
group I	11	.36	.030
group II	7	1.00	.364
D. willistoni[b]			
group I	10	—	.112
group II	18	—	.223
Mus musculus[c]			
group I	17	.24	.089
group II	11	.45	.106
Homo sapiens[d]			
group I	24	.21	.048
group II	47	.32	.077

[a]Data are from Kojima, Gillespie, and Tobari (1970).
[b]Data are from Ayala et al. (1972).
[c]Data are from Selander and Yang (1969).
[d]Data are from Harris and Hopkinson (1972).

up to the beginning of the citric acid cycle, including the pentose "shunt." All other enzymes are in Group II. It does indeed appear that, in varying degrees, the enzymes of glycolysis are less variable than the others, but it should also be noted that glycolytic enzymes are overrepresented in the samples relative to the proportion of all enzymes known in metabolism. On this basis, the average genetic heterogeneity over the whole set of enzyme-specifying genes has been underestimated. So neither the distinction between particle-bound and soluble enzymes nor the distinction between functional groups of enzymes leads us to revise our estimates of heterozygosity downward. If anything, they are too conservative.

What of nonenzymatic proteins? There exists a large body of data on genetic polymorphism for human nonenzymatic proteins which at first sight is of no use to us. These are the data on blood groups. It is a cornerstone of our method that we choose our sample of loci without reference to how variable they are. Unfortunately, blood groups are detected from antibody-antigen reactions, so a blood group locus remains undetected unless there is at least one variant in the population. Moreover, since blood groups are detected in routine cross-matching, the greater the variation, the more likely a locus is to be detected. This bias would disappear, however, if enough bloods could be tested, because presumably every blood group locus is mutable and therefore would eventually be detected as a rare variant. At first, when only a small number of bloods have been cross-matched, only the most polymorphic loci will appear, but as time goes on and a progressively larger number of tests have been run, the cumulative estimate of genetic variation should fall as more and more relatively invariant loci show up.

This historical method is applied to the English population in table 25 and figure 12 (Lewontin, 1967a). As expected, the first blood groups discovered were highly polymorphic, and the estimates of proportion of loci polymorphic and heterozygosity per individual were very high. As the number of individuals tested increased, and as sophisticated methods of analysis were developed, more and more nearly monomorphic loci were found, so that the estimates of polymorphism and heterozygosity reached an asymptote while the number of loci discerned rose steeply. The asymptotic values are approximately 36 percent for the proportion of loci polymorphic and 16 percent for the proportion of an individual genome that is heterozygous. To compare this heterozygosity with that estimated from electrophoretic variation, we must take into account that the antigens are not themselves the primary products of gene action but reflect changes in glycosyl transfer enzymes, and are not restricted to charge changes. On the assumption that electrophoretic changes constitute only about one-third of all amino acid substitution, we get from Harris and Hopkinson an estimate of total heterozygosity for enzymes of $3 \times 0.067 = 0.201$, which is in reasonable agreement with the value of 0.162 for the blood groups. Of course, the factor of 3 is somewhat arbitrary and cannot be taken too seriously. The point is that heterozygosity estimated from electrophoretic variation of enzymes is once more seen to be conservative.

While all the available evidence reinforces the conclusion that a very large fraction of genes in most sexually reproducing organisms is polymorphic, and that each individual in a population is heterozygous at a large fraction of his loci, there will always remain a

TABLE 25
Information on human blood groups

	Blood group	*Year discovered*	*Frequency of most common allele*	*Hetero-zygosity at locus*	*Cumulative hetero-zygosity*	*Proportion poly-morphic*
1	ABO	1900	0.437	0.512	.512	1.00
2	MNS	1927	0.389	0.700	.606	1.00
3	P	1927	0.540	0.497	.569	1.00
4	Se	1930	0.523	0.499	.552	1.00
5	Rh	1940	0.407	0.662	.574	1.00
6	Lu	1945	0.961	0.075	.491	1.00
7	K	1946	0.936	0.122	.438	1.00
8	Le	1946	0.815	0.301	.421	1.00
9	Levay	1946	~1.00	~0	.374	.889
10	Jobbins	1947	~1.00	~0	.337	.800
11	Fy	1950	0.549	0.520	.353	.818
12	Jk	1951	0.514	0.500	.366	.833
13	Becker	1951	~1.00	~0	.337	.769
14	Ven	1952	~1.00	~0	.313	.714
15	Vel	1952	~1.00	~0	.292	.667
16	H	1952	~1.00	~0	.274	.625
17	Wr	1953	0.999	0.002	.258	.588
18	Be	1953	~1.00	~0	.244	.556
19	Rm	1954	~1.00	~0	.231	.526
20	By	1955	~1.00	~0	.219	.500
21	Chr	1955	0.999	0.002	.209	.476
22	Di	1955	~1.00	~0	.199	.454
23	Yt	1956	0.995	0.010	.191	.434
24	Js	1958	~1.00	~0	.183	.417
25	Sw	1959	0.999	0.002	.176	.400
26	Ge	1960	~1.00	~0	.169	.384
27	Good	1960	~1.00	~0	.163	.370
28	Au	1961	0.576	0.489	.175	.393
29	Lan	1961	~1.00	~0	.168	.380
30	Bi	1961	~1.00	~0	.163	.366
31	Xg	1962	0.644	0.458	.173	.387
32	Sm	1962	~1.00	~0	.167	.375
33	Tr	1962	~1.00	~0	.162	.364

Note: From Lewontin (1967a). Data refer to the English population.

nagging doubt when we contemplate the few scores of genes examined in man in contrast with the 3 million genes that might be coded by the 3×10^9 nucleotides in the DNA of each of his sperm.

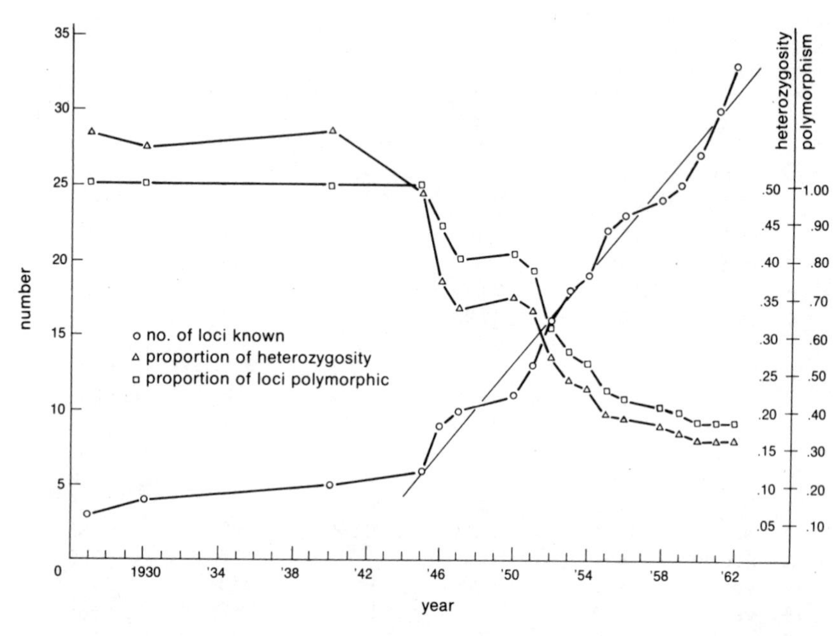

FIGURE 12

Relations between year of discovery and blood group polymorphism. From Lewontin (1967a).

A CLOSE LOOK AT *DROSOPHILA PSEUDOOBSCURA*

After the initial and somewhat unsatisfactory survey of lines of *D. pseudoobscura*, the Chicago group began a systematic survey of a large number of populations in order to see in detail the pattern of allelic variation over the range of the species. *Drosophila pseudoobscura* is distributed over western North America and Central America from southern British Columbia to Guatemala and from the Pacific coast to the Rocky Mountains and central plateaus of Texas and Mexico (figure 13). In addition, there is a large, disjunct population in the highlands of Colombia around Bogotá. The species is typically found in cool forests, being restricted in higher elevations in the more southerly or arid parts of the range. It is found essentially at sea level in northern California, but not below 5500 feet in Guatemala or below 7000 feet in Colombia. The species occurs in the semi-

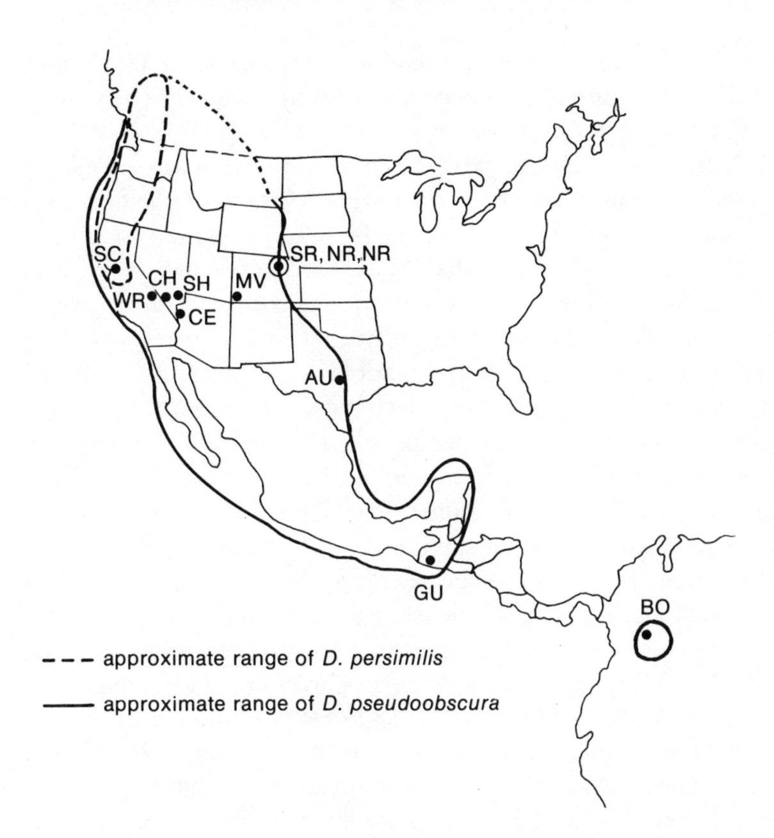

FIGURE 13

Distribution range of *Drosophila pseudoobscura* and *D. persimilis*, and location of populations sampled. Abbreviations are those used in the text.

isolated mountain ranges of the Basin and Ranges region of the western United States, but in the spring it can be found around small oases even in Death Valley and the Mojave Desert. Whether such populations persist, even as aestivating forms, during the summer heat is doubtful. The general impression given is of a widespread species that is almost continuous in its distribution during the favorable spring and fall seasons but that contracts markedly to isolated, cool, moist foci in the hot and dry seasons, at least in the temperate part of the distribution. Since the flies are easy to entice into baited traps when they are active, but extremely elusive in their free state, nothing is known about their actual distribution in cold winter months, nor have their food sources or breeding sites been found, although they certainly feed on yeast and bacteria (Dobzhansky and Epling, 1944).

Collections designed to span the biogeographic and ecological range of the species have been made from a variety of populations. These are shown on the map in figure 13. One population, SC, is at the center of abundance of the species and has a rich inversion polymorphism on the third chromosome. Five populations, WR, CH, SH, CE, and MV, are from isolated mountain ranges and plateaus in the more arid part of the distribution, forming an east-west transect through the distribution. Two population groups, the cluster SR, HR and NR, and the population AU, are from the extreme eastern ecological boundary of the species range, and the GU sample is from the extreme southern boundary of the main distribution, in the highlands of Guatemala. The last population, BO, is isolated from the rest of the species range by 1500 miles.

Of the 24 loci examined (Prakash, Lewontin, and Hubby, 1969), 11 are monomorphic and homogeneous over the entire species range. The allelic frequencies of the 13 polymorphic loci are given in tables 26A–J and 27A–E. The 10 loci represented in table 26 show three general features. The first, and most obvious, is that heterozygosity at these loci is spread throughout the entire species range. Except for the 2 local polymorphisms, *acid phosphatase-4* and *aldehyde oxidase*, polymorphic loci are generally polymorphic. Second, the allelic frequencies at these loci show a remarkable uniformity over the main contiguous distribution of the species. There is no evidence of polytypy or of geographical races. When one allele is strongly predominant in frequency, this predominance is universal. Moreover, when there are two frequent alleles, as in *xanthine dehydrogenase* (table 26H) with roughly a 3:1 ratio of the two major alleles, or *pt-8* (table 26I) with a 1:1 polymorphism of the two major alleles, these ratios are characteristic of all the population. Even the *esterase-5* locus (table 26J), for which as many as 10 different alleles may be segregating, has a rough similarity, with allele *1.00* being the most frequent in all populations, allele *1.07* being the second most frequent in almost all cases and allele *0.95* being in third or fourth place in all cases but one.

The uniformity of order and of relative frequency of alleles from populations as distant as California, Colorado, and Texas, including

TABLE 26

Frequency of alternative alleles of ten polymorphic loci in various populations of *Drosophila pseudoobscura*

	Populations											
Allele	*SC*	*WR*	*CH*	*SH*	*CE*	*MV*	*SR*	*HR*	*NR*	*AU*	*GU*	*BO*
A. Larval acid phosphatase-4, X chromosome												
0.93	—	—	—	—	—	—	—	—	—	.028	—	—
1.00	1.000	1.000	1.000	1.000	1.000	1.000	1.000	1.000	1.000	.860	1.000	1.000
1.05	—	—	—	—	—	—	—	—	—	.112	—	—
B. Acetaldehyde oxidase-2, chromosome II												
0.90	.010	—	—	—	—	—	—	—	—	—	—	—
0.93	.030	—	—	—	—	—	—	—	—	—	—	.050
1.00	.940	1.000	1.000	1.000	1.000	1.000	1.000	1.000	1.000	1.000	1.000	.830
1.02	.020	—	—	—	—	—	—	—	—	—	—	.120
C. Malic dehydrogenase, chromosome IV												
0.80	—	—	—	—	.011	—	.019	—	—	—	—	—
1.00	.969	1.000	.936	.929	.954	.948	.962	.882	.904	.957	.727	1.00
1.20	.031	—	.064	.071	.034	.052	.019	.118	.096	.043	.273	—
D. Octanol dehydrogenase-1, chromosome II												
null	—	—	.014	—	—	—	—	—	—	—	—	—
0.75	—	.026	—	—	—	—	—	—	—	—	—	—
0.86	—	—	—	.056	.020	.013	—	—	—	—	—	—
1.00	.977	.871	.951	.902	.939	.961	.965	.972	.885	1.000	1.000	1.000
1.05	—	.077	.014	.028	.010	—	.035	—	.115	—	—	—
1.22	.023	.026	.021	.014	.031	.026	—	.028	—	—	—	—
E. Leucine aminopeptidase, autosome												
0.83	—	.012	—	.012	.009	—	—	—	—	—	—	—
0.90	.008	—	.016	—	—	.025	—	—	.019	.043	.036	—
0.95	.050	.024	.039	.024	.018	.008	—	—	—	.022	—	—
1.00	.892	.916	.897	.893	.954	.940	.875	1.000	.923	.870	.964	.947
1.10	.050	.048	.048	.071	.018	.025	.125	—	.058	.054	—	.054
1.12	—	—	—	—	—	—	—	—	—	.011	—	—

Allele	Populations											
	SC	*WR*	*CH*	*SH*	*CE*	*MV*	*SR*	*HR*	*NR*	*AU*	*GU*	*BO*
F. Pt-7, chromosome II												
0.68	.005	—	—	—	—	—	—	—	—	—	—	—
0.73	.005	—	.021	.014		.009	.040	.025	—	.012	—	.050
0.75	.954	.950	.979	.971	.987	.955	.960	.950	1.000	.966	1.000	.925
0.77	.036	.050	—	.014	.013	.036	—	.025	—	.023	—	.025
G. Pt-13, autosomal												
1.23	.057	—	.082	—	.045	.025	.058	—	.022	.022	—	—
1.30	.943	1.000	.918	1.000	.940	.975	.942	1.000	.935	.978	1.000	.725
1.37	—	—	—	—	.015	—	—	—	.043	—	—	.275
H. Pt-8, chromosome II												
0.80	.014	.025	.008	—	—	.009	.019	.027	—	.011	—	.870
0.81	.472	.450	.600	.514	.473	.410	.539	.595	.480	.441	.625	.093
0.83	.514	.525	.392	.472	.527	.576	.442	.378	.480	.512	.375	.037
0.85	—	—	—	.014	—	—	—	—	.040	.035	—	—
I. Esterase 5, X chromosome												
null	—	—	—	—	.016	—	—	—	—	—	—	—
0.85	—	.013	.007	.047	—	.035	—	—	—	—	—	—
0.90	—	.027	.030	—	—	—	—	—	—	.015	—	—
0.95	.123	.149	.096	.140	.114	.113	.237	.216	.114	.031	.158	.026
0.97	—	.027	.015	—	—	—	—	—	—	.031	—	—
1.00	.424	.460	.356	.419	.317	.365	.474	.486	.341	.292	.579	.974
1.02	.014	.013	.022	—	—	.048	—	.108	.182	.108	.053	—
1.03	.080	—	—	—	—	.039	—	—	—	—	—	—
1.04	.004	.041	.193	.198	.211	.104	.017	.135	.045	.154	—	—
1.07	.193	.243	.200	.174	.260	.196	.271	.054	.273	.262	.210	—
1.09	.009	—	.007	—	—	—	—	—	—	—	—	—
1.12	.132	.027	—	.023	.081	.100	—	—	.045	.046	—	—
1.16	.019	—	.073	—	—	—	—	—	—	.062	—	—

Allele	SC	WR	CH	SH	CE	MV	SR	HR	NR	AU	GU	BO
J. Xanthine dehydrogenase, chromosome II												
0.90	.053	—	.007	—	—	.016	.035	—	—	.018	—	—
0.92	.074	—	.030	.040	—	.073	.089	.026	.020	.036	—	—
0.97	—	.133	.098	.077	.012	—	—	—	—	—	—	—
0.99	.263	.200	.188	.173	.131	.300	.286	.289	.220	.232	.278	—
1.00	.600	.667	.647	.710	.857	.581	.555	.633	.720	.661	.722	1.000
1.02	.010	—	.030	—	—	.030	.035	.052	.040	.053	—	—

Note: Data are from Prakash, Lewontin, and Hubby (1969) and Prakash, Lewontin, and Crumpacker (1973).
SC = Strawberry Canyon, Calif.; WR = Wild Rose, Calif.; CH = Charleston, Nev.; SH = Sheep Range, Nev.; CE = Cerbat, Ariz.; MV = Mesa Verde, Colo.; SR = State Recreation, Colo.; HR = Hardin Ranch, Colo.; NR = Nelson Ranch, Colo.; AU = Austin, Tex.; GU = Guatemala; BO = Bogotá, Colombia.

TABLE 27

Allelic frequencies of three polymorphic loci on chromosome III of *Drosophila pseudoobscura* from the same populations as in table 26

	Populations											
Allele	*SC*	*WR*	*CH*	*SH*	*CE*	*MV*	*SR*	*HR*	*NR*	*AU*	*GU*	*BO*
A. *Pt-10*												
1.02	.005		.007	.014	.015	.022	—	—	—	.010	—	—
1.04	.615	.898	.945	.943	.985	.970	.770	.694	.308	.935	—	—
1.06	.380	.102	.048	.043	—	.008	.230	.306	.692	.054	1.000	1.000
B. *α-Amylase*												
.74	.030	—	—									
.84	.290	.206	.090	.172	.194	.211	.380	.391	.548	.125	1.00	1.00
1.00	.680	.794	.910	.828	.806	.789	.620	.609	.452	.875	—	—
C. *Pt-12*												
1.18	.550	.736	.750	.792	.733	.940	1.000	1.000	.972	.900	1.000	1.000
1.20	.450	.264	.250	.208	.267	.060	—	—	.028	.100	0	0

central and peripheral populations, prosperous and ecologically marginal ones, is a fact that must figure critically in any explanation of the vast heterozygosity in the species Different explanatory hypotheses about the origin and fate of genetic variation make different predictions about the similarity between ecologically and geographically diverse populations, as we shall see in chapter 5.

The third feature of the observation is that Bogotá (BO), the isolated Andean population, is an exception to the widespread polymorphism and uniformity of gene frequencies seen in the other populations. For the more polymorphic loci (table 26D–J) the BO population is markedly less heterozygous than the average population from the main part of the distribution, strikingly so for the species' three most polymorphic loci (tables 26H–J). For the *pt-8* locus an allele that is rather rare elsewhere is the preponderant one in Bogotá, but for the other loci it is the most common allele in the species that predominates in Bogotá as well. The most striking case is *esterase-5*, which is immensely polymorphic everywhere except in Bogotá, where allele *1.00* is virtually monomorphic. The depauperate genetic variation in the large, prosperous, but isolated Andean population must also figure strongly in any theory of variation. That it is indeed the isolation of Bogotá, rather than its tropical environment, that is responsible for its unusual characteristics is shown by the data from Guatemala (GU). An extremely small sample was captured there, so the lack of observed polymorphism among the less variable loci is to be expected. For the more variable loci, however, GU is quite as heterozygous as the rest of the main distribution (table 26E–I), in contrast to the homogeneity at these loci in Bogotá.

A very different picture emerges when we look at the three loci of table 27. In place of the uniformity of polymorphism and gene frequency over the main continental distribution of the species, with BO being homozygous for the most common species allele, there is considerable variation in allele frequency from population to population. For two of the three loci, BO is homozygous for an allele that is in a minority elsewhere, and for all three loci GU is identical with BO. There are apparently five geographic races: California, Death Valley, eastern Colorado, Texas, and tropical. The three loci of table 27 are on the third chromosome of *D. pseudoobscura*. This chromosome element is highly polymorphic for a series of overlapping inversions that, when heterozygous, reduce recombination on the third chromosome to less than 1 percent of its normal value (Dobzhansky and Epling, 1948). The frequencies of the inversions show strong geographical differentiation (Dobzhansky and Epling, 1944), so that if the different alleles at the *pt-10, pt-12*, and *α-amylase* loci were closely associated with different inversions, the geographical variation in allelic frequencies might be explained.

Prakash and Lewontin (1968, 1971) determined the allelic composition of a variety of gene arrangements sampled from several populations, and their results are given in table 28. There is a very strong association at the *pt-10* locus between the allele *1.04* and the inversion types Standard, Arrowhead, and Pikes Peak, whereas allele *1.06* characterizes the other arrangements. At the *α-amylase* locus one of the alleles, *1.00*, is associated with the three arrangements, Standard, Arrowhead, and Pikes Peak, but the other arrangements are predominantly *0.84*, although the association is not as strong as at the *pt-10* locus. For *pt-12* the variation is much simpler. The Standard gene arrangement is 80 percent *pt-12*$^{1.20}$, while that allele is practically absent from all other inversions. These patterns of association are even more striking when the evolutionary relations between the chromosome types are considered. Because the arrangements differ from each other by single overlapping inversions, it is possible to deduce their phyletic relationships. Figure 14 shows that Standard, Arrowhead, and Pikes Peak belong to a cluster of related inversions, the "Standard phylad," connected by single inversion steps but separated from the "Santa Cruz" phylad of inversions by a hypothetical arrangement that has never been found in nature. Moreover, the Standard arrangement is shared between *D. pseudoobscura* and its sibling species, *D. persimilis*, which has its own cluster of inversions, also part of the Standard phylad. Thus it appears that the alleles *pt-10*$^{1.04}$ and *α-amylase*$^{1.00}$ characterize the Standard phylad of gene arrangements while their alternate alleles are associated with the Santa Cruz phylad. These connections are further borne out by the fact that *D. persimilis* is homozygous for *pt-10*$^{1.04}$ and is segregating for *α-amylase*$^{1.00}$ and for a number of alleles that are unique to *D. persimilis,* but is virtually without *α-amylase*$^{0.84}$, as we expect for a species whose inversions belong entirely to the Standard phylad (Prakash, 1969). The association between specific inversions must then date back to the original split

TABLE 28
Allelic frequencies in different gene arrangements of chromosome III of *Drosophila pseudoobscura*

Gene arrangement	Allele	Populations MA	SC	MV	AU	CE	BO
A. Pt-10							
Standard	*1.04*	1.00	1.00	×	1.00	1.00	—
	1.06	—	—	—	—	—	—
Arrowhead	*0.94*	0.10	—	0.02	—	—	—
	1.02	—	—	0.97	—	0.02	—
	1.04	0.90	1.00	0.01	1.00	0.98	—
	1.06	—	—	—	—	—	—
Pikes Peak	*1.02*	—	—	—	0.01	—	—
	1.04	1.00	—	1.00	0.99	1.00	—
Santa Cruz	*1.06*	—	1.00	—	—	—	1.00
Chiricahua	*1.04*	0.50	—	—	—	—	—
	1.06	0.50	1.00	—	×	×	—
Treeline	*1.04*	—	—	—	0.33	—	—
	1.06	1.00	1.00	—	0.66	—	1.00
Olympic	*1.06*	—	—	—	×	—	—
Estes Park	*1.06*	—	—	—	—	1.00	—
B. α-Amylase							
Standard	*0.84*	0.05	0.12	—	0.80	—	—
	0.92	0.10	—	—	—	—	—
	1.00	0.85	0.88	—	0.20	×	—
Arrowhead	*0.84*	0.05	0.30	0.21	0.29	0.22	—
	1.00	0.95	0.70	0.79	0.71	0.78	—
Pikes Peak	*0.84*	—	—	—	—	0.04	—
	1.00	1.00	—	—	1.00	0.96	—
Santa Cruz	*0.84*	—	1.00	—	—	—	1.00

between the two phylads and must predate the divergence of *D. persimilis* from *D. pseudoobscura* and the current geographic distribution of these species, which at the most conservative estimate is Arcto-Tertiary, about one million years ago.

The associations of the alleles of *pt-10*, *α-amylase*, and *pt-12* do indeed explain the frequency differences between the populations in table 27. For example, both Guatemala and Bogotá are segregating

Gene arrangement	Allele	Populations					
		MA	*SC*	*MV*	*AU*	*CE*	*BO*
α-Amylase (cont).							
Chiricahua	*0.84*	0.36	0.36	—	×	—	—
	1.00	0.64	0.64	—	—	—	—
Treeline	*0.74*	—	0.14	—	—	—	—
	0.84	0.90	0.79	—	1.00	1.00	1.00
	1.00	0.10	0.07	—	—	—	—
Olympic	*0.84*	—	—	—	—	—	×
Estes Park	*0.84*	—	—	—	—	1.00	—
C. Pt-12							
Standard	*1.18*	0.20					
	1.20	0.80					
Arrowhead	*1.18*	0.95					
	1.20	0.05					
Pikes Peak	*1.18*	1.0					
	1.20	—					
Chiricahua	*1.18*	1.0					
	1.20	—					
Treeline	*1.18*	0.95					
	1.20	0.05					

Note: Data are from Prakash and Lewontin (1968, 1971) and Prakash, Lewontin, and Crumpacker (1973). When only one chromosome was examined, an X indicates presence of an allele.

MA = Mather, Calif.; SC = Strawberry Canyon, Calif.; MV = Mesa Verde, Colo.; AU = Austin, Tex.; CE = Cerbat, Ariz.; BO = Bogotá, Colombia.

only for inversions of the Santa Cruz phylad and so are expected to be virtually homozygous for $pt\text{-}10^{1.06}$, $\alpha\text{-}amylase^{0.84}$, and $pt\text{-}12^{1.18}$, in contrast to the low frequency of $pt\text{-}10^{1.06}$ and $\alpha\text{-}amylase^{0.84}$ in the rest of the species, in which the Standard phylad is much more common. In general there is an excellent and detailed match between the observed frequencies of alleles for these three loci in different populations and the frequencies predicted from the relations in table

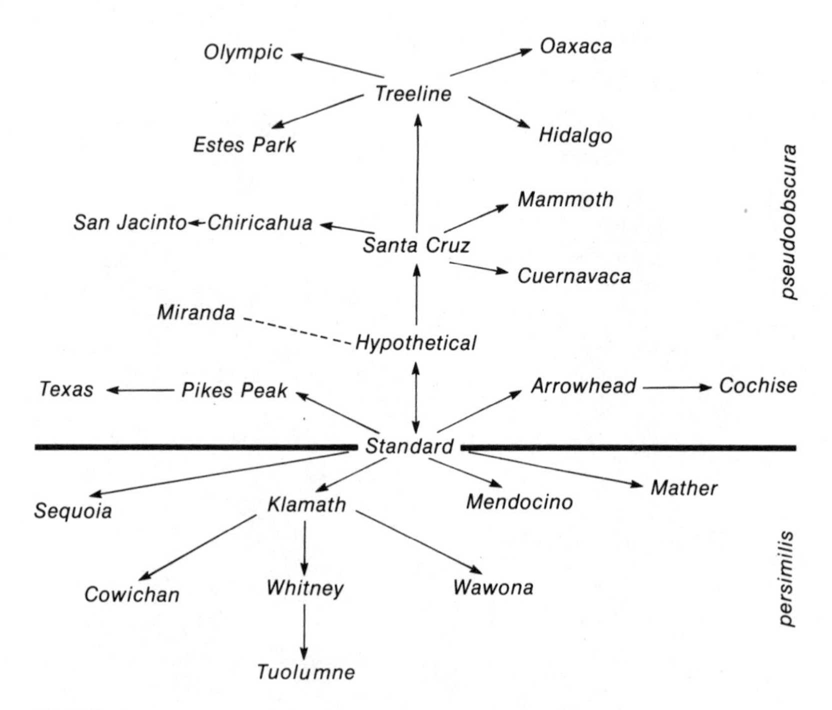

FIGURE 14

Phylogenetic relationships among the arrangements on the third chromosome of *Drosophila pseudoobscura* and *D. persimilis*. Adapted from Dobzhansky and Epling (1944).

28 together with the known frequencies of inversions given in table 29 (Prakash and Lewontin, 1968, 1971; Prakash, Lewontin, and Crumpacker, 1973). Note especially the striking similarity between the total frequency of the Standard phylad in table 29 with the frequency of *pt-10*$^{1.04}$ in table 27A.

There is some suggestion, although it has not been critically tested, that variation among populations in the frequency of alleles at the sex-linked *esterase-5* locus (see table 26J) may be the result of similar association between alleles and the "sex-ratio" inversions of the X chromosome (Prakash, Lewontin, and Crumpacker, 1973).

The association of specific alleles with inversions also has the consequence that the alleles of two different loci are not at random with respect to each other. For example, the frequency of gametes of the genotype $\underline{pt\text{-}10^{1.06},\ \alpha\text{-}amylase^{1.00}}$ in the Strawberry Canyon population should be (0.320) (0.680) = 0.258, according to the

TABLE 29
Frequencies of different gene arrangements of chromosome III in populations of *D. pseudoobscura*

	Populations											
Gene arrangement	*SC*	*WR*	*CH*	*SH*	*CE*	*MV*	*SR*	*HR*	*NR*	*AU*	*GU*	*BO*
Standard	.47	.34	.28	.35	.38	.01	.03	—	.09	.06	—	—
Arrowhead	.09	.44	.67	.60	.52	.98	.42	.40	.28	.08	—	—
Pikes Peak	.03	.06	.02	—	.05	.01	.27	.21	.18	.78	—	—
Total Standard phylad	.59	.84	.97	.95	.95	1.00	.75	.61	.55	.92	—	—
Chiricahua	.19	.11	.03	.05	.04	—	—	.02	—	.01	—	—
Treeline	.14	.04	—	—	.01	—	—	.04	.17	.06	.65	.37
Estes Park	.06	.01	—	—	—	—	.25	.33	.24	—	—	—
Santa Cruz	.01	—	—	—	—	—	—	—	—	—	.30	.63
Olympic	.01	—	—	—	—	—	—	—	.04	.01	—	—
Oaxaca	—	—	—	—	—	—	—	—	—	—	.05	—
Total Santa Cruz phylad	.41	.16	.03	.05	.05	—	.25	.39	.45	.08	1.00	1.00
Proportion of heterokaryotypes	.71	.67	.47	.52	.58	0.04	.69	.69	.79	.38	.49	.47

Note: Data are from Mayhew et al. (1966), Strickberger and Wills (1966), and Prakash, Lewontin, and Crumpacker (1973). The abbreviations are the same as in table 26.

allelic frequencies in table 27. However, because of the complete association of *pt-10*$^{1.06}$ with the Santa Cruz phylad of inversions and the strong association of *α-amylase*$^{1.00}$ with the Standard phylad, the frequency of the gamete is only 0.131. If we consider the triple combination *pt-10*$^{1.06}$, *α-amylase*$^{1.00}$, *pt-12*$^{1.20}$, the frequency would be $(0.380)(0.680)(0.450) = 0.116$ if the loci were assorted at random, but the real frequency is only 0.005; that is, practically absent.

In chapter 6 I shall discuss the importance of these associations between alleles at different loci as indications of the forces controlling the genetic variation, and as elements in the dynamic theory of genetic change. Clearly, if such associations are frequent, then a sufficient set of dimensions for a description of evolutionary processes must be related to gametic types rather than to allelic frequencies at individual loci.

All the data on *D. pseudoobscura* are summarized in table 30. Once again the lack of any difference among the populations of the main continental distribution is clear. What differences do exist in average heterozygosity per locus disappear completely when inversion heterozygosity is discounted. The two extremes, Strawberry Canyon with 16 percent heterozygosity and Guatemala with 8 percent, turn out to be essentially the same if third-chromosome genes are removed from the calculation. Only the isolated population of Bogotá is clearly differentiated from the rest, departing more than ten standard deviations from the average heterozygosity of the main continental distribution. Yet these continental populations include the complete biogeographical range of the species, from the year-round abundant population of Strawberry Canyon through the populations living at the eastern ecological margin of the species, detectable only in favorable seasons, to the very sparse population of the highlands of Guatemala, where a week of collecting produced only eight individuals of *D. pseudoobscura* among numerous Drosophila, including large numbers of its close relative, *D. azteca*. Moreover, the similarity among the continental populations lies not only at the gross level of average heterozygosity per locus, but at the fine level of close similarity of allelic frequencies.

This uniformity of genetic composition contrasts sharply with the considerable variation in frequencies of third chromosome arrangements among populations (table 29). Since the extensive work of Dobzhansky and his colleagues has shown that these gene arrange-

TABLE 30

Proportion of loci polymorphic out of 24 examined, average number of alleles per locus with frequency greater than 1 percent, and proportion of genome estimated as heterozygous in 12 populations of *Drosophila pseudoobscura*

Population	*Number of polymorphic Loci*	*Proportion of loci polymorphic*	*Average number of alleles per locus*	*Proportion of genome heterozygous*	
				including chromosome III	*excluding chromosome III*
Strawberry Canyon, Calif.	12	.50	2.29	.161	.116
Wild Rose, Calif.	9	.38	1.92	.129	.105
Charleston, Nevada	11	.46	2.21	.126	.113
Sheep Range, Nevada	10	.42	1.96	.125	.108
Cerbat, Arizona	11	.46	1.92	.112	.093
Mesa Verde, Colorado	11	.46	2.12	.117	.109
State Recreation, Colo.	10	.42	1.71	.131	.110
Hardin Ranch, Colo.	8	.33	1.62	.123	.098
Nelson Ranch, Colo.	10	.42	1.79	.140	.113
Austin, Texas	11	.46	2.21	.126	.119
Guatemala	(5)[a]	(.21)[a]	(1.29)[a]	.081	.092
Bogotá, Colombia	6	.25	1.37	.051	.058
Mean excluding Bogotá and Guatemala	10.3	.43	1.97	.129 ± .0058	.108 ± .0034
Grand Mean	9.5	.40	1.86	.119 ± .0081	.103 ± .0048

Note: Data are from Prakash, Lewontin, and Hubby (1969) and Prakash, Lewontin, and Crumpacker (1973).
[a]Sample size was so small that it produced spuriously low values.

ments respond to selection (see Dobzhansky, 1970, chapter 5, for a review), it must be that the populations we have sampled are living under physical and biotic conditions that differ from population to population in a way that is significant to the physiology of the species. Our interpretation of the genic variation we have seen must account for this divergence in the apparent action of natural selection between chromosomal arrangements and allelic substitutions at individual loci not associated with inversions.

GEOGRAPHICAL PATTERNS IN OTHER SPECIES

The pattern of geographical variation in *Drosophila pseudoobscura* has four features: (1) monomorphic loci are identically monomorphic over all populations; (2) most polymorphic loci show no pronounced geographical differentiation; (3) loci associated with inversions show patterns of geographic differentiation characteristic of the inversions; and (4) a completely isolated population is more homozygous than populations in the contiguous range of the species. Several other species show patterns that are consonant with the situation in *D. pseudoobscura*, but there are some interesting differences as well.

Drosophila willistoni is widely distributed in the islands of the Caribbean, in tropical Central and South America as far south as northern Argentina, and as far north as southern Florida. A comparison of four continental populations from Colombia with six Caribbean island populations for 20 loci shows the same kind of overall similarity as in *D. pseudoobscura* (Ayala, Powell, and Dobzhansky, 1971). Of 16 polymorphic loci, 14 have closely similar allele frequencies in all populations, but for 2 there is a marked and consistent difference in frequencies between the islands and the mainland. The average heterozygosity on the islands is 0.162 as compared to 0.184 on the mainland but, because of the small number of localities tested, this difference, while suggestive, is not significant.

One of the most interesting features of the genetics of *D. willistoni* is the rich inversion polymorphism on all of its chromosome arms. Unlike the large overlapping inversions that form the third chromosome polymorphism in *D. pseudoobscura*, the inversions of *D. willistoni* are small, numerous, and often nonoverlapping (da Cunha, Burla, and Dobzhansky, 1950). Unfortunately this has made it impossible to estimate the frequencies of individual inversions in various populations, homozygotes being virtually indistinguishable from each other in polytene chromosomes, but heterozygotes can be counted. Average inversion heterozygosity varies widely over the species, with central populations heterozygous for many inversions (9.36 per female in Mojolinho, Goiás, Brazil) but marginal and island populations nearly homozygous (0.27 heterozygous inversions per female on the island of St. Vincent). Although the continental populations in the study of Ayala, Powell, and Dobzhansky (1971) were not examined for inversions, we can judge from their position in the distribution that they are heterozygous for about 7 inversions per female, in contrast to the islands, which vary from 0.23 inversions per female on tiny St. Kitts to 3.18 per female on the very large and proximate island of Trinidad (Dobzhansky, 1957).

The lack of a major difference in genic heterozygosity between the mainland and the islands seems at first in contradiction to the pattern of inversion variation, especially since the inversions are uniformly spread over all the chromosomes and so, presumably, include all or most of the gene loci. However, the connection between inversion heterozygosity and genic heterozygosity only becomes powerful when there is an extreme association between genes and inversions. For example, suppose alleles *A* and *a* were in frequencies 0.8 and 0.2, respectively, in one chromosomal type, whereas they were in reverse frequencies, 0.2 and 0.8, in an inverted chromosome. Then populations completely homozygous for either chromosomal arrangement would have a genic heterozygosity of $2(0.8)(0.2) = 0.32$, while a population with equal frequencies of the two chromosomal arrangements would have a genic heterozygosity of 0.50. Moreover, the effect of chromosomal heterozygosity is not cumulative. That is, if there were 18 inversions covering 18 different parts of the genome, each with a gene having the same 0.8:0.2 association, then a population in which all 18 were segregating would still only have an average genic heterozygosity of $(18)(0.50)/18 = 0.50$, although the average number of heterozygous inversions per individual would be 9. The completely structurally monomorphic population, with zero inversions heterozygous per individual, would, nevertheless, be genically heterozygous for 0.32 of its genome. In the absence of detailed matching of gene alleles with inversions in *D. willistoni*, it is therefore impossible to say whether associations like those in *D.*

pseudoobscura exist. The fact that two loci show marked and consistent frequency differences between the islands and the continent suggests that these two may be strongly correlated with particular inversions, some of which become unusually frequent on the islands. For example, inversion II*L-a* is absent on the continent but 40 percent of flies on Trinidad are heterozygous for it.

The extensive survey of *Drosophila obscura* from Finland replicates the pattern of *D. pseudoobscura*. Populations in southern and central Finland and marginal populations in Lapland were studied for 30 loci (Lakovaara and Saura, 1971a), 16 of which were polymorphic. The 14 monomorphic loci were identical in all populations, and there was little variation in allele frequency among the polymorphic loci; for one esterase locus the most common allele in central Finland was absent in Lapland, replaced in northern Lapland by an allele unique to the north, and for a second esterase locus the commonest allele in central Finland was rare in the south. There was no local differentiation in average heterozygosity, and the marginal populations of Lapland were just as heterozygous as other populations.

Lakovaara and Saura's study of *Drosophila subobscura* gave similar results (1971b). Of 15 polymorphic loci, 9 showed no differences among populations, 2 showed strong differentiation from population to population, and 3 were homozygous in an island population. In these latter three cases an allele that is otherwise rare was fixed on the islands, a curious circumstance and quite different from the Bogotá population of *D. pseudoobscura*, in which the common allele in the rest of the species usually was common or fixed in the isolated population. The three fixed loci in *D. subobscura* do not show any geographical variation outside the islands and are among the most heterozygous loci in the study.

Drosphila subobscura is distributed from Persia to Finland. The Lakovaara and Saura study included only the northern margin and thus more extensive differentiation might exist in the species range as a whole. Extreme southern and northern populations are monomorphic for different inversions, whereas Central European populations are highly polymorphic, so there is a good chance that the species as a whole may be more differentiated in allelic composition (Sperlich and Feuerbach, 1969).

A study of five loci in *Mus musculus*, the house mouse, from 16

North American regions, including an island population from Jamaica, again showed the island population to be virtually monomorphic as compared with the mainland ones (Selander, Yang, and Hunt, 1969). There is fairly general uniformity of pattern over the continental populations, but some large regional similarities can also be detected. Thus, although the major allele at the *est-2* locus is in about the same frequency everywhere, the minor component consists of a single allele in the Northeast but three low-frequency alleles in the Southwest. As another example, there is a 1:1 polymorphism for the hemoglobin β–chain locus in California and Arizona, while for the rest of the continent the polymorphism is about 4:1. Thus there is some correlation between neighboring regions and some differentiation between distant places, but no major variation and no obvious pattern.

Two cases are known in which much more extensive and regular differentiation of gene frequencies occurs. The first is in *Peromyscus polionotus*, the beach mouse, from the Gulf Coast of the United States (Selander et al., 1971). The results of the survey of 32 loci are shown in table 31 and figure 15. Each circle, representing a population, is filled in in proportion to the heterozygosity per individual, a full circle denoting a heterozygosity of 0.10. There is no apparent reason why the population of peninsular Florida should be 75 percent more heterozygous than the main continental part of the distribution. The western beach populations, however, are geographically isolated from the rest of the species and are living in an atypical habitat, grassy beach dunes, as opposed to

TABLE 31

Average heterozygosity per individual in mainland and insular populations of *Peromyscus polionotus*

Locality	*Heterozygosity*
Western beach	
Santa Rosa Island (4, 5)	.0195
Peninsulas (1–3, 6)	.0323
Florida panhandle (7–12)	.0523
South Carolina and Georgia (13–21)	.0506
Peninsular Florida (23–28)	.0810
Eastern beach islands (29, 30)	.0825

Note: Data are from Selander et al. (1971). Population numbers in parentheses refer to figure 15.

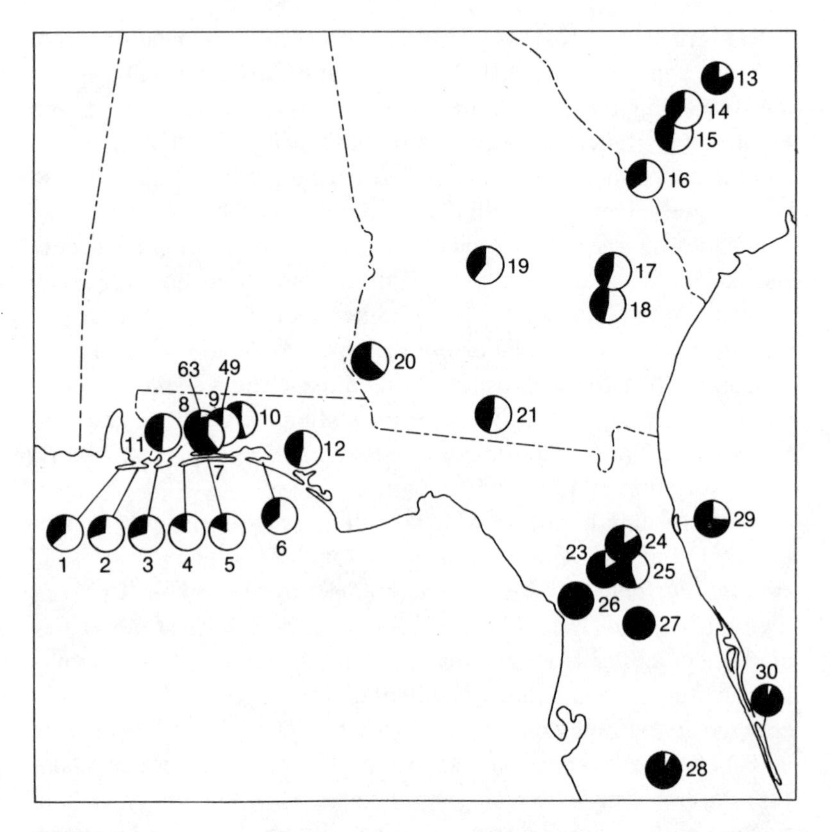

FIGURE 15

The average heterozygosity per population in *Peromyscus polionotus.* The dark area of each circle is proportional to 10× the heterozygosity per individual, a completely darkened circle representing a heterozygosity of 0.10. Adapted from Selander et al. (1971).

the usual abandoned fields that are characteristic of *P. polionotus.* Yet both of these features are characteristic also of the two Florida beach island populations, although isolation from the mainland may be less and population sizes are probably larger. A second characteristic of the western beach island and peninsular populations is a considerably greater heterogeneity of allelic frequencies between populations than for the mainland. Both this heterogeneity and the lower heterozygosity point to isolation and random fluctuation in gene frequencies as important agents in determining the genetic variation of the beach mice.

The second case shows even greater variation among populations than does *Peromyscus*. Stone and colleagues (1968) and F. M. Johnson (1971) studied samples of *Drosophila ananassae* and *D. nasuta* from island groups that stretched 4500 miles across the western Pacific, from Samoa to the Philippines. The island groups are separated by 200 to 2000 miles of open water, whereas islands within a group are 50 to 60 miles apart at most or, in the case of the two Philippine samples, separated by 800 miles of almost continuous land surface.

As usual, there were no loci that were fixed for different alleles in different island groups, but there were many polymorphic loci at which large and consistent difference in allele frequencies occurred between island groups. The largest differences observed were for *acid phosphatase* in *D. ananassae*, in which one of the five alleles had a frequency of 0.81–0.93 in Samoa but only 0.05–0.08 in Palau and 0.025 in Ponape. Somewhat less extreme was *esterase-F* in *D. nasuta*. Here an allele with a frequency of 0.48–0.55 in Samoa had a frequency of 0.07–0.16 in Fiji, only 450 miles away. Not all polymorphic loci were variable, however. The major allele of *leucine aminopeptidase* varied only from 0.79 to 0.96 over the entire set of islands, with most of the variation being contributed by two island groups. The two Philippine localities were remarkably similar in each case despite the 800-mile separation.

Once again we see that isolation seems to be a necessary although not sufficient condition for marked differentiation in allele frequencies. On the other hand, there were no appreciable differences in heterozygosity per individual when averaged over loci, nor were the differences that did occur related to island size or isolation. The highest heterozygosities were found on the vast Philippine archipelago, only 400 miles from the Asian mainland, and on the miniscule island of Yap in mid-Pacific.

The most extreme case of geographical divergence is the polytypic species *Acris crepitans* (Dessauer and Nevo, 1969) that I discussed in the preceding section. An average of only 12 percent of its loci were polymorphic per population, but 60 percent of the loci showed significant genetic variation over the whole species. Four out of 20 loci were completely homozygous within populations yet differed between populations. Three geographical races could be clearly demarcated—an Appalachian, a Gulf, and a Midwestern race—and the question is open whether these geographical races are indeed exchanging genes in nature, or whether they are in some degree sexually isolated from each other.

An analogous situation exists for the two subspecies *Mus musculus musculus* and *M. musculus domesticus* in northern and southern Denmark, studied by Selander, Hunt, and Yang (1969). Both subspecies are heterozygous (see table 22) but there are profound differences in gene frequencies between them. Twenty loci were identically monomorphic in the two subspecies, but every variable locus was clearly differentiated between the groups. It seems almost certain that there is strong, perhaps complete, reproductive isolation between the two neighboring subspecies so that, as in *Acris*, we are dealing with separately evolving entities rather than a single species.

MARGINAL POPULATIONS

It is often observed that populations at the margins of a species' range are less phenotypically variable than those near the center. Mayr (1963) notes that a study of phenotypically polymorphic species "reveals almost invariably that the degree of polymorphism decreases toward the border of the species, and that the peripheral populations are not infrequently monomorphic."

The accuracy of the generalization is improved if a distinction is made between geographically peripheral and ecologically marginal populations. The area inhabited by a species is a spatio-temporal mosaic of favorable and unfavorable habitats. As the frequency in space and time of the favorable "patches" grows less and less along some geographical gradient, there finally comes a point at which the species is unable to maintain a population. This is a species border, and it will be ill-defined and fluctuate as a result of fluctuations in the environment and in the numbers of immigrants that reach it from the center of the species distribution. As well as being geographically peripheral, populations at this border are also ecologically marginal. Yet the boundaries of the species need not all be of the ecologically marginal sort. For a terrestrial species the seashore is always a periphery, but the populations living there are not necessarily living on an ecological margin in our sense. Thus the Pacific Ocean is the western boundary of the distribution of *Drosophila pseudoobscura*, but the population living in the San Francisco Bay area is one of the most prosperous and polymorphic in the entire species range.

Alternatively, a species may have an internal "margin," a region of low frequency of favorable habitats surrounded by very favorable regions. An isolated lowland surrounded by continuously increasing elevations is an example. For *Drosophila pseudoobscura* the Isthmus of Tehuantepec in southern Mexico is such an internal margin, and the species is rare or absent there although present in reasonable numbers on either side. Of course there is a danger of circularity in this reasoning, since the ecological requirements of most species are not well known, and a border may be judged an ecological margin if the species is monomorphic there but merely an abrupt border if the species populations are locally polymorphic. The best escape from this circularity is to make the judgment from average population size and its temporal variation, compared with the center of the species.

The greater monomorphism of marginal populations extends to chromosomal as well as morphological variation. I have already pointed out that inversion heterozygosity in *Drosophila willistoni* is very high in central Brazil, with more than 9 inversions heterozygous per female, whereas the species is virtually monomorphic in the islands of the Lesser Antilles. There are few inversions at the northernmost end of the distribution in Florida (2.06 per female) and at the southern end in extreme southern Brazil and northern Argentina, as well as along the southeastern Atlantic coast of Brazil and the western Pacific coast of Ecuador. The degree of chromosomal polymorphism correlates well with an index of environmental diversity based on climate, vegetation, and closely related competitors (da Cunha and Dobzhansky, 1954; da Cunha et al., 1959). We have seen, however, that genic polymorphism in *D. willistoni* is only slightly less on the Caribbean islands than on the mainland of South America so that there seems to be no effect, or at most a slight effect, of marginality on variation at the genic level.

Drosophila robusta is a species of the eastern half of the United States and southern Canada, following closely the distribution of the American elm, *Ulmus americanus*. A rich inversion polymorphism is spread throughout all the major chromosome arms. Populations at the center of the distribution, in Missouri and Tennessee, are heterozygous for 8 to 9 inversions per female, while the southernmost extension of the range (in central Florida) has only about 1 inversion heterozygous per individual, and the westernmost extension in Nebraska is monomorphic (Carson, 1958). A survey by Prakash (1973) of 20 enzymes and proteins from Nebraska, Missouri, and Florida populations gave heterozygosities per individual of 0.13, 0.14, and 0.15, respectively. Thus, as in *D. willistoni*, there is no relation between genic polymorphism and marginality of the population, or between average genic heterozygosity and average inversion heterozygosity. This does not, however, rule out association between specific alleles and specific inversions except those associations that are nearly complete. There is, in fact, evidence of such associations in *D. robusta* (Prakash, 1973).

The relation between geography and inversion heterozygosity in *Drosophila pseudoobscura* is not simple and does not clearly follow any rule about marginal and central populations. The most chromosomally polymorphic populations are in California, as might be expected from the abundance of the species there, and at the eastern margin in Colorado. The only populations that approach monomorphism are those of Arizona, Utah, and western Colorado. All other populations, including the eastern margin in Texas, the central plateau of Mexico, and the southern margin in Guatemala, have an equal and intermediate degree of inversion heterozygosity. The bottom line in table 29 gives illustrative examples of these generalities. Strawberry Canyon, California, has 71 percent heterokaryotypes, Mesa Verde in Colorado only 4 percent, and Austin, Texas, and Guatemala have 38 and 49 percent, respectively. Nor can we invoke ecological marginality to explain these figures, since the population density of *D. pseudoobscura* is vastly greater in Mesa Verde than in Guatemala. These wide variations in the extent of heterokaryotypy in different populations make a sharp contrast with the relative uniformity of genic heterozygosity shown in table 28. Of course only one chromosome arm is involved in the chromosomal polymorphism of this species. Nevertheless, there is certainly no relationship between marginality or centrality of a population and its average *genic* heterozygosity in *D. pseudoobscura,* especially if the third chromosome is discounted.

There is some disagreement about the explanation for the greater inversion polymorphism seen in central populations of species like *D. willistoni* and *D. robusta*. Dobzhansky (1951) believes the inversions are differentially adapted to various environmental conditions, so that central populations living in a more diverse environment will hold many inversions in stable equilibrium by natural selection, while in marginal environments one arrangement will be most fit because the range of environments is narrow. Carson (1959) believes, however, that special combinations of alleles must be selected in the stringent marginal environments, unlike those that are found in high frequency at the center. Therefore recombination is necessary at the margins in order to promote the formation of these new combinations, which are then driven to homozygosity by natural selection once they have arisen ("homoselection"). Both Dobzhansky and Carson agree that marginal environments demand more specialization than central ones. Their disagreement is only on whether the required specialized genotypes are already in existence, tied up in one or another of the inversions (Dobzhansky), or are novelties that must be produced by recombinational events (Carson).

My own view (Lewontin, 1957) has been quite different. First, in contrast to both Carson and Dobzhansky, I emphasize the *temporal instability* of marginal environments, so that the variation in selection at the margins is at least as great as it is at the center. Second, I have been skeptical (although on no evidence) whether *chromosomal* hetero- and homozygosity can be equated with *genic* hetero- or homozygosity. On the evidence that has now come to light from electrophoretic studies, it appears that such skepticism was justified. Marginal populations are as heterozygous as central ones, despite great differences in chromosomal polymorphism and despite associations of alleles with certain inversions. This does not mean that the chromosomal polymorphism is irrelevant. Quite the contrary. Dobzhansky seems to me correct in his view that in central populations with predictable, spatially diverse environments, a small number of distinct and diverse physiological and developmental modes will be selected. These modes are determined by coadapted genotypes tied up in inversions that prevent recombination. Carson seems to me right when he emphasizes the necessity of recombination in marginal populations in order to produce combinations of alleles that are not represented in the normal modes. But it is not some particular, specialized, homozygous genotype that is being selected in the marginal environment. In the highly unstable and unpredictable environment of the margin, quite different genotypes are being selected at different times. It is not surprising, then, that genic heterozygosity is high, and remains high, because no particular genotype is favored for very long. The metaphor of the laboratory is sometimes used to describe marginal populations. They are thought of as performing genetic "experiments." If so, these are frustrating experiments, the very opposite of Hershey's Heaven.

RACIAL VARIATION IN MAN

It is a commonplace that the human species is divided into more or less easily distinguishable racial groups. After all, everyone can tell a black African from a Chinese from a European, while to an African, one lank-haired, gray-faced European must look pretty much like another. But we must be careful. The apparent homogeneity within races as compared to the "obvious" difference between them stems partly from the fact that our consciousness of racial differences is constantly being reinforced socially because racial distinctions serve economic and political ends, and partly because the very characters we use to distinguish races—skin color and texture, hair form, eye, nose, and lip shape—are those to which we are most keenly attuned for the purpose of distinguishing individuals. When

TABLE 32
Allelic frequencies at seven polymorphic loci in Europeans and black Africans

	Europeans			*Africans*		
Locus	*allele 1*	*allele 2*	*allele 3*	*allele 1*	*allele 2*	*allele 3*
Red cell acid phosphatase	.36	.60	.04	.17	.83	—
Phosphoglucomutase-1	.77	.23	—	.79	.21	—
Phosphoglucomutase-3	.74	.26	—	.37	.63	—
Adenylate kinase	.95	.05	—	1.00	—	—
Peptidase A	.76	—	.24	.90	.10	—
Peptidase D	.99	.01	—	.95	.03	.02
Adenosine deaminase	.94	.06	—	.97	.03	—
Average heterozygosity per individual		.068 ± .028			.052 ± .023	

Note: From Harris (1970), modified by later data.

TABLE 33
Examples of extreme differentiation and close similarity in blood group allele frequencies in three racial groups

Gene	*Alleles*	*Caucasoid*	*Negroid*	*Mongoloid*
Duffy	*Fy*	.0300	.9393	.0985
	Fy^a	.4208	.0607	.9015
	Fy^b	.5492	—	—
Rhesus	R_0	.0186	.7395	.0409
	R_1	.4036	.0256	.7591
	R_2	.1670	.0427	.1951
	r	.3820	.1184	.0049
	r'	.0049	.0707	0
	others	.0239	.0021	0
P	P_1	.5161	.8911	.1677
	P_2	.4839	.1089	.8323
Auberger	Au^a	.6213	.6419	
	Au	.3787	.3581	
Xg	Xg^a	.67	.55	.54
	Xg	.33	.45	.46
Secretor	*Se*	.5233	.5727	
	se	.4767	.4273	

Note: From the summary of Cavalli-Sforza and Bodmer (1971).

we examine allele frequencies at randomly chosen loci, we get a rather different picture.

Harris (1970) studied 27 enzyme loci in Europeans and black Africans. Of these, 20 loci were identically monomorphic in both groups. The 7 polymorphic loci are shown in table 32. There is no strong differentiation between the groups with exception of the *PGM-3* locus, at which opposite alleles are in the majority in whites and blacks.

Among the 34 known blood group genes, again no case is known in which fixation for different alleles has occurred in different races, but some of the polymorphic genes show a good deal of geographical variation. Summaries of the 15 polymorphic loci are given for "Caucasoids," "Negroids," and "Mongoloids" by Cavalli-Sforza and Bodmer (1971), of which 6 show marked racial differentiation. Table 33 gives the three most extreme examples of differentiation together with the three most homogeneous cases. The ABO system, which shows some large divergences, especially in groups like the

American Indians (some of which are nearly pure O), nevertheless betrays a strong clustering of frequencies (Brues, 1954). Figure 16 shows a representative sample of diverse human groups plotted on a trilineal diagram, showing the strong clustering of ABO allele frequencies around 20 percent I^A, 15 percent I^B and 65 percent *i*.

It is possible to partition the total genetic variation observed in man into components ascribable to different levels of population aggregation. We can calculate the heterozygosity within national groups from allele frequencies. Let us call the average heterozygosity within national groups H_o. Suppose we now consider a

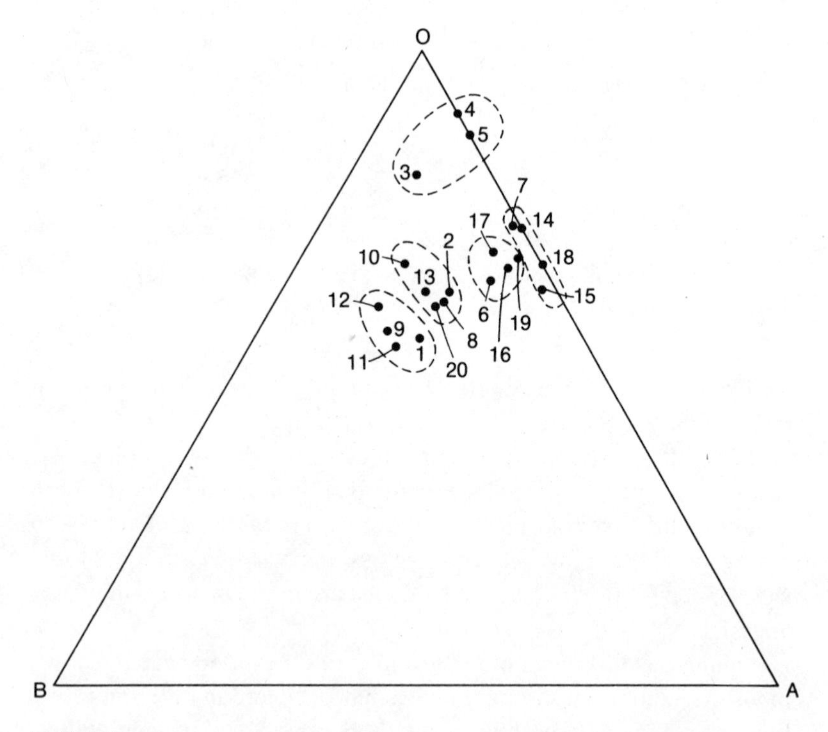

FIGURE 16

Trilineal diagram of the ABO blood group allele frequencies for human populations. Each axis represents one of the alleles (no distinction is made between A_1 and A_2) Each point is a human population. *1–3:* Africans; *4–7:* American Indians; *8–13:* Asians; *14, 15:* Australian Aborigines; *16–20*: Europeans. Dashed lines enclose arbitrary classes based on gene frequency, irrespective of "race." From Jacquard (1970).

hypothetical population made up by merging all the tribal or national populations within one race, for example, Ewe, Kikuyu, Pygmies, and others, making up one large African "race," and calculate the heterozygosity H_1 in this composite. If all the national groups were identical in allele frequencies, then the merger would have no effect and H_1 would be equal to H_0, but if there were some differentiation in allelic frequencies between nations or tribes, the merged group would be more heterozygous. In exactly the same way we could (on paper) merge all the world's races into a single large species group and calculate H_2, the total human species genetic diversity. The relative difference $(H_1 - H_0)/H_2$ is a measure of the proportion of the total genetic diversity that is due to differences between national or tribal groups, and $(H_2 - H_1)/H_2$ is the proportion of the total diversity that arises from racial differentiation.

At present 17 polymorphic genes have been well enough studied over a variety of human populations to make a reasonable calculation possible. Table 34 gives the values of H_0, H_1, and H_2 for these genes, together with the partition of the average heterozygosity. The actual values in the table are not heterozygosities but a nearly identical measure, the Shannon information, $H = -\Sigma\, p \ln p$ (Lewontin, 1972a). Although there is variation between loci in their relative contributions, the average values show that 85 percent of human genetic diversity is *within* national populations and only 7.5 percent between nations within races and 7.5 percent between major races. The relative division between the last two categories depends in part on how national groups are assigned to races and how many "races" are constructed. Are Hindi-speaking peoples of India Caucasoids or a separate group? Do they belong together with Vedic speakers? Where do the Turks or the Lapps belong? There are many possible ways of dividing groups, depending upon linguistic, morphological, or genetic criteria. But these decisions in no way affect the 85 percent variation within groups and only alter in a small way the relative sizes of the between-race and between-nation components.

In any case, the 85 percent within-population component is an underestimate, since all groups were equally weighted in the calculation. But small, isolated groups like the American Indians, Basques, Eskimos, and Australian Aborigines are usually the most deviant from the world average in allele frequencies. Then these very small

TABLE 34
Measure of human diversity within populations, H_o, within races, H_1, and for the whole species, H_2, for 17 loci

Locus	H_0	H_1	H_2	$\frac{H_0}{H_2}$	$\frac{H_1 - H_0}{H_2}$	$\frac{H_2 - H_1}{H_2}$
Haptoglobin	.888	.938	.994	.893	.050	.057
Lipoprotein Ag	.829	—	.994	.834	—	—
Lipoprotein Lp	.600	—	.639	.939	—	—
Xm	.866	—	.869	.997	—	—
Red cell acid phosphatase	.917	.977	.989	.927	.061	.012
6-Phosphogluconate dehydrogenase	.286	.305	.327	.875	.058	.067
Phosphoglucomutase	.714	.739	.758	.942	.033	.025
Adenylate kinase	.156	.160	.184	.848	.022	.130
Kidd	.724	.930	.977	.741	.211	.048
Duffy	.597	.695	.938	.636	.105	.259
Lewis	.960	.993	.994	.966	.033	.001
Kell	.170	.184	.189	.899	.074	.027
Lutheran	.106	.139	.153	.693	.215	.092
P	.949	.978	1.000	.949	.029	.022
MNS	1.591	1.663	1.746	.911	.041	.048
Rh	1.281	1.420	1.900	.674	.073	.253
ABO	1.126	1.204	1.241	.907	.063	.030
Average				.849 within populations	.075 between populations within races	.075 between races

Note: From Lewontin (1972a).

populations make a greatly excessive contribution to the between-group comparisons. Since most of the world's population is made up of Chinese, Indians, Europeans, and the recently hybridized populations of South America, who vary less from each other than do the small isolated groups, the correct proportion of human genetic variation that is within nations or tribes is closer to 95 than to 85 percent.

The taxonomic division of the human species into races places a completely disproportionate emphasis on a very small fraction of the total of human diversity. That scientists as well as nonscientists nevertheless continue to emphasize these genetically minor differences and find new "scientific" justifications for doing so is an indication of the power of socioeconomically based ideology over the supposed objectivity of knowledge.

Indeed the whole history of the problem of genetic variation is a vivid illustration of the role that deeply embedded ideological assumptions play in determining scientific "truth" and the direction of scientific inquiry. Those who, like Monod (1971), think that facts speak for themselves will suppose that the struggle between the classical and balanced schools is over, having been decisively concluded by the hard observations of the new molecular population genetics. But they will be wrong. The classical hypothesis has been developed in extended form, feeding upon, digesting, assimilating, and waxing fat on the very facts that were meant to give it fatal indigestion. It is not the facts but a world-view that is at issue, a divergence between those who, on the one hand, see the dynamical processes in populations as essentially conservative, purifying and protecting an adapted and rational *status quo* from the nonadaptive, corrupting, and irrational forces of random mutation, and those, on the other, for whom nature is process, and every existing order is unstable in the long run, who see as did Denis Diderot that "Tout change, tout passe, il n'y a que le tout qui reste."

CHAPTER 4 / THE GENETICS OF SPECIES FORMATION

It is an irony of evolutionary genetics that, although it is a fusion of Mendelism and Darwinism, it has made no direct contribution to what Darwin obviously saw as the fundamental problem: the origin of species. I do not mean to say that the theory and observations of population genetics have not influenced and even permeated theories of species formation. One has only to read Dobzhansky's *Genetics and the Origin of Species* (1951) or Mayr's *Animal Species and Evolution* (1963) to see how population genetics informs modern ideas about speciation. Mayr's whole thesis, that speciation is the tearing apart of a group of genetically interconnected populations, normally held together by strong cohesive forces of migration and natural selection, comes directly out of the balance school of population genetics, which he has assimilated into the theory of geographic speciation. But it is a long way from describing speciation in general genetic terms to constructing a quantitative theory of speciation in terms of genotypic frequencies. While it is a question of elementary population genetics to state how many generations will be required for the frequency of an allele to change from q_1 to q_2, we do not know how to incorporate such a statement into a speciation theory, in large part because *we know virtually nothing about the genetic changes that occur in species formation.*

How much of the genome is involved in the early steps of divergence between two populations, causing them to be reproductively isolated from each other? We do not know. Mayr writes of a "genetic revolution" in speciation, but we cannot put quantitative limits on this revolution (which may after all turn out to be only a minor reform) until we begin to characterize the *genetic* differences between populations at various stages of *phenotypic* divergence. The whole concept of a genetic revolution arises out of the undoubted truth that every gene affects every character, that "no gene frequency can be changed, nor any gene added to the gene pool, without an effect on the genotype as a whole, and thus indirectly on the selective value of any other genes" (Mayr, 1963, p. 269).

But it does not follow that every gene substitution really matters. It may be true "that thou canst not stir a flower without troubling of a star," but the computer program for guiding a space capsule does not, in fact, have to take my gardening into account. General principles are not the same as quantitative relationships.

The problem of making quantitative statements about the multiplication of species has been that we have been unable to connect the phenotypic differentiation between populations, races, semispecies, and species with particular genetic changes. It is our old problem in a new context. The "stuff of evolution," the subtle changes in cell physiology, developmental pattern, behavior, and morphology that lead to reproductive isolation and ecological differentiation are the observables, but the only entities for which we have been able to construct a dynamic theory are genotypes. Until we are able to specify the genotypic differences between populations at various levels of phenotypic differentiation, we will not have the beginnings of a quantitative genetic theory of speciation. Even when we are able to measure genotypic differentiation, however, it will be only a beginning, since in the end we must know in what way particular genetic differences are related to the particular reproductive and ecological properties that separate two species.

The general theory of geographic speciation postulates a multistage process after an initial geographical isolation. Populations must become geographically isolated from each other because even a small amount of migration will prevent genetic differentiation between populations unless some extraordinarily strong selection virtually fixes different alleles in different populations. Although it is theoretically possible for disruptive selection to lead to reproductive isolation without geographical isolation, Mayr makes a strong and convincing case that such a phenomenon must be rare in nature.

Thoday and Boam (1959) and Thoday and Gibson (1962) succeeded in producing stable sexual isolation by disruptive selection without isolation, but some unusual genetic condition must have underlain their result, since several attempts to produce the phenomenon again have failed (for example, Scharloo, 1971). Attempts to produce sexually isolated populations within a species by selecting against hybrids between them have succeeded, but the isolation that appeared during the course of the selection disappears if the populations are again allowed to interbreed (Wallace, 1954; Knight, Robertson, and Waddington, 1956). If there is any element of the theory of speciation that is likely to be generally true, it is that geographical isolation and the severe restriction of genetic exchange between populations is the first, necessary step in speciation.

The first stage of speciation is the appearance, after some period of isolation, of genetic differences that are sufficient to restrict severely the amount of gene exchange that can take place between the populations if they should again come into contact. In addition, there is some divergence in ecological niche and, indeed, this divergence might be the direct cause of the reproductive isolation. Changes in feeding preference, pattern of diurnal activity, breeding season, and so on could lead to micro-temporal or micro-spatial overlapping. The question of the relation between ecological divergence and reproductive isolation is open. How often does reproductive isolation arise "accidentally," even in the absence of any other differentiation? How often is it simply an indication of a general genetic divergence for basic developmental patterns? How often is it the direct consequence of ecological divergence? We do not know, because the genetic analysis and ecological study necessary to answer these questions have not been, and for the most part cannot be, carried out on numerous populations in the early phases of the speciation process.

The second stage in speciation occurs if and when the isolated populations come into contact again. If sufficient reproductive isolation of the proper sort developed while the populations were allopatric, there may be a reinforcement of the reproductive barriers by natural selection. If, during isolation, physiological differences have arisen that cause hybrid offspring to be less viable or fertile, then individuals that mate heterospecifically will leave fewer genes to future generations, and there will be selection for characters that reduce the amount of heterospecific contact. There is an interesting quantitative problem here. How strong must the reproductive barrier be between the two newly sympatric populations in order that the differences they have accumulated in isolation from each other not be swamped out by the crossing between them? The theory of geographic speciation rests on the assumption that even a little migration between populations is enough to prevent them from speciating in the first place. Hence the requirement for geographical isolation. Does this mean that even a little gene flow between populations newly come together will destroy their differentiation, so that successful speciation demands the virtually *complete* rejection of genes flowing from one population to the other?

During the second stage there will also be selection for ecological divergence between the forming species, at least until they have the minimum niche overlap necessary for stable coexistence with each other and with the other species in the community. Although the principle that newly formed species must diverge ecologically from each other has long been asserted and was given an explicit form in Brown and Wilson's concept of character displacement (1956), the foundations of a quantitative and predictive theory of the limiting similarity between coexisting species have only recently been laid by MacArthur and Levins (1967). It is important to realize that selection for niche divergence and selection for secondary sexual isolation are not necessarily independent phenomena, but that, on the contrary, divergence in life habits may in itself usually lead to sexual isolation by reducing temporally and spatially the opportunity for the union of gametes from the two newly formed species.

There is yet a third stage in the evolution of species, which is not generally thought of as part of the speciation process. The newly formed species continue to evolve, but not in response to each other. Each becomes simply a part of separate communities of species undergoing phyletic evolution, splitting, and extinction, with no special relation to each other as the offspring of a common ancestor. Then the degree of genetic similarity between contemporaneous representatives of the two new phyletic lines reflects only the total passage of time since the original split, and the average speed of evolution of a more or less long series of evolutionary episodes.

The importance for us of distinguishing the three stages of species divergence is that the genetic questions involved at each stage are very different and will be totally confounded if we ask only What is the genetics of speciation? or What is the genetic difference between two species? At the first stage we want to know how much and what kind of genetic differentiation is required for primary mechanisms of reproductive isolation to arise. Is it true in general, as Mayr believes, that a genetic revolution occurs as a result of a reduction in population size, followed by selection of quite new genotypes, and that reproductive isolation is the incidental concomitant of this revolution? Or is a very small fraction of the genome involved, so that at the end of the first stage much the greatest part of the genome is unchanged? Even if there are substantial changes in the greater part of the genome, is the reproductive isolation a result of differentiation of a few loci only?

At the second stage we want to know how much more genetic divergence must occur to produce ecologically differentiated, stable members of the species community. If only minor differentiation occurs during geographical isolation, the major portion of genetic diversification during speciation may occur during the second, sympatric stage when natural selection is operating to push the new entities apart. Alternatively, stage two may be only a fine-tuning process in which the finishing touches are put on an already accomplished speciation.

Finally, the questions about the third stage are really questions of genetic systematics. That is, if we consider species judged by the criteria of evolutionary systematics to be more closely and less closely related, and therefore more recently and less recently diverged from a common ancestral population, how much genetic similarity is there between them? What is the rate of independent genetic divergence in absolute and "taxonomic" time?

For historical and pedagogical reasons it is better to take up these questions in reverse order—to begin with the genetic differences between species, about which we know something, and to end with the first steps in speciation, which are almost totally obscure.

GENETIC DIFFERENCES BETWEEN SPECIES

Once again we are faced with the methodological contradiction. Species are, by definition, reproductively isolated from each other, but genetic investigations are carried out by making crosses. The contradiction is not absolute, because there do exist some pairs of species which, although completely isolated reproductively in nature, can be made not only to hybridize, but even to yield F_2 generations and backcrosses in the laboratory or garden. It is nevertheless true that as long as doing genetics meant making successful crosses, investigations of the genetics of species differences had to be restricted to very closely related, partly compatible species. The problem is less severe in plants than in animals because isolating mechanisms in plants are more often meiotic than developmental, so that if enough crosses are made, a few seeds will be set and these can be grown for further analysis.

Most of the analyses of genetic differences between species (see Mayr, 1963, p. 543; Dobzhansky, 1970, pp. 261–63) have concentrated on the morphological characters that are used to differentiate them. In a few cases a single gene substitution with some modifiers may account for a clear-cut species difference. More often there appear to be many, but an unknown number of, gene differences involved. In any event, it is impossible to estimate what proportion of the genome differentiates the species.

A second major line of investigation has been on the nature of the sterility in hybrids between species (see Dobzhansky, 1951, chapter 8). Nearly all of these analyses have been at the chromosomal level, showing that the haploid sets from the two parental species fail to pair properly at meiosis, or that there is an incompatibility between the chromosomes of one species and the cytoplasm of another, or that sex determination is disturbed by mixtures of sex chromosomes or autosomes from different species.

Perhaps the case that was best worked out, given the limitations of the methods available, was Dobzhansky's analysis of male sterility in the hybrids between *Drosophila pseudoobscura* and *D. persimilis* (1936). Crosses between these two sibling species are easy to make and produce large numbers of fully fertile F_1 females and completely sterile F_1 males. When the cross is made of *D. persimilis* ♀ × *D. pseudoobscura* ♂, the sterile F_1 males have testes about one-fifth normal size, while the reciprocal cross gives sons with normal-sized testes, although these males too are completely sterile. Backcrosses of F_1 females to either parental species are possible, and the backcross sons have testes of varying size and are variously fertile, testis size and fertility being correlated. By means of marker chromosomes, Dobzhansky was able to show that there are at least two genes on each of the large chromosomes influencing testis size and that an interaction between the sex chromosome of one species and the autosomes of the other was a predominant effect.

Whether they are concerned with morphological differences or sterility, such studies are essentially attempts to work out the genetics of complex quantitative characters. Even under the best conditions, within a species that has an excellent array of marker genes and chromosomal aberrations available as tools, the genetic dissection of a quantitative character is a difficult and somewhat slippery affair, and the results, although framed in terms of "genes," are really in terms of chromosome segments.

For species crosses in which generations beyond the F_1 are difficult to produce, in which there is a paucity of mutant markers, and in which many of the critical genotypes may barely survive, if at all, one can usually do very little beyond assigning influences to different chromosomes. Moreover, no matter how fine the analysis, the dissection of a particular morphological difference between species, or even of the sterility barrier between them, is not really an answer to the question of how much genetic difference there is between two species, of what the relative genetic divergence is between taxa of different evolutionary relationship.

The ambiguity of continuously varying phenotypes can be overcome by studying simple Mendelizing differences between species; by analogy with the same problem in the study of genetic variation within a species, antigenic differences come immediately to mind. If nonspecific antigenic tests are made—for example, by preparing antiserum against tissue extracts of one species and measuring the intensity of reaction of this antibody against other species—nothing has been gained since we are dealing again with a genetically unanalyzable quantitative character. It is necessary to make antibodies against single antigens specified by single genes and then to survey species for the possession of homologous antigens. This technique was utilized by Irwin (1953) to study several genera of doves (Columbidae). He first found nine antigenic differences between the blood cells of *Streptopelia chinensis* and *S. resoria*, establishing by species crosses that these were the result of nine separate gene differences. He then tested the blood cells of 23 other species against the nine specific antisera, with results shown in table 35. Antigen d-1, for example, seems widely distributed over the

TABLE 35
Number of species in different genera of the Columbidae that possess red blood cell antigens homologous to nine antigens of *Streptopelia chinensis*.

	Genus Streptopelia		*Genus Columba*		*Other genera*	
Antigen	*present*	*absent*	*present*	*absent*	*present*	*absent*
d-1	6	2	5	6	5	1
d-2	3	5	4	7	0	6
d-3	4	4	2	9	0	6
d-4	5	3	3	8	3	3
d-5	2	6	5	6	0	6
d-6	2	6	4	7	0	6
d-7	2	6	9	2	0	6
d-11	6	2	8	3	0	6
d-12	4	4	4	7	0	6
Total	34	38	44	55	8	46

Note: From Irwin (1953).

family whereas d-3 is nearly restricted to *Streptopelia*. Congeners of *S. chinensis* have slightly more similarity to it than do species of *Columba*, and the two genera together are more similar than are other genera. One species, *Streptopelia senegalensis,* was homologous to *S. chinensis* for all nine antigens.

Irwin could further test whether the homologies were indeed identities by saturating an antiserum with the cells of one species and then seeing whether the antiserum still retained any power against the cells of its own species. If not, then presumably the cells of the two species had identical antigens. Except for *S. senegalensis*, four of whose antigens were identical with *S. chinensis*, nearly all other cases of homology failed to show identity. The failure of this approach to the genetic similarity between species is, as Irwin points out, that we do not know how many red cell antigens the species are identical, because the tests depended upon finding an antigenic difference between *S. chinensis* and *S. risoria* in the first place. As I pointed out in chapter 3, this is the problem in general of using antigens to measure genetic differences. There has to be a difference to begin with for a gene to be detected at all.

The methodological problems of measuring the differences between the genomes of different species are the same as for comparing different individuals within a species except that there is the added problem that most species cannot be crossed with each other.

There is all the more reason, then, to have a method which does not depend totally on classical genetic analysis and in which the invariant genes can be detected as well as differentiated ones. We are again led to the study of a random sample of specific enzyme and protein molecules by means of a comparison of their electrophoretic mobilities. By use of proteins, especially enzyme proteins, whose genetics have been established by intraspecific study, species can be sampled even when they cannot be crossed, although crosses should be made whenever possible to establish gene homologies.

The pioneering work in this field was the study of *Drosophila virilis* and its relatives by Hubby and Throckmorton (1965). They compared the electrophoretic mobility of the soluble proteins from ten species of the *virilis* group. Different numbers of bands appeared in different species, from 29 in *D. lacicola* and *D. flavomontana* to 42 in *D. texana*. Because no genetic work had then been done, protein bands could not be assigned to loci, so that only presence or absence of a band at a particular location in the gel could be scored. If, for example, a locus had ten electrophoretically different alleles, each fixed in a different species, a total of ten potential band positions would be scored and each species would show "presence" for one and absence for the other nine. If, on the other hand, each species had a unique locus specifying a protein, not shared by the other species, there would again be ten potential positions in the total study and each species would have a band at one and lack it at the nine other positions. There is thus a confounding between different alleles at one locus and the presence or absence of the products of different genes.

A second problem is that with so many protein bands, a change in electrophoretic mobility of the protein from one locus could simply move the band into a position superimposed on a protein from another locus.

Third, it is not certain that each band within a species is coded by a different gene. Indeed, if any of the strains tested were polymorphic (only one strain per species was tested), a difference between species might be assumed that was only a chance difference between strains.

Finally, if the proteins form complexes with small molecules that affect the net charge of the complex, then a whole group of bands from one species might be shifted relative to another species, although the gene products were the same.

All of these difficulties make the analysis imprecise, but they do not create any major biases (except for the last problem, which, if widespread, would be serious). Each band is, to a first approximation, a gene product, and a different mobility means an elementary genetic difference. Unfortunately, as in all electrophoretic work, a lack of difference is ambiguous since many gene changes are not reflected in charge. *This last fact means that all estimates of genetic differences between species based on electrophoretic studies of proteins are underestimates*, by perhaps a factor of three.

Table 36 is a summary of Hubby and Throckmorton's findings. As little as 2.6 percent of the proteins of a species may be unique to it (*D. virilis*) or as many as 28.2 percent, with an average of 14.3 percent. This does not mean, however, that 14 percent of the proteins of these species have arisen since they diverged from their common ancestor, or even since the common ancestor of two very closely related forms. In any pair of species that have derived from a common ancestor, one of the unique alleles may be ancestral.

The probable phylogenetic tree of the ten species in Hubby and Throckmorton's study, based on chromosomal rearrangements and protein similarities, shows that the ten extant species most probably

TABLE 36

Percentages of larval proteins that are unique to a species, shared with other species in the phylad, and shared with some member of the species group as a whole, for ten species of the *Drosophila virilis* group

Species	*Total bands*	*% Unique*	*% Common to phylad*	*% Common to species group*
Virilis phylad				
D. americana	38	5.3	23.7	71.1
D. texana	42	21.4	16.7	61.9
D. novamexicana	38	7.9	21.1	71.1
D. virilis	38	2.6	21.1	76.3
Montana phylad				
D. littoralis	39	28.2	25.6	46.2
D. ezoana	35	8.6	25.7	65.7
D. montana	37	18.9	37.8	43.2
D. lacicola	29	20.7	20.7	58.6
D. borealis	42	19.0	28.6	40.6
D. flavomontana	29	10.3	37.9	51.7
Average	36.6	14.3	25.9	59.8

Note: Data are from Hubby and Throckmorton (1965).

trace back to four immediate ancestral forms, so that 40 percent of unique proteins could nevertheless be ancestral. Making this maximum allowance, we conclude that at a minimum 8.5 percent of the proteins in the extant species have arisen since their speciation. How much of this has been "phase 3" evolution, and how much stems from the initial speciation events, we cannot tell. A further 25.9 percent of the proteins are unique to one of the two phylads that make up the species group but common to species within the phylad. By the same reasoning, half of these could have been present in the ancestral form, so that as little as 15 percent of the protein species may have arisen since the origin of the two phylads and, adding this to the minimum estimate from unique species proteins, we get a minimum estimate of 23.5 percent for the proportion of proteins that have changed from the ancestral form. Again we must bear in mind that this is doubly a minimum, since it takes into account only the electrophoretically identifiable changes.

Recently J. Hubby, L. Throckmorton, and R. Singh have shown that "alleles" identified by electrophoresis are, in fact, heterogeneous classes in some cases. By determining the heat stability of *xanthine dehydrogenase* and *octanol dehydrogenase* alleles in *Drosophila virilis* and its close relatives, these workers have detected amino acid substitutions within electrophoretic classes, thus multiplying the number of different alleles segregating within a species by a factor of 2.5 and in the species group as a whole by a factor of 3.8. In turn, some species thought to be identical with respect to allelic composition at the *octanol dehydrogenase* or *xanthine dehydrogenase* loci turn out to be carrying different alleles that are detected by the heat-stability criterion.

The second study of Hubby and Throckmorton (1968) was an advance in precision because, in place of the large number of proteins of unknown genetic control, it was restricted to enzymes and larval hemolymph proteins, each of which is the product of a separate gene. Nine triads of species were chosen, two members of each triad being sibling species, the third being a nonsibling member of the same species group. A list of the species is given in the first column of table 37, the first two species in each group being the siblings. The table shows the percentage of genes with identical alleles between siblings (I), the percentage of loci with an allele identical between one of the siblings and the nonsibling (II), and the percentage of loci with identical alleles for all three species. Because

TABLE 37
Percentages of loci with identical alleles between sibling species (I), between a sibling and nonsibling in the same species group (II), and among all three species of a triad (III) for nine triads of species in the genus *Drosophila*

Triads	I	II	III	I + III	II + III
arizonensis *mojavensis* *mulleri*	42.1	6.3	6.3	48.4	12.6
mercatorum *paranaensis* *peninsularis*	55.0	11.8	11.8	66.8	23.6
hydei *neohydei* *eohydei*	43.8	3.2	6.3	50.4	9.5
fulvimaculata *fulvimaculoides* *lemensis*	50.0	13.2	15.8	65.8	29.0
melanica *paramelanica* *nigromelanica*	26.3	10.0	5.3	31.6	15.3
melanogaster *simulans* *takahashii*	52.9	7.9	0.0	52.9	7.9
saltans *prosaltans* *emarginata*	36.8	7.7	10.5	47.3	18.2
willistoni *paulistorum* *nebulosa*	7.1	11.6	15.4	22.5	27.0
victoria *lebanonensis* *pattersoni*	64.3	0.0	21.4	85.7	21.4
Average	42.0	7.9	10.4	52.4	18.3

Note: Data are from Hubby and Throckmorton (1968).

only a single strain was tested for each species, the question of polymorphism within species does not enter into the observations (except in one case), although it does enter into the interpretation of the results, as I shall show.

The percentage of identity between siblings (I + III) varies considerably from group to group. *Drosophila victoria* and *D. lebanonensis*, for example, are nearly identical siblings, with the same allele at 85.7 percent of their loci, whereas *D. willistoni* and *D. paulistorum* are identical only at 22.5 percent of their loci. There is rather less variation for the comparison between nonsiblings (II + III). On the average, sibling species differ at 47.6 percent of their loci and nonsiblings at 81.7 percent. This study thus shows a much greater degree of genetic divergence than the results on *D. virilis* suggested. There is such immense variation between triads, however, that the *D. virilis* results are well within the range observed.

It is sometimes supposed that sibling species are species *in statu nascendi*, entities in the process of speciation. But as Mayr (1963) explains, that is certainly not the case. Sibling species are those for which the taxonomist is unable, or was unable at one time, to find reliable morphological differences despite clear-cut reproductive isolation. But similarity is in the eye of the beholder (the flies seem to have no difficulty in doing their own taxonomy!), and subsequent more careful study has sometimes led to the discovery of distinctions. However, the close morphological similarity is not without meaning, or the entire practice of morphological systematics would have resulted in a meaningless jumble. Total morphological similarity is, if not an infallible guide, at least a reliable indication of genetic similarity, and the data of Hubby and Throckmorton are an elegant confirmation of this proposition. Sibling species have three times as much genetic similarity as nonsiblings, and only in one case out of nine (*willistoni-paulistorum-nebulosa*) is the genetic similarity greater for the nonsiblings than the siblings

A certain care must be exercised in interpreting the large genetic differences between closely related species in the Hubby and Throckmorton data. Only a single strain, virtually always homozygous as it turned out, was examined for each species. Suppose that two species differ only in allele frequency at a polymorphic locus, one with alleles *A* and *a* in frequency 0.8 and 0.2, respectively, the other with the reverse frequencies. Then a little more than two-thirds of the time, single strains from each species will be fixed for opposite alleles, and the two species will be judged different. But how great, biologically, is the difference between a 0.8:0.2 and a 0.2:0.8 polymorphism? It is certainly not a basis for reproductive or ecological isolation. Is it possible that much of the difference between the species is variation in frequencies of polymorphic loci? Of the species listed in table 37, only four, *D.*

TABLE 38
A comparison of allele frequencies at 13 polymorphic loci in *D. pseudoobscura* and *D. persimilis*

Locus	*Alleles*	*D. persimilis*	*D. pseudoobscura*
Pt-7	*0.75*	1.00	.95
	others	—	.05
Pt-12	*1.18*	1.00	.55 (.20)[a]
	1.20	—	.45 (.80)
Pt-13	*1.23*	—	.06
	1.30	1.00	.94
Malic dehydrogenase	*1.00*	1.00	.97
	1.20	—	.03
Octanol dehydrogenase	*1.00*	1.00	.97
	1.22	—	.03
Acetaldehyde oxidase-2	*1.00*	1.00	.94
	others	—	.06
Pt-8	*0.80*	.02	.02
	0.81	.88	.47
	0.83	.10	.51
Glucose-6-phosphate dehydrogenase	*1.00*	.79	1.00
	1.10	.21	—
Leucine amino peptidase	*1.00*	.71	.89
	1.10	.18	.05
	1.12	.11	—
	others	—	.06
Pt-10	*1.02*	—	.005
	1.04	1.00	.615 (1.00)[a]
	1.06	—	.380

melanogaster, D. simulans, D. willistoni, and *D. paulistorum*, have been tested subsequently for electrophoretic polymorphism; between 42 and 86 percent of their loci are polymorphic. Clearly we need more refined information on the array of alleles present in natural populations of a pair of species before we can have a full appreciation of their specific differences.

The first detailed study of allelic frequency patterns in two closely related species was Prakash's (1969) comparison of natural populations of *D. pseudoobscura* and its sibling species, *D. persimilis*. The range of *D. persimilis* is, as far as is known, completely included within that of *D. pseudoobscura*, being confined to higher and moister localities in California, Oregon, Washington, and British Columbia. There is a broad spatial and temporal overlap in the dis-

Locus	Alleles	D. persimillis	D. pseudoobscura
Xanthine dehydrogenase	*0.99*	—	.26
	1.00	.22	.60
	1.02	.68	.01
	1.04	.10	—
	others	—	.13
Amylase	*0.74*	—	.03
	0.84	.02	.29
	0.92	.08	—
	1.00	.54	.68 (1.00)[a]
	1.05	.06	—
	1.09	.03	—
Esterase-5	*0.95*	—	.12
	1.00	—	.43
	1.02	—	.02
	1.03	—	.08
	1.07	.02	.19
	1.09	—	.01
	1.12	.04	.13
	1.16	.06	.02
	1.20	.74	—
	1.29	.02	—
	1.33	.12	—

Note: Data are from Prakash (1969) and Prakash, Lewontin, and Crumpacker (1973).

[a]Frequencies in the Standard gene arrangement of the third chromosome. *D. persimilis* arrangements belong to the Standard phylad of inversions.

tribution of the two species. They can be crossed, and the chromosomes are completely homologous, differing only in a few large inversions. Both species have extensive inversion heterozygosity on chromosome III and share one arrangement, Standard (see figure 14). There is no question about the homology of loci in the species, and it has been shown, for example, that hybrid enzyme is formed between esterase monomers produced by the two species, both in species hybrids and in *in vitro* dimerization (Hubby and Narise, 1967).

Prakash studied 24 loci in *D. persimilis* from Mather, California. Of these, 11 are monomorphic in *D. pseudoobscura*; they turned out to be *identically* monomorphic in *D. persimilis*. The remaining 13 loci are shown in table 38, compared with allelic frequencies in the

Strawberry Canyon population of *D. pseudoobscura*. For third chromosome loci, gene frequencies in the Standard arrangement of *D. pseudoobscura* are also given, since *D. persimilis* is related to the Standard phylad of inversions.

The results are remarkable. Except for the *esterase-5* locus, where the overlap in gene frequencies is small, there is very little differentiation in gene frequencies. Indeed, in 10 out of the 13 cases the major allele is the same in both species. If we add the 13 cases where both species are identically monomorphic, there is only slight or no gene frequency differentiation at 21/24 or 88 percent of loci, clear quantitative differentiation in allele frequencies at 2 loci, *pt-8* and *xanthine dehydrogenase,* and something approaching qualitative differentiation at 1 locus, *esterase-5*. There are sporadic occurrences of alleles that are unique to one species or the other, but except for the *esterase-5* locus no unique allele reaches a frequency greater than 25 percent. Most important, there is not a single case of fixation or even near-fixation for alternative alleles in the two species.

If there are "species-distinguishing" genes as indeed we suppose there must be, since these species are ecologically differentiated and reproductively isolated, they have not been picked up in a random sample of 24 loci. Thus, even if such species-differentiating genes are large in absolute number, they must be a small fraction of the whole genome, almost surely less than 10 percent of it. An alternative is that there are no such species-distinguishing genes but that the difference between species lies in the accumulation of quantitative differences in allelic frequencies, as in the case of the *esterase-5* locus. This latter hypothesis is not particularly attractive because it assumes that species differences simply represent very low probabilities of total genetic identity between individuals. Yet with the degree of polymorphism within species that has been revealed, the probability of genetic identity within a species is already essentially zero. For example, using only the 20 most polymorphic genes known at present in man, the probability of genetic identity between two Englishmen is already less than 10^{-6}. It seems more reasonable to suppose that *D. persimilis* and *D. pseudoobscura* do indeed differ completely at certain loci, like those found by Dobzhansky in his tudy of sterility in their hybrids, but that only a special part of the genome is involved, while most of the genome remains undifferentiated. The discovery of considerable genetic heterogeneity within electrophoretic classes, through heat denaturation studies (see p. 169), might alter this conclusion significantly.

Ayala and Powell (1972) have asked a somewhat different question about such data. If we look at the species with the eyes of the systematist, we can ask whether a locus such as the *esterase-5* gene would be a good diagnostic character for the species. This can be done by calculating the frequency of the diploid genotypes at a locus in each species and then adding up the overlap in the distributions. If unknown individuals were assigned to that species in which their genotype was the more frequent, the error of assignment would be half the overlap. We can call a locus "diagnostic" if the probability of error is small, 0.01 or less. On this criterion only the *esterase-5* locus is diagnostic in the Prakash sample, the next best locus being *xanthine dehydrogenase,* with a 5 percent overlap in diploid genotypes.

The value of Ayala and Powell's approach is that it draws attention to the diploid genotypic distributions, which are, after all, what matter to the organisms, rather than to the gene frequency distributions, which are more of an abstraction. Ayala and Powell report finding 3 more loci, not examined by Prakash, that are diagnostic for *D. pseudoobscura* and *D. persimilis* although they are not as differentiated as *esterase-5*, and 12 more loci that do not differentiate the species. Thus there are 4 diagnostic loci out of 39.

Another pair of sibling species, *D. melanogaster* and *D. simulans*, were examined by Kojima, Gillespie, and Tobari (1970). The result of studying 17 enzyme loci was much the same as for *D. pseudoobscura*. No loci were fixed for alternative alleles in the two species, although one case, *aldehyde oxidase,* was highly polymorphic in *D. simulans* but monomorphic in *D. melanogaster* for an allele with frequency less than 0.01 in the former species. Several polymorphic loci had different major alleles in the two species, but only 2 (11 percent) were diagnostic at the 0.01 level. On the whole there is more differentiation of allelic frequencies at the highly polymorphic loci than is the case for *D. pseudoobscura* and *D. persimilis*, and unique alleles are a more common occurrence. The general picture, however, is not different in the two cases.

The most extensive and, in many ways, the most interesting, detailed comparison of gene frequencies among species is the work of

TABLE 39
Frequencies of alleles at 12 "diagnostic" loci in four species of *Drosophila*. Several alleles that occur with low frequencies are not included

Gene	Alleles	*Drosophila willistoni*	*Drosophila tropicalis*	*Drosophila equinoxialis*	*Drosophila paulistorum*
Lap-5	*0.98*	.09	.02	—	—
	1.00	.29	.19	—	—
	1.03	.50	.63	.004	.004
	1.05	.09	.15	.21	.08
	1.07	.007	.01	.71	.86
	1.09	—	—	.07	.04
Est-5	*0.95*	.03	—	.03	.03
	1.00	.96	—	.94	.84
	1.05	.01	—	.02	.13
Est-7	*0.96*	.02	.02	—	—
	0.98	.16	.11	—	—
	1.00	.54	.62	—	.002
	1.02	.23	.23	—	.08
	1.05	.05	.03	—	.78
	1.07	.003	.001	—	.09
Aph-1	*0.98*	.02	—	—	—
	1.00	.84	.05	.02	.01
	1.02	.08	.90	.92	.93
	1.04	.06	.04	.06	.03
Acph-1	*0.94*	.05	.95	.01	—
	1.00	.92	.03	.17	—
	1.04	.02	.006	.81	.16
	1.06	—	—	—	.21
	1.08	—	—	.004	.62
Mdh-2	*0.86*	.001	.994	.003	.001
	0.94	.02	.005	.994	.993
	1.00	.97	—	.004	.006

Ayala and Powell (1972) on the four sibling species of the *Drosophila willistoni* group, *D. willistoni, D. paulistorum, D. equinoxialis*, and *D. tropicalis*. These tropical South and Central American species differ very much in their ecological latitude and geographical range. *D. willistoni* is the most inclusive species, ranging from southern Florida and Mexico to northern Argentina. *Drosophila paulistorum* is absent from the northern part of this range but is very common in South America, being dominant in the drier and moderately humid areas, whereas *D. willistoni* is the major species in the superhumid tropics. *Drosophila tropicalis* and *D. equinoxialis* are absent both in southern Brazil and in Florida and

Gene	Alleles	Drosophila willistoni	Drosophila tropicalis	Drosophila equinoxialis	Drosophila paulistorum
Me-1	*0.90*	—	.03	—	—
	0.94	—	.91	—	.004
	0.98	.02	.06	—	.99
	1.00	.95	—	.005	.005
	1.04	.02	—	.99	—
Tpi-2	*0.94*	.003	.01	—	.02
	1.00	.98	.98	.02	.98
	1.06	.01	.01	.98	—
Pgm-1	*0.96*	.04	—	.01	.02
	1.00	.87	.01	.35	.94
	1.04	.08	.98	.62	.04
Adk-2	*0.96*	.01	.05	—	—
	0.98	.05	—	—	—
	1.00	.88	.92	.04	.98
	1.02	.05	—	—	—
	1.04	.004	.03	.94	.02
Hk-1	*0.96*	.04	.02	.08	—
	1.00	.95	.96	.91	.01
	1.04	.006	.02	.005	.97
	1.08	—	.001	.002	.02
Hk-3	*1.00*	.98	.97	.95	.07
	1.04	.006	.01	.04	.92

Note: Data are from Ayala and Powell (1972). A dash indicates that the allele has not been found in the species.

Lap = leucine aminopeptidase; *Est* = esterase; *Aph* = alkaline phosphatase; *Acph* = acid phosphatase; *Mdh* = malic dehydrogenase; *Me* = malic enzyme; *Tpi* = triose phosphate isomerase; *Pgm* = phosphoglucomutase; *Adk* = adenylate kinase; *Hk* = hexokinase.

Mexico and have their highest proportions in the Greater Antilles and on the Caribbean coast of Colombia and Panama, where they are a major and sometimes dominant proportion of the four species (Burla et al., 1949). On the basis of chromosomal evidence, *D. paulistorum* and *D. equinoxialis* are more closely related to each other than they are to the other two species and, as we will see, this is borne out by the genic evidence.

The data of Ayala and Powell are worth displaying, both for their value in the present context and for our future discussion of the meaning of polymorphism (chapter 5). Table 39 gives the allelic frequencies at 12 out of the 28 loci examined. The loci displayed are those that are "diagnostic" for at least one species in each case, and so represent the loci at which there is significant differentiation. Even among these diagnostic loci there are some extraordinary similarities, as for example *esterase-5*, which is included, presumably, because no activity at this locus was detected in *D. tropicalis*, but which is nearly identical in frequency distribution in the other three species. Although there is only a single case in which the distributions have no overlap at all (*D. tropicalis* and *D. equinoxialis* at the *me-1* locus), there are many instances among the last 7 loci in the table where species are nearly fixed for alternative alleles, for example *D. tropicalis* and *D. equinoxialis* for *mdh-2*, and *D. equinoxialis* and *D. paulistorum* for *me-1*. Indeed the *me-1* locus almost completely distinguishes all four species from each other.

On the other hand, there are some impressive similarities between pairs of species at highly polymorphic loci. A particularly interesting case is the near-identity of *D. willistoni* and *D. tropicalis* for the 6 alleles at the *esterase-7* locus, in view of their considerable differentiation from *D. paulistorum* at that locus. A similar case is at the *aph-1* locus, for which three species are identically polymorphic and the fourth differs but is still polymorphic. One gains a general impression from the data that *D. willistoni* and *D. tropicalis* form a related pair and *D. equinoxialis* and *D. paulistorum* form a second.

In interpreting the data of table 39, one must not lose sight of the 16 loci not shown, for which all four species were similar or identical in their gene-frequency distributions. When these are taken into account, the proportions of diagnostic loci distinguishing various species pairs are those shown in table 40. The fraction of diagnostic genes varies between 14 and 35 percent.

Generally, the differentiation among the four species of the *D. willistoni* group is higher than for the *melanogaster-simulans* pair or the *pseudoobscura-persimilis* pair. The feature held in common is

TABLE 40

Percentages of loci that are "diagnostic" at the 1 percent level for pairs of species in the *Drosophila willistoni* group

	D. tropicalis	*D. equinoxialis*	*D. paulistorum*
D. willistoni	17.9	21.4	25.0
D. tropicalis		21.4	35.7
D. equinoxialis			14.3

the general absence of alleles that are fixed in one species and lacking in another. Even in the *willistoni* group there are only 5 cases out of 112 in which a dominant allele in one species is not found in at least one other species at a frequency of 0.01 or more. Indeed, in only 45 of 336 cases is the dominant allele in one species not found in *all three* other species at a frequency of 0.01 or greater. There is not a single case of a unique dominant allele.

These observations taken together mean that where species are highly differentiated in their alleles there is at least a low-level polymorphism in one species for the genes that characterize the other. There is then a potential genetic transition between species that does not require the chance occurrence of new variation by mutation. That is, *the overwhelming preponderance of genetic differences between closely related species is latent in the polymorphisms existing within species.* Obviously this generalization becomes less and less true as species diverge farther and farther in the course of their phyletic evolution. We find, for example, no allelic commonality between Drosophila species belonging to different species groups, such as *D. melanogaster* and *D. pseudoobscura*. But our evidence does tell us that this greater differentiation requires only the occasional input of mutational novelties and that the early stages of phyletic divergence make use of an already existing repertoire of genetic variation.

SPECIES *IN STATU NASCENDI*

All of the evidence given so far has concerned species that have long since completed their speciation and are underdoing their third stage of evolutionary differentiation. Evidence for species in their second stage, the reinforcement period when formerly allopatric populations have come together again, is much more difficult to gather. The chief difficulty is one of identification. How are we to know sympatric elements in the second stage of speciation when we see them? Unless there is some morphological or cytological differentiation between the entities, they will not be recognized as distinct. And if they are distinct enough to be recognized, there must be evidence of hybridization between them, hybridization that does not result in a swamping of their differences.

At present I know of only two cases that have been surveyed for patterns of allelic variation. One is the pair of subspecies of *Mus musculus* in Denmark (Selander, Hunt, and Yang, 1969), and the other is the complex of "semispecies" of *Drosophila paulistorum* in South America (Richmond, 1972).

In Denmark there is a light-bellied northern race, *Mus musculus musculus,* which meets a southern, dark-bellied race, *M. m. domesticus,* in a narrow zone of overlap and hybridization. Intermediate populations occupy an east-west belt only a few kilometers wide. Selander, Hunt, and Yang surveyed 41 enzyme loci in four populations of *M. m. musculus* and two of *M. m. domesticus,* with the results shown in table 41. Only 16 loci are given, the other 25 being identically monomorphic in both races. The first three *M. m. musculus* samples were from the northern part of the main Jutland peninsula, but the fourth, although morphologically *musculus,* was sampled from three islands, half the sample coming from an island very close to the mainland and far south of the mainland borderline between the subspecies. In several instances the allele frequencies in this population show evidence of introgression from *domesticus;* the most striking examples are *esterase-1*, in which an allele usually missing from *musculus* is present in 5 percent, and *hexose-6-phosphate dehydrogenase,* in which two alleles have been introduced from *domesticus,* in equal and low frequency. Thus the genetic evidence confirms that *domesticus* and *musculus* are genetically differentiated entities with some hybridization and a small amount of introgression in their overlap zone, but not sufficient gene flow to destroy the difference between them. These are exactly the characteristics of populations in stage two of their speciation process.

Table 41 shows a great deal of differentiation. Discounting the introgressed population of *musculus*, there are two cases of fixation for alternative alleles in the two subspecies, *esterase-1* and *hexose-6-phosphate dehydrogenase*, and six additional cases in which opposite alleles are clearly in the majority in the different entities. Thus 8 loci out of 41, or 20 percent of loci, are differentiated between the forming species, and of these, 5, or 12 percent, are diagnostic at the 0.01 level. The very small sample sizes tend to exaggerate the lack of overlap so that the two loci fixed at alternative alleles may not, in fact, have disjunct frequency distributions. Nevertheless, the degree of divergence is not different from that of the fully formed species of the *willistoni* complex in *Drosophila* and, if

TABLE 41
Allele frequencies for 16 loci in samples of *Mus musculus musculus* and *M. m. domesticus*

Locus	Allele	*M. m. musculus samples* 1	2	3	4	*M. m. domesticus samples* 1	2
Esterase-1	a	1.00	1.00	1.00	.95	—	—
	b	—	—	—	.05	1.00	1.00
Esterase-2	a	—	.17	.07	.05	1.00	1.00
	b	1.00	.83	.93	.95	—	—
Esterase-3	a	.30	.67	.47	.32	.40	.30
	b	.70	.33	.53	.68	.60	.40
Esterase-5	a	1.00	.86	1.00	1.00	.78	.69
	b	—	.14	—	—	.12	.31
Alcohol dehydrogenase	a	.83	.70	.90	.60	.97	.97
	b	.17	.30	.10	.40	.03	.03
Lactate dehydrogenase regulator	a	.95	.64	.69	—	.38	.29
	b	.05	.36	.31	—	.62	.71
Supernatant malic dehydrogenase	a	.37	.37	.37	.47	1.00	1.00
	b	.63	.63	.63	.53	—	—
Mitochondrial malic dehydrogenase	a	.95	.93	.80	.71	1.00	1.00
	b	.05	.07	.20	.29	—	—
Hexose-6-phosphate dehydrogenase	a	—	—	—	.03	.50	.53
	b	—	—	—	.03	.50	.47
	c	1.00	1.00	1.00	.94	—	—
6-Phosphoglu-conate dehydrogenase	a	.33	.83	.93	.76	1.00	1.00
	b	.67	.17	.07	.24	—	—
Isocitrate dehydrogenase	a	.17	.23	—	.08	.93	.98
	b	.83	.77	1.00	.92	.07	.02
Indophenol oxidase	a	.02	.13	—	.42	1.00	1.00
	b	.98	.87	1.00	.58	—	—
Phosphoglu-comutase-1	a	.75	.43	.40	.37	1.00	1.00
	b	.25	.57	.60	.63	—	—
Phosphoglu-comutase-2	a	.03	—	.23	.13	1.00	1.00
	b	.97	1.00	.77	.87	—	—
Phosphoglu-cose isomerase	a	1.00	1.00	1.00	1.00	.93	.93
	b	—	—	—	—	.07	.07
Hemoglobin	a	.10	.10	.63	.50	—	.23
	b	.90	.90	.37	.50	1.00	.77

Note: Data are from Selander, Hunt, and Yang (1969).

anything, is greater than for the sibling species *melanogaster-simulans* and *pseudoobscura-persimilis*, although, of course, a comparison of evolutionary rates in such divergent forms as Diptera and mammals is bound to be shaky.

Drosophila paulistorum is distributed through Central and South America from Costa Rica and the Greater Antilles to southern Brazil. Over that range it is divided into six semispecies that overlap in distribution more or less, depending upon geography. Thus the northernmost, "Centro-American," and the southernmost, "Andean-Brazilian," occupy large exclusive areas but overlap at their margins with the four central semispecies, which are much more broadly overlapping with each other. The "Interior" semispecies, for example, inhabits the overlap zone between the "Amazonian" and the Andean-Brazilian.

The semispecies are not morphologically distinguishable although there has been cytological differentiation and reproductive isolation. Even in laboratory conditions females of one semispecies usually refuse to accept males of another although no choice of mate is offered, and when crosses do occur the F_1 hybrid males are sterile. Gene passage between semispecies happens, however, through the "Transitional" race, which will cross with other races and give fertile sons (see Dobzhansky, 1970, pp. 369–72, for a review and references concerning this fascinating group).

A study of 17 enzyme loci in the six semispecies from a great variety of localities has been made by Richmond (1972). The species as a whole turns out to be as polymorphic as its sibling, *D. willistoni*, with about two-thirds of its loci polymorphic in any population and about 20 percent heterozygosity per individual. Of the 17 loci only 4 show any differentiation between semispecies, and even in those cases the differentiation is small. For 1 locus, three of the semispecies (C, T, and AB) are segregating for two alleles in a ratio of about 0.9:0.1 while the other three (O, I, and A) are in the ratio 0.3:0.7. In a second case the semispecies C, T, and AB are segregating in varying proportions for four alleles while O, I, and A are nearly monomorphic for one of them. At a third locus that has 7 alleles segregating, the most frequent allele in the A race is different from the one that is common in the other races, but there is a broad overlap in the frequency distributions.

Finally, a locus that is nearly monomorphic in the rest of the semispecies is polymorphic in the AB race, with about 25 percent of a unique allele. Moreover, the O race, represented unfortunately only by laboratory stocks, is monomorphic for an allele that is found segregating only in the C race. With the possible exception of this last case, then, there are no diagnostic loci for any of the semispecies comparisons. It is not surprising that in two of the cases it is the Centro-American, Andean-Brazilian, and Transitional races that are similar to each other but dissimilar to the other three groups. It is precisely the Centro-American and Andean-Brazilian semispecies that are cross-fertile with the Transitional one.

The lack of genetic differentiation among the semispecies of *D. paulistorum*, entities that are strongly isolated from each other ethologically and by sterility barriers, is in sharp contrast to the considerable differentiation of the group of four sibling species of which *D. paulistorum* is a part. If the semispecies can be regarded as a model for the second step in speciation, which, in the past, led to the formation of the present-day sibling species, then we see that most of the differentiation of the sibling species has occurred since speciation was completed. The differences we observe among the sibling species of the *willistoni* group are best explained as the result of phyletic evolution after successful speciation, whereas the speciation process itself resulted in very little differentiation.

Although we cannot prove that the events in *D. paulistorum* today are a repetition of the history of past speciation in the *D. willistoni* complex, a better comparison of stage two of speciation will never be found. All reconstructions of evolutionary processes that have left no fossil record depend upon the assumption that successive stages in the evolution of a system, in time, can be seen at present in different systems in the various stages of their own evolution. This principle of ergodicity lies at the base not only of biological systematics, but of a great deal of general historical reconstruction as well.

IN FLAGRANTE DELICTO

If so little genetic divergence characterizes races in their second stage of speciation, what will we find during the first stage, when populations have newly acquired reproductive barriers in isolation from each other? At present only one case has been studied, and it is very illuminating. By a rare combination of luck and intuition, Prakash (1972) has discovered that the Bogotá population of *Drosophila pseudoobscura* is in the first stage of becoming a new species!

When females from Bogotá are crossed with males from any other locality, the F_1 males are completely sterile. The reciprocal cross produces perfectly normal sons. In accord with Haldane's rule (Haldane, 1922), in *Drosophila* male sterility is over and over again one of the primary reproductive isolating mechanisms. Moreover, as we saw on p. 164, the cross between *Drosophila pseudoobscura* and *D. persimilis* produces F_1 males with very small testes when the cross is made in one direction but not in the reciprocal. Thus, the sterility of sons of the cross between Bogotá females and "mainland" males but not from the reciprocal cross, as a first step in speciation, is quite in accord with similar occurrences in the past. Since the Bogotá population is only in the first stage of speciation we do not expect, and Prakash did not find, an ethological isolation between the Bogotá population and the rest of the species. Mating preferences presumably develop in the second stage, after the incipient species come into contact again.

From the evidence in chapter 3, the Bogotá population has all the earmarks of a recent colonization from a small number of propagules. Bogotá is only half as heterozygous as the rest of the species populations; it is usually monomorphic or nearly so for the allele that is most common in the rest of species, except for one locus at which it has a very high frequency of an allele that is usually low; and it is segregating for the two most frequent chromosomal rearrangements found in its nearest neighbors in Guatemala.

Extensive collections of *Drosophila* were made in Colombia in 1955 and 1956 by the expert University of Texas group, and 114 different species were found, but not a single individual of *D. pseudoobscura*. Much of the collecting effort was expended, however, in tropical localities where *D. pseudoobscura* is normally absent. In 1960 *D. pseudoobscura* appeared in traps in the city of Bogotá, and by 1962 this species comprised between 1 and 50 percent of the species trapped in various localities. Although it is impossible to rule out absolutely the presence of *D. pseudoobscura* at a very low level in Colombia for a very long time, the evidence is strong that the species was introduced not much before 1960 and found an environment to which it was preadapted but from which it had been barred by its previous failure to cross the gap of 1500 miles from Guatemala. The sample of flies on which Prakash's experiments were done was collected in Bogotá in late 1967, so a reasonable guess is that ten years passed between the colonization and the sampling. During those 10 years the first step in speciation was taken, probably as a result of the initial colonization.

The first step in speciation, reproductive isolation, has occurred in the Bogotá population, yet there has been no genetic differentiation at the 24 loci examined by Prakash, Lewontin, and Hubby (1969). If anything, Bogotá is supertypical of *D. pseudoobscura*, since it is often homozygous or nearly homozygous for the most common alleles in the species (see table 26), with the exception of the *pt-8* locus, for which it has a high frequency (87 percent) of an allele that is found at only 1 or 2 percent in other localities. The first step in speciation has been taken, not by a wholesale reconstruction of the genome, but by the chance acquisition of an isolating mechanism that probably has a very restricted genetic basis. Total male sterility in only one of the two reciprocal population crosses would probably not be a sufficient mechanism in itself to cause speciation to be completed if the Bogotá population came in contact with the rest of the species now. But it is a large step toward speciation in a very short time. To what extent such a "quantum" loss of reproductive compatibility without genic divergence is the rule in speciation we cannot know until we accumulate many more cases of the early stages of speciation. The tools are readily available.

THE GENETICS OF SPECIES FORMATION

The evidence is sparse, and the closer we get to the beginning of the speciation process, the sparser the evidence. The most solid and coherent argument comes from the comparison of the sibling species of the *willistoni* group with the semispecies of *D. paulistorum*. The data on the other closely related species, on the races of *Mus*, and on the remarkable case of *D. pseudoobscura* from Bogotá "lend versimilitude to an otherwise bald and unconvincing tale."

The first stage of speciation, the acquisition of primary reproductive isolation in geographical solitude, does not require a major overhaul of the genotype and may result from chance changes in a few loci. During the second stage, when the isolated populations have again come into contact, there is some genetic differentiation, perhaps as much as 10 percent of the genome having marked differences in the distribution of allelic frequencies. Complete divergence with fixation of alternative alleles is rare, and loss of old or acquisition of new functions rarer still. Given that the allelic classes detected by electrophoresis are themselves heterogeneous and that there are allelic differences between species that are detectable by techniques like heat denaturation, complete divergence is undoubtedly more common than has so far been estimated, probably by a factor of three.

It is only in the third stage of species evolution, the open-ended phyletic change that occurs more or less independently of the evolution of the sister species, that the major divergence occurs. The speed of this divergence is very variable relative to morphological change, at least in the early stages, so that sibling species may differ markedly in as little as 10 percent of their genome or as much as 50 percent, but even then the divergence is predominantly quantitative. It is rare for a species to differ by alleles that are not polymorphic in its very closely related species, so that most of the species divergence in the early stages of phyletic evolution makes use of the already available repertoire of genetic variants and is not limited by the rate of appearance of novelties by new mutation. This last point, that considerable evolutionary change (including speciation and divergence of new full species) occurs without being limited by the rate of appearance of novel genes is the chief consequence, for the process of speciation, of the immense array of genetic variation that exists in populations of sexually reproducing organisms.

PART III / THE THEORY

CHAPTER 5 / THE PARADOX OF VARIATION

For many years population genetics was an immensely rich and powerful theory with virtually no suitable facts on which to operate. It was like a complex and exquisite machine, designed to process a raw material that no one had succeeded in mining. Occasionally some unusually clever or lucky prospector would come upon a natural outcrop of high-grade ore, and part of the machinery would be started up to prove to its backers that it really would work. But for the most part the machine was left to the engineers, forever tinkering, forever making improvements, in anticipation of the day when it would be called upon to carry out full production.

Quite suddenly the situation has changed. The mother-lode has been tapped and facts in profusion have been poured into the hoppers of this theory machine. And from the other end has issued—nothing. It is not that the machinery does not work, for a great clashing of gears is clearly audible, if not deafening, but it somehow cannot transform into a finished product the great volume of raw material that has been provided. The entire relationship between the theory and the facts needs to be reconsidered.

THE MODULATORS OF VARIATION

The genotypic distribution in a population is subject to a complex of forces that act separately and in interaction to increase, decrease, or stabilize the amount of variation. We will need to look closely at these forces and some of their interactions, but we first need a more superficial overview.

Variation is introduced into a population by either mutations or the immigration of genes from other populations with different alleles. In some cases significant variation may be introduced by the introgression of alleles from other subspecies or species, if reproductive isolation breaks down somewhat in a zone of contact. For example, there is direct evidence from the gene frequencies in table 41 (page 181) that genes from *Mus musculus domesticus* are entering *Mus musculus musculus* at a substantial rate in their zone of overlap in Denmark. Loci that are usually monomorphic in *M. m. musculus* have a demonstrable polymorphism in the border population, and the average heterozygosity in that population is 0.108 as compared with 0.083 in populations far from the overlap zone. The same kind of increase of genetic variation from hybridization has often been suggested on the basis of morphological and physiological characters (see Mayr, 1963, chapter 6, for a review) and might even serve as an important source of new variation for adaptation (Lewontin and Birch, 1966). Direct evidence from allele frequencies such as for *Mus* is not common but ought to be quite easy to obtain from electrophoretic studies.

Once allelic variation has been introduced by mutation or immigration, the total array of genotypes may be considerably increased by recombination, but of course this source of genetic variance is completely limited by the allelic variation that is being recombined. If loci are nearly monomorphic, then although recombination will produce some new unique combinations of alleles at different loci and although such unique combinations may serve as the basis for new adaptations, they will be generated at a very low frequency. For example, if there are two nearly monomorphic loci, each with a variant allele at a frequency of 0.001, it might easily be that the double mutant gamete *ab* is completely absent. But the rate of production of such a gamete by recombination is the frequency of recombination multiplied by the frequency of double heterozygotes, or, at maximum, $(0.5)(0.002)(0.002) = 2 \times 10^{-6}$. Thus, recombination, which can produce new variation between two loci at a rate no greater than the product of their heterozygosities, becomes effective only when heterozygosities are already high.

The third force that can increase or at least conserve variation is natural selection. If an initially rare allele is favored by selection, either because the environment has changed, or because the allele has only recently been introduced into the population, the increase in allelic frequency toward 0.5 means an increase in genetic variation. However, if the allele is unconditionally more fit and is eventually fixed in the population, or at least driven to a very high frequency, then variation again decreases as the allele passes from a frequency of 0.5 to near 1. Selection for the replacement of an old "wild type" by a new one is then a temporary producer of heterozygosity. As we shall see, this transient heterozygosity cannot account for much of the genetic variation in a population.

There is one form of selection, however, that is a powerful preserver of genetic variation: balancing selection. In its simplest form, balancing selection arises from an unconditional superiority in fitness of heterozygotes over homozygotes, so-called overdominance of fitness* or single-locus heterosis. If the heterozygote *Aa* is superior in fitness to the homozygotes *AA* and *aa*, there will be a stable intermediate equilibrium of allelic frequencies, and a stable heterozygosity. For example, if there are only two alleles and we assign the fitnesses

AA	*Aa*	*aa*
$1 - s$	1	$1 - t$

then the equilibrium allelic frequency of *a* is

$$q_a = \frac{s}{s + t} \tag{1}$$

and the equilibrium heterozygosity is

$$H = \frac{2st}{(s + t)^2} \tag{2}$$

The greater the asymmetry in fitness between the homozygotes, the closer the equilibrium allelic frequency is to zero or one and the less

*Overdominance in fitness is often referred to simply as "heterosis" but this is an extremely confusing usage because it fails to make the critical distinction between the *observation* that crosses between lines produce more fit progeny and the *explanation* that this higher fitness is the result of superior heterozygotes at particular loci. Indeed this confusion is what chapter 2 of this book is all about.

the heterozygosity. This can best be seen by rewriting the heterozygosity as

$$H = \frac{1}{1 + \left(\frac{s/t + t/s}{2}\right)} \tag{3}$$

So if the two homozygotes are equal in fitness, $s = t$ and $H = 0.5$, whereas if one homozygote suffers a loss in fitness ten times as great as does the other homozygote,

$$H = \frac{1}{1 + \frac{10.1}{2}} = 0.165$$

Unconditional overdominance is not the only form of balancing selection that will maintain stable heterozygosity. Selection operating in opposite directions in the two sexes or in gametic and zygotic stages, and a number of cases in which the fitnesses of genotypes are functions of their relative frequencies in the population may also lead to a stable equilibrium (Wright, 1969). A variety of situations in which fitnesses vary in space or time can also lead to stable equilibria, even though the heterozygote is not superior to the homozygotes in any of the individual spatial or temporal modes. No general criterion has been developed for distinguishing those kinds of varying selection that preserve variation from those that reduce it, but a rather broad range of models predicts the maintenance of variance (Levene, 1953; Dempster, 1955; Kimura, 1955; Haldane and Jayakar, 1963).

Genetic variation is removed from populations by both random and deterministic forces. Every population eventually loses variation because every population is finite in size and therefore subject to random fluctuation in allele frequencies. Such random fluctuation, if unopposed by the occurrence of new variation, leads ineluctably to the fixation of one allele in the population, although the loss of heterozygosity may be extremely slow if population size is very large. But most populations of organisms, especially free-living terrestrial species, undergo periodic severe reductions in population size either regularly with the change of seasons, or in episodes of colonization, or aperiodically when a particular conjunction of adverse environmental factors creates a demographic catastrophe. At such times the loss of heterozygosity may be great, as only a few individuals survive to reproduce.

Irrespective of population size, newly arisen alleles are in constant danger of loss from the Mendelian mechanism itself. Even if every individual in a bisexual population left exactly two offspring, so that the population was stable in size and there was no opportunity for differential reproduction among genotypes, a heterozygote for a new mutation would have a 25 percent chance of failing to pass on that mutation to either of its offspring. If we add to this the fact that there is always some variation in family size, the probability of loss becomes even greater. If, for example, family size has a Poisson distribution, the probability of loss of a new mutant in one generation is $1/e \cong .37$.

Random fluctuations in selection intensity may also reduce genetic variation. It is true that some kinds of variation in selection will act as balancing selection, even in the absence of overdominance, yet other patterns of fluctuation will hasten homozygosity by driving allele frequency close enough to zero or one for random loss to be important, although the average selection for or against the allele may be very small or even zero (Wright, 1948; Kimura, 1955).

The strongest force reducing genetic variation is, of course, selection against recessive or partly dominant deleterious genes. The effect of such selection on the species as a whole depends upon whether the same allele is favored in all populations. If there is one wild type over the whole of the species range, then the result will be uniformity over all populations. If there is local adaptation, however, with different alleles selected in different localities, the species will be polytypic but individual populations will be homozygous, as in *Acris crepitans*. There is, in this case, storage of variation in the species as a whole, rather than in the heterozygosity of individual populations, although migration between such populations will result in their heterozygosity as well.

Another not uncommon form of selection that reduces heterozygosity is selection against heterozygotes. When the heterozygote is less fit than the homozygote, there is an unstable equilibrium and the allele frequency is driven to zero or one, depending upon which side of the unstable point it finds itself on at the time the selection process begins. In effect this means that the introduction of new mutations is strongly resisted. Like selection for locally adapted genotypes, selection against heterozygotes could lead to polytypy. Indeed selection for locally adapted alleles and selection against heterozygotes are the two forms of natural selection postulated for the process that converts locally differentiated populations into those ecologically and reproductively isolated populations we call species. Maternal-fetal incompatibility is an example of selection against heterozygotes, so the maintenance of the apparently stable polymorphism for human blood groups is rather paradoxical.

A complete listing of the forces that preserve or remove genetic variation, with representative references, is given by Karlin and McGregor (1972).

CLASSICAL AND BALANCE THEORIES

That natural selection could be both the preserver and the destroyer of intrapopulation variation would have been a surprise to Darwin. For him, natural selection was the converter of intrapopulation variation into temporal and spatial differentiation. Lacking a correct theory of genetics, and especially one that included segregation of alleles from hybrids, Darwin could not have imagined forms of selection that would actually stabilize inherited variation. Without Mendelism, the theory of natural selection is inevitably one that predicts that a more fit type will completely replace the less fit types if the trait is at all heritable.* Classical Darwinism saw evolution as the passage from one more or less uniform state to another and in this sense was no different from pre-Darwinian ideas of evolution. What Darwin added was the realization that the variation present in populations was the source of the eventual variation between species, but the ontogeny of the variation itself was a mystery about which Darwin changed his mind in the course of his life. The lack of a satisfactory explanation for the origin and replenishment of variation that was constantly being

*It is interesting that biometrical genetics has preserved the literal truth of this statement, even with Mendelism, by a semantic convention. *Heritability*, in modern technical usage, is precisely that fraction of the total genetic variance in a population that will lead to genetic change under selection. At gene frequency equilibrium, the heritability is, by definition, zero!

reduced by "survival of the fittest" was a serious flaw in evolutionary theory, a flaw that was not repaired until the reappearance of Mendelism. Classical Darwinian theory held that heritable variations arose from a variety of sources and that natural selection sorted through these variations, rejecting all but the most fit type. Natural selection was seen as *antithetical* to variation—"Many are called, but few are chosen."

Those who support the classical theory of population structure are the direct inheritors of this pre-Mendelian tradition. For them, variation arising from mutation is constantly being removed by the purifying force of directional selection and, to some extent, by random genetic drift. Although it seems strange to say, the classical theory owes virtually nothing to Mendelism. The fact is that almost the entire theoretical apparatus of random genetic drift and directional selection can be derived from a *haploid* model of the genome and that the introduction of diploidy and sexual recombination makes no qualitative change and only trivial quantitative changes in the predictions of evolution under these forces. For random drift, the introduction of diploidy simply alters by a factor of 2 the rate of loss of genetic variation in a population except in the totally unrealistic case of no variance in family size. For selection, so long as the heterozygote has a fitness somewhere within the range spanned by the homozygotes, the process of allelic frequency change can be adequately represented by the model of "genic" selection, and the genetic load (or variance in fitness) in a population at equilibrium between mutation and selection is virtually the same for both models. Indeed, if we accept Muller's contention that severely deleterious mutations contribute only a small proportion of the total human genetic load, the entire argument of his article "Our load of mutations" (1950) would follow if sex had never been invented. A search through textbooks and technical literature on the theory of random drift and directional selection shows that a haploid model is often used for simplicity of derivation, diploidy being added later as a refinement.*

*A chapter in Moran's *The Statistical Processes of Evolutionary Theory* (1962) has the remarkable title "Progress to homozygosity in finite populations," although it deals entirely with a *haploid* model, but the anomaly disappears if we read "homogeneity" for "homozygosity."

The balance theory, as its name suggests, emphasizes that aspect of natural selection which would have been foreign to Darwin and which is the unique contribution of Mendelism to evolutionary theory, the possibility that natural selection preserves and even increases the heritable variation within populations. In the absence of special symbiotic relations between genotypes, natural selection can stabilize variation only when there are heterozygotes and segregation of genes. Only then can there be heterosis, but also the stabilization of variation by temporally fluctuating selection is critically dependent upon a diploid sexual model (Dempster, 1955). The balance school sees the maintenance of variation within populations and adaptive evolution as manifestations of the same selective forces, and therefore it regards adaptive evolution as immanent in the population variation at all times. Because the alleles that are segregating in a population are maintained in equilibrium by natural selection, they are the very alleles that will form the basis of adaptive phyletic change or speciation.

The reverse is true for the classical theory, in which variation is present *faute de mieux* and in which the genetic basis for further evolution is either lacking or extremely rare most of the time in the history of a population because natural selection is efficiently sweeping out any variation that might otherwise accumulate.

Nowhere is the contrast between the classical and balanced views of diploidy more apparent than in the attitudes toward the importance of sexual reproduction in evolution. If the standing variation in populations is indeed under natural selection and is the source of future evolution, then sexually reproducing individuals can form an immense array of new genotypes through recombination in each generation and so increase the adaptive range of their offspring. On the other hand, Muller, who was one of the first to deal with the question of the advantage of sex (1932), and more recently Crow and Kimura (1965) claim that the chief advantage of sexual reproduction is in bringing together the rare adaptive mutations that are the real basis of evolutionary change. This assumption, that the alleles that are eventually to be fixed by natural selection are rare at the beginning of the selection process, is critical to the Muller-Crow-Kimura argument (Maynard Smith, 1968; Eshel and Feldman, 1970).

But why this continued juxtaposition of classical and balance theories? Have not the evident facts of vast quantities of polymorphism and heterozygosity firmly established the balance theory? In the face of the evidence given in chapters 3 and 4, how can the classicists hold their ground? The answer is, Easily.

THE NEOCLASSICAL THEORY

If we take it as given that balancing selection is rare and that natural selection is nearly always directional and "purifying," how can we explain the observed polymorphism for electrophoretic variants at so many loci? We can do so by claiming that the variation is only apparent and not real. That is, we can suppose that the substitution of a single amino acid, although detectable in an electrophoresis apparatus, is in most cases not detectable by the organism. If it makes no difference to the physiological function of an enzyme whether it has a glutamine or a glutamic acid residue, for instance, on the surface of the folded molecule far away from the enzyme's active site, then the variations detected by electrophoresis or by any method that is sensitive to amino acid substitutions may be completely indifferent to the action of natural selection. They are "genetic junk," revealed by the superior technology of the laboratory but redundant physiologically. From the standpoint of natural selection they are *neutral mutations.*

The suggestion that most, if not all, of the molecular variation in natural populations is selectively neutral has unfortunately led to widespread use of the terms "neutral mutation theory" and "neutralists" to describe the theory and its proponents (see almost any discussion of the problem of genic heterozygosity since 1968). But these rubrics put the emphasis in just the wrong place and obscure both the logic of the position and the historical continuity of this theory with the classical position. It is not claimed that nearly all mutations are neutral or that evolution proceeds without natural selection, chiefly by the random fixation of neutral mutations. Both these statements are patently untrue and both are foreign to the spirit of the proposition that is being made. On the contrary, the claim is that many mutations are subject to natural selection, but these are almost exclusively deleterious and are removed from the population.

A second common class is the group of redundant or neutral mutations, and it is these that will be found segregating when refined physicochemical techniques are employed. In addition the theory allows for the rare favorable mutation, which will be fixed by natural selection, since after all adaptive evolution does occur. But it supposes this event to be uncommon. Finally, it also allows that occasional heterotic mutants might arise but that these do not represent a significant proportion of all the loci in the genome.

Thus the so-called neutral mutation theory is, in reality, the classical Darwin-Muller hypothesis about population structure and evolution, brought up-to-date. It asserts that when natural selection occurs it is almost always purifying, but that there is a class of subliminal mutations which are irrelevant to adaptation and natural selection. This latter class, predictable from molecular genetics and enzymology, is what is observed, they claim, when the tools of electrophoresis and immunology are applied to individual and species differences.

The neoclassical theory cannot be refuted by erecting a neutralist strawman and refuting that. So, for example, the demonstration that single amino acid substitutions can in some instances make big differences in physiology is irrelevant. The range of effects of single amino acid substitutions can be illustrated by human hemoglobin. Of 59 variant α and β hemoglobin chains listed by Harris (1970), 43 are without known physiological effects at least in the heterozygous state in which they are found, 5 cause methemoglobinemias because they are near the site of the heme iron and are therefore mildly pathological, and 11 cause instability of the hemoglobin molecule that results in various degrees of hemolytic anemia. Although most of this last group result from the substitution of noncharged by noncharged amino acids on the inside of the three-dimensional structure, two are caused by charge changes at position 6 on the outside of the molecule. One, a substitution of lysine (+) for the normal glutamic acid (–) causes the benign hemoglobin C disease, but the other, a substitution of valine (0), is the famous hemoglobin S, causing sickle-cell anemia (Perutz and Lehman, 1968).

We cannot make anything of the relative proportion of pathological (16) to asymptomatic (43) substitutions, since the former are detected from the pathology they cause, whereas the latter turn up in routine electrophoretic screening. We do not know, for example, how many neutral substitutions on the surface or in the interior of the molecule go undetected because of their lack of detectable physiological effect. Nor is it certain, conversely, that the 43 surface charge changes are absolutely neutral. Most have never been seen in homozygous condition. The point is that single amino acid substitutions, charged and uncharged, on the surface or the interior of the molecule, run the gamut of effects from apparently neutral to severely pathological, and this range is in no way contradictory to the neoclassical theory.

Second, the neoclassical theory is not refuted by occasional observations of overdominance for fitness, because the theory does not deny that cases exist but only that they are common and explain a significant proportion of natural variation. So it is no use trotting out that tired old Bucephalus, sickle-cell anemia, as a proof that single-locus heterosis can exist. Anyone who has taught genetics for a number of years is tired of sickle-cell anemia and embarrassed by the fact that it is the only authenticated case of overdominance available. "If balancing selection is so common," the neoclassicists say, "why do you always end up talking about sickle-cell anemia?"

Finally, the neoclassical theory cannot be disposed of by pointing to the elephant's trunk and the camel's hump. The theory does not deny adaptive evolution but only that the vast quantity of molecular variation within populations and, consequently, much of the molecular evolution among species, has anything to do with that adaptive process.

The neoclassical theory is supported by taking a diverse collection of theoretical results on genetic load from gene substitution (Haldane, 1957), genetic load from balanced polymorphism (Crow, 1958), the amount of heterozygosity that will be maintained by mutation in a finite population (Kimura and Crow, 1964), the probability of fixation of favorable mutations (Haldane, 1927) and of neutral mutations (Kimura, 1962), and the equilibrium rate at which new mutations will be fixed in populations (Kimura, 1968), and substituting into these theoretical results present estimates of average mutation rates, average genome size, heterozygosity per locus, and population size, together with the estimates of the rate of amino acid substitution in evolution from numerous well-studied polypeptides (King and Jukes, 1969). These diverse strands have been brought together by Kimura and Ohta in a series of overlapping papers, two of which (1971b, 1971c), together with their excellent book (1971a), contain all the calculations and arguments. Nowhere, however, is the entire structure of the argument specifically displayed.

The neoclassical argument is made up of two complementary parts. First, it attempts to show that the balance theory is untenable because it involves internal contradictions, and then attempts to show that the neoclassical theory is compatible with the data. Both parts of the argument are essential. To show only that the neoclassical view is compatible with the facts, to show that it is a sufficient theory, is not enough, for then the two theories stand side by side with no way to choose between them. It is a cornerstone of the neoclassical argument that the balance view is irreconcilable with all the important known facts.

This two-sided argument is applied to two different sets of facts, the amount of heterozygosity in populations and the rate of substitution of alleles in evolution. The neoclassicists maintain that the amount of allelic variation and the rate of amino acid substitution in proteins during evolution are both too large to be accounted for by selection but can be satisfactorily explained by assuming that the genetic variation for amino acid substitutions is neutral and that the differences in amino acid composition of most proteins are the result of random fixation of these neutral alleles during evolution.

Note again that the theory does not say that all amino acid variants are neutral or that the amino acid sequence of proteins in evolution is completely free to vary. On the contrary it holds that many amino acid variants will have deleterious effects on function, but these will not be seen segregating in populations at high frequencies and *ipso facto* will not be seen as substitutions in evolution. The hypothesis of neutrality applies only to the bulk of genic variation that is actually present.

There is no indissoluble link between the theory as it applies to standing variation within populations and to differences between species. It is entirely possible that virtually all intrapopulation variation for enzymic forms is neutral, yet all the differences between species accumulated in evolution might be adaptive. Conversely, it is entirely possible that most of the variation in populations is held there by balancing selection, but that speciation involving initially small, isolated populations could cause considerable random fixation of genetic differences nonadaptively. In fact, Mayr's theory of the genetic revolution assumes that precisely such random divergence may be the first step in speciation. It is important to separate the application of the neoclassical theory to population variation from its application to gene replacement in evolution, because they do not stand or fall together.

EVIDENCE FROM TOTAL HETEROZYGOSITY

Let us consider the argument, first given in detail by Kimura and Crow (1964), that the amount of variation within populations is too great for balancing selection. If we consider a locus with two alleles and with fitnesses

AA	Aa	aa
$1-s$	1	$1-t$

then at gene frequency equilibrium the heterozygosity is

$$H = \frac{2st}{(s+t)^2} \tag{4}$$

and the mean fitness of the population is

$$\bar{W} = 1 - \frac{st}{(s+t)} \tag{5}$$

or, substituting equation (4) into (5),

$$\bar{W} = 1 - H\left(\frac{s+t}{2}\right) = 1 - H\bar{s} \tag{6}$$

where $(s+t)/2 = \bar{s}$ is the average fitness advantage of the heterozygote.

As a low estimate, the heterozygosity per locus is 10 percent and the proportion of polymorphic loci is 30 percent. If we take a conservative estimate of the size of the genome, which will favor the balance theory, there are 10,000 genes coding for enzymes and proteins in Drosophila. Then there are 3000 loci segregating, with an average heterozygosity per segregating locus of $0.10/0.30 = 33$ percent. If the effects on fitness of the various loci are independent, then the mean fitness of the population will be the product of the fitnesses at the separate loci and, for the genome as a whole,

$$\bar{W} = (1 - 0.33\bar{s})^{3000} \cong e^{-1000\bar{s}}$$

if $\bar{s}$ is small. Suppose that the overdominance were only 10 percent at each locus. Then

$$\bar{W} = e^{-100} \cong 10^{-43}$$

Relative to the fitness of a completely heterozygous individual, the average fitness of the population is thus 10^{-43}. In a population that is neither increasing nor decreasing rapidly over long periods, the average reproductive rate per individual must be around 1, so the reproductive ability of a complete heterozygote would be 10^{43}, an absurdity. One can hardly imagine a Drosophila female, no matter how many loci she was heterozygous for, laying 10^{43} eggs.

We could ask, reciprocally, how large a value of $\bar{s}$ could be postulated and still allow a reasonable fitness for the hypothetical fittest genotype. If we were generous and allowed that genotype a fitness of 100, relative to the mean, then $\bar{s} \cong 0.005$. But if the average heterosis at a locus is so low, less than 0.5 percent, then the normal fluctuation of the environment and random genetic drift become vastly more important than the average selection coefficient. A locus with an average heterosis of 0.005 is likely to experience long sequences of generations when there is no heterosis at all. The balance hypothesis is in serious difficulties if it must rely on postulated average selection of such magnitude.

This argument about genetic load, used by Lewontin and Hubby (1966) in connection with their observations on allozyme heterozygosity, was immediately criticized as being naive and misleading by three independent sources (King, 1967; Milkman, 1967; Sved, Reed, and Bodmer, 1967). All pointed out that the excessive fitness of the hypothetical multiple heterozygote was an artifact of the multiplicative model of fitness and that it was equally reasonable to suppose that fitness reached some plateau asymptotically not far above the mean heterozygosity. The contrast between these two models is shown in figure 17. The imposition of the upper threshold has no effect on selection within actual populations because individuals with very high heterozygosities compared with the mean do not, in fact, exist. If there are 3000 loci with an average heterozygosity of 0.33 per locus, then the mean number of loci heterozygous per individual is 1000 and the standard deviation (assuming a binomial distribution) is 25.7; thus 99.9 percent of population will be heterozygous for fewer than 1080 loci, which, under the multiplicative models with $\bar{s} = 0.1$ and $H = 0.33$, would corre-

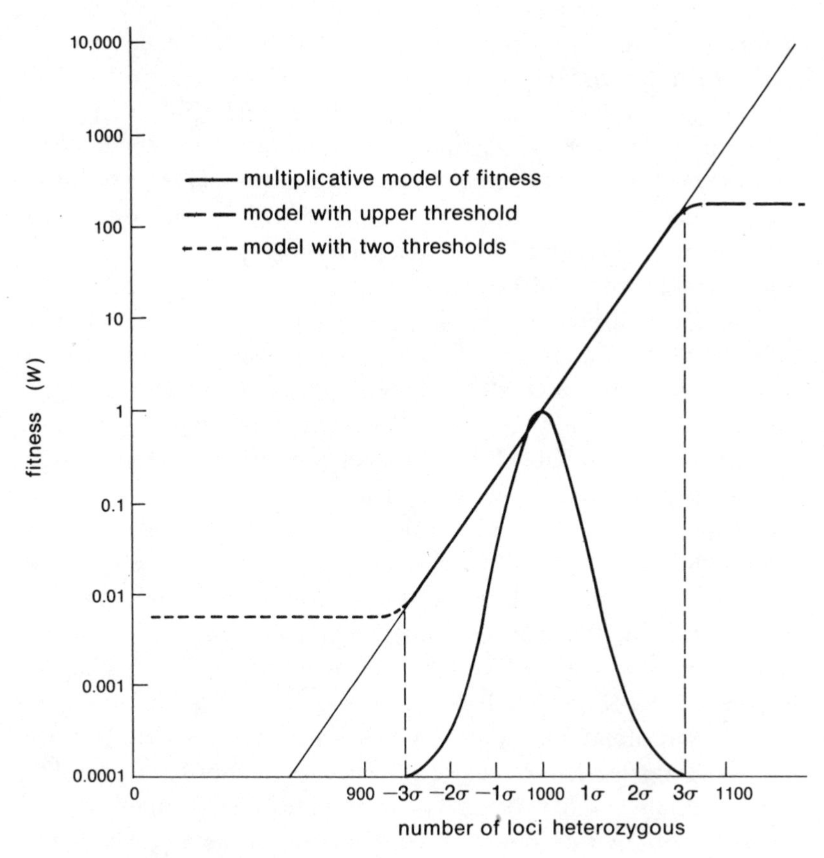

FIGURE 17

Diagrammatic contrast between the multiplicative model of fitness; a model with an upper threshold, to accommodate the problem of the optimum genotype; and a model with two thresholds, to accommodate the problem of inbreeding depression.

spond to a maximum fitness 220 times the mean of the population. Figure 17 has been drawn so as to place the upper asymptote at 220.

A reasonable model of selection that generates such a plateau was suggested by King (1967). Let us suppose that, because of limited resources, only a proportion W_0 of the fertilized zygotes can survive and reproduce. W_0 is then analogous to the mean fitness. Let us further suppose that survival depends upon an adaptive trait that has some variance in the population, with both a genetic and an environmental component. Finally, we will suppose that the adaptive trait increases linearly with the number of loci that are heterozygous. When selection occurs, individuals survive if they have a value of the trait greater than some threshold, the threshold being determined by the proportion W_0 of the total population that is to remain. The advantage of heterozygotes in this model arises from their having a higher value of the adaptive trait and therefore a greater probability that they will be included within the surviving group. Table 42 gives a few sample values of $\bar{s}$, the heterosis arising in this model on the assumption of different survival proportions of the population and different degrees of genetic determination of the adaptive trait. A heterosis of 10 percent can be achieved if the adaptive trait is largely determined genetically and only 1 percent of the zygotes survive, and in general reasonably large values of heterosis arise from moderate values of selection and heritability.

Although the plateau models of selection cope adequately with the problem of mean fitness, they do not resolve the paradox that there is too much polymorphism to explain by selection. Another consequence of the balance model, inbreeding depression, poses serious quantitative problems, as foreseen by Sved and his colleagues and by King. While we can claim that the excessive fitness of hypothetical multiple heterozygotes is a pseudo-problem because such heterozygotes never exist anyway, we cannot be so cavalier about multiple homozygotes. By use of inbreeding techniques, populations with high degrees of homozygosity can be and have been produced. For example, the marker scheme of chapter 2, when applied to the second chromosome of *Drosophila melanogaster*, makes 40 percent of the genome homozygous; when applied to both second and third chromosomes simultaneoulsy, as in the experi-

TABLE 42

Heterosis per locus, $\bar{s}$, that would arise from King's truncation model of selection for a given proportion of the variance of the adaptive trait that is genetic, r, and a given survival proportion, W_0, on the assumption of 3000 loci with an average heterozygosity of 0.33

	r		
W_0	*1.00*	*0.1*	*0.01*
0.50	0.030	0.010	0.003
0.10	0.065	0.038	0.006
0.01	0.101	0.060	0.013

ments of Temin and co-workers (1969), it results in populations that are homozygous for 80 percent of the entire genome and 98 percent of the autosomal genome. Thus males are essentially complete homozygotes. Keeping to our model of independently acting loci with fitnesses $1 - s$, 1, and $1 - t$ for the genotypes AA, Aa, and aa, respectively, we find that the fitness of an inbred line *relative to the random-mating population from which it was derived* is, for small s and t, approximately

$$\frac{W_{\text{inbred}}}{\bar{W}_{\text{random}}} = (1 - FH\bar{s})^n = e^{-FHn\bar{s}} \tag{7}$$

where F is the inbreeding coefficient and n is the number of segregating loci. For our numerical case, $H = 0.33$ and $n = 3000$, so

$$\frac{\bar{W}_{\text{inbred}}}{\bar{W}_{\text{random}}} = e^{-1000F\bar{s}}$$

which will be a very small number when $\bar{s}$ is any considerable size. For example, if the second chromosome of *D. melanogaster* is made homozygous, $F = 0.40$ (0.50 if the X chromosome is not counted) so that

$$\frac{\bar{W}_{\text{inbred}}}{\bar{W}_{\text{random}}} = e^{-400\bar{s}} = \begin{cases} 10^{-17} & \text{when } \bar{s} = 0.1 \\ 0.018 & \text{when } \bar{s} = 0.01 \\ 0.14 & \text{when } \bar{s} = 0.005 \end{cases}$$

Recall that Sved and Ayala (1970) found a 14 percent net fitness for second chromosome homozygotes in *D. melanogaster* when measured in competition with heterozygotes, which then implies heterosis on the order of 0.5 percent. We are once again in the position that we must postulate average heterosis of a fraction of a percent, this time to be consonant with observed inbreeding depression. Heterosis of the order of 10 percent is clearly out of the question.

The dilemma of inbreeding depression could be avoided if we were willing to assume a *lower* threshold for fitness as well as an upper one, as shown in figure 17. There is no evidence for such a threshold and indeed the experiments of Spassky, Dobzhansky, and Anderson (1965) and Temin and co-workers (1969) show that, if anything, the reverse is true and that multiple homozygotes are less fit than predicted from the multiplicative model. Nor does King's threshold model help, since it predicts about the same inbreeding depression as the multiplicative model.

The second quantitative problem of the balance hypothesis is presented by the genetic variance of fitness to be expected for the balance model. Natural selection implies not only an average loss of fertilized zygotes, but also a variation in the number of offspring from genotype to genotype. If there were no genetic variance in fitness there could be no selection. Under the model of fitness that we have been discussing, the variance in fitness in the equilibrium population is

$$\sigma^2 = \left[1 + \left(\frac{st}{s + t - st}\right)^2\right]^n - 1 \tag{8}$$

which is, to very close order of approximation,

$$\sigma^2 \cong e^{nH\tilde{s}^2/2} - 1 \tag{9}$$

where $\tilde{s}$ is the geometric mean of the fitnesses of the homozygotes, which, in the asymmetrical cases we have considered will be somewhat smaller than $\bar{s}$. With $nH = 1000$ as usual and $s = 0.1$,

$$\sigma^2 \cong e^{500\tilde{s}^2} - 1 \cong 24$$

Because this value depends upon $\tilde{s}^2$ in a positive exponent, it is highly sensitive to assumptions about the intensity of selection. So, for example, with $\tilde{s} = 0.05$, $\sigma^2 = 1.2$. But it is also very sensitive to the assumption about the number of loci. Suppose there were 20,000 loci, rather than 10,000 specific enzymes. Then $N = 6000$ and $\sigma^2 = 600$. Thus the genetic variance in fitness is extremely sensitive to the various guesses of parameters, much more so than other predictions of our theory.

To compare these predicted values with observed variances, we note that in a sexual population that is just replacing itself, with a Poisson distribution of offspring, the *total* variance in offspring number is 2, so that the genetic variance in fitness is likely to be less than 1. In human populations, which are growing, and which have more than Poisson variance in offspring number, the total variance ranges from about 3 for stable Great Britain to 21 for rapidly growing Brazil (Cavalli-Sforza and Bodmer, 1971). The genetic component of the variance certainly does not exceed 25 percent, from the available evidence, and is more likely to be on the order of 5 percent, so that again we should expect the genetic variance in fitness to be 1 or less. From equation (9), such a value of genetic variance in fitness means that if the average heterozygosity over all loci including monomorphic genes is 10 percent, then

$$n\bar{s}^2 \sim 10$$

Thus if the genome were made up of 100,000 genes of the kind sampled by electrophoresis, s could not be greater than 0.01.

To sum up, the most telling evidence against the balance hypothesis as the explanation for the observed standing variation in populations is that the predicted inbreeding depression under the heterotic model is vastly greater than what is observed, unless the postulated overdominance is less than 1 percent. This contradiction would be even greater if a less conservative estimate of the genome size were used. In addition, the observed genetic variance in fitness appears rather too small for the theory, although this is not certain.

The second half of the neoclassical argument is that the facts of intrapopulation variation are completely compatible with a classical picture, and indeed that the classical hypothesis predicts some observations that are not predicted (or contradicted) by the balance theory.

First, the observed proportion of heterozygosity can be entirely explained as allelic variation of no effect at all on fitness. Each locus is capable of mutating to a large number of forms, some 10^{200} for a cistron of ordinary length. Of course a very large but unknown number of substitutions would change the enzyme so as to destroy or reduce its activity and so would be selected against. Many may be neutral, however, and these will mostly be lost within a few generations of their occurrence. Some mutations, although they are eventually lost, may rise temporarily to intermediate gene frequencies by random drift. Still others, about $1/2N$ of new neutral mutations, will eventually become fixed in the population, and some of these may be in intermediate or high frequency. At any moment, the majority of loci will be represented by only one allele, but decreasing proportions of loci will be represented by 2, 3, 4, . . . n alleles at varying frequencies. When this process has gone on for some time, a steady state will be reached as a kind of dynamic balance between the input of new mutations, the random increase of these mutations by drift, and the random loss of variation. We expect that the higher the mutation rate and the larger the population size, the more of this neutral variation will accumulate without being lost. In fact, at the steady state, the heterozygosity H will be

$$H = 1 - \frac{1}{4N\mu + 1} \tag{10}$$

where N is population size and μ the mutation rate per locus to neutral alleles. For example, suppose $N = 3 \times 10^4$ and $\mu = 10^{-6}$; then H will be 11 percent, quite close to what we actually observe. Indeed, no matter how much heterozygosity is observed, it can be explained by suitable choices of N and μ. The form of equation (10) is both the strength and the weakness of the neutral theory. It contains two parameters, one that is a very large but unknown number, N, and one that is a very small but unknown number, μ, appearing only as their product $N\mu$. Any value of $N\mu$ is then "reasonable," and so every observed heterozygosity is consonant with the classical theory. It is important to note that the agreement between observation and theory in this case is not intended to be a proof of the neoclassical theory but only to show that the observed values of heterozygosity are not at variance with it because no level of heterozygosity would be.

Despite the unchallengeable agreement between the heterozygosity observed in any given case and the predictions of neutral theory, the application of neutral theory to *every* case of heterozygosity turns out to lead to an absurd result. This absurdity arises from the peculiar behavior of H as a function of $N\mu$, illustrated in figure 18. Heterozygosity is extremely sensitive to $N\mu$ over a very short range, while if $N\mu$ is smaller than about 0.01, heterozygosity is nearly zero no matter how small $N\mu$ is, and if $N\mu$ is greater than about 10, heterozygosity is virtually complete no matter how large $N\mu$ may be.

The observed range of heterozygosities over all the species listed in table 22 lies in the sensitive region, between 0.056 and 0.184. This range corresponds to values of $N\mu$ between 0.015 and 0.057. Since there is no reason to suppose that mutation rate has been specially adjusted in evolution to be the reciprocal of population size for higher organisms, we are required to believe that higher organisms including man, mouse, Drosphila and the horseshoe crab all have population sizes within a factor of 4 of each other.* Moreover,

*The reader should not try to match the heterozygosities of particular species in table 22 with particular population sizes. The standard errors of individual estimates of H make this a useless procedure.

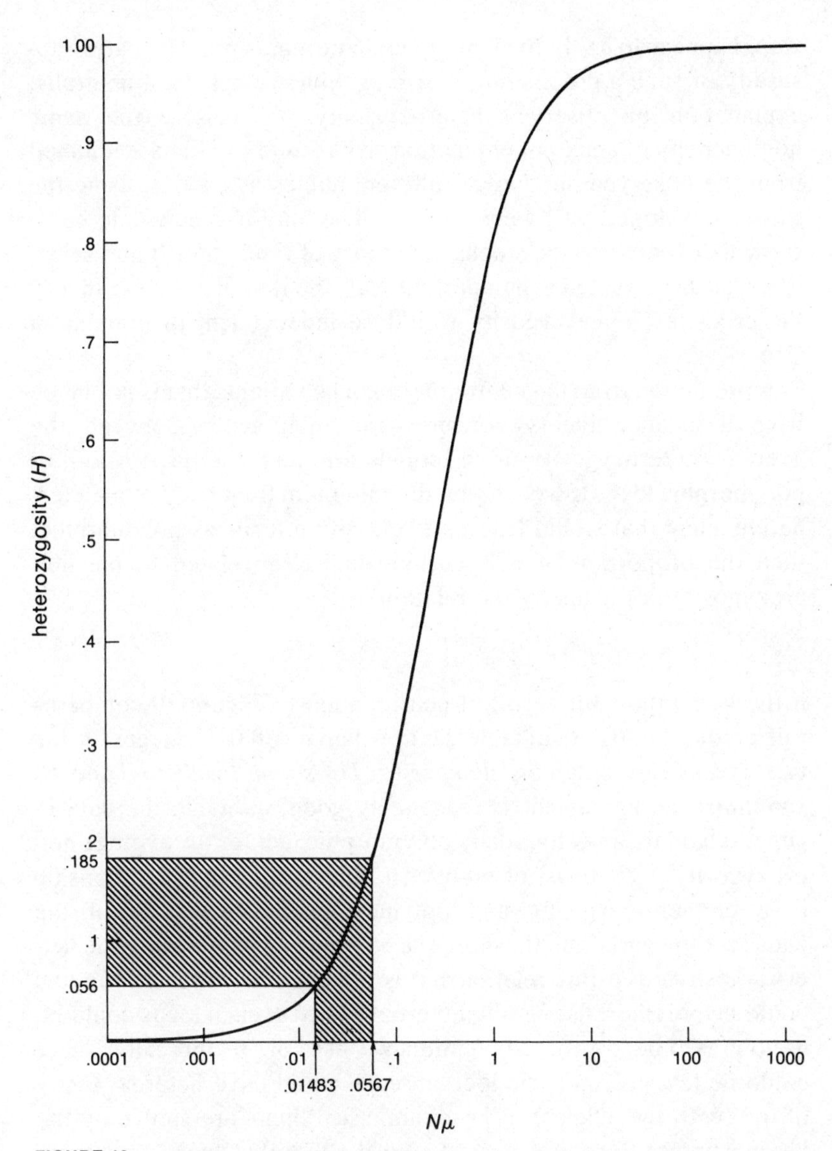

FIGURE 18

The heterozygosity H predicted by a neutral mutation hypothesis, as a function of the product of population size N and mutation rate μ. The heavy horizontal and vertical lines enclose the range of observed heterozygosities and inferred values of $N\mu$ in various species.

other organisms, less well studied, including eels, wild grasses, chickens, and Pogonophora, give values of heterozygosity in the same range, so this extraordinary invariance of population size would appear to apply to all multicellular organisms. The patent absurdity of such a proposition is strong evidence against a neutralist explanation of observed heterozygosity. Precisely the same homogeneity of apparent population size results if $N\mu$ is estimated from the observed number of different alleles at a locus, using the theory developed by Ewens (1972). The only escape would be to show that somehow the stochastic theory of random drift and selection was incomplete in an important way and that in a correct theory the predicted heterozygosity would be independent of population size.

A prediction from the neutrality of allelic variants that is not made from a balance theory concerns the relationship between the average heterozygosity in a population and the proportion of polymorphic loci. If we let q be the minimum frequency of an alternative allele that would lead us to classify a locus as polymorphic, then the proportion of polymorphic loci P is related to the heterozygosity per locus by the relation

$$P = 1 - q^{H(1 - H)}$$

if the variation is the result of neutral alleles. Figure 19 compares this prediction to data of table 22, in which $q = 0.01$. Except for the two excessively polymorphic species, *Drosophila willistoni* and *D. simulans*, the agreement is reasonably good, although the data in general tend to show too many polymorphic loci for the average heterozygosity. There is, of course, a purely numerical relationship between heterozygosity and polymorphism: irrespective of the causes of the variation, the more the polymorphism the more the heterozygosity. But this relationship is a weak one, since all the loci could be polymorphic yet the heterozygosity at each locus could be as low as 0.02 by the convention we use, or, reciprocally, there could be few polymorphic loci but each could have heterozygosity of 0.5 (with two alleles). These numerical limits are shown by the heavy straight lines in figure 19, and the fact that there are no observed points in the upper left region especially, is corroborative although not compelling evidence that the variation is unselected.

A similar kind of relationship, however, gives evidence against the neoclassical theory (Johnson and Feldman, 1972). If there were k alleles at a locus and their frequencies were exactly equal, then each would have a frequency $1/k$, and the total homozygosity at the locus

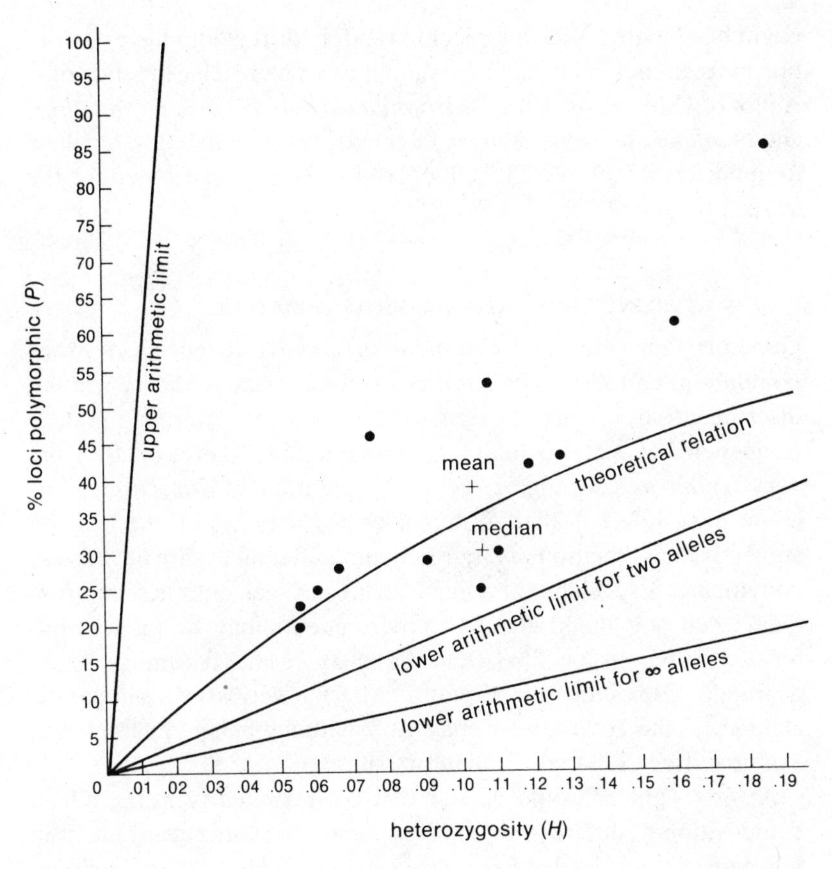

FIGURE 19

The observed relation between the proportion of loci polymorphic P and the average heterozygosity per locus H for the data of table 22, compared with the theoretical line under a neutral hypothesis, if a locus is defined as polymorphic at the 1 percent level.

would be

$$\sum p_i^2 = \left(\frac{1}{k}\right)^2 k = \frac{1}{k}$$

and would be a minimum for a given k. If there were any variation in the allele frequencies, then Σp_i^2 would be greater than $1/k$ and at the extreme, when all the p_i were nearly zero except one very common allele, Σp_i^2 would be equal 1. Then $k\Sigma p_i^2$ would vary from 1 to k, depending upon how uneven the distribution of allele frequencies might be. Neutral alleles subject to random drift would have a variation in frequency such that $k\Sigma p_i^2$ should be an increasing function of k. A plot of $k\Sigma p_i^2$ against k for *Drosophila pseudoobscura, D. simulans,* and *D. affinis,* however, shows a decreasing relationship. The allele frequencies at multiple allelic loci are too evenly distributed for the neutral theory.

EVIDENCE FROM GEOGRAPHICAL VARIATION

General similarities or differences in allele frequencies between populations can all be taken as support for a selective theory without discrimination. Thus if populations are very different in allele frequencies we can postulate local adaptation, whereas if they are very similar we can suppose general adaptation. *A priori* arguments about how different allele frequencies ought to be, on the basis of subjective appreciation of environmental differences, are never very convincing. Especially for animals, who can seek out microenvironments that suit them, effective environments may be far less different in different localities than they may seem from the gross appearance of the climate and biota. Alternatively a vital but subtle element of the environment may vary withoug being apparent; for example, trace minerals in plant nutrition.

At first sight, it would appear that close similarity in the allelic composition of different populations would be a strong argument in favor of balancing selection, because such similarity is at variance with our expectation from a random process. If allelic frequencies are the result of occasional mutations being spread by random drift, we expect two populations to have about the same average heterozygosity, but the polymorphic loci will not be the same and the frequencies of particular alleles at any one locus will have no relationship to each other in different populations. It is the essence of the drift process that the particular allele that is in high frequency in one population is unrelated to the allele that is dominant in a different locality. Yet the observations of chapter 3 show clearly that the same alleles have the same frequency in population after population. Except for the Bogotá population and for the genes associated with third chromosome inversions, the allelic frequencies in *Drosophila pseudoobscura* are remarkably alike in every population examined, from California to Texas to Guatemala. The same is true for *D. willistoni* and in varying degrees for the other organisms surveyed, including man.

The homogeneity of allele frequencies over broad areas seems to be in direct contradiction to the random differentiation among populations that is predicted by a hypothesis of neutrality. But not so, because we must not forget migration between populations. A surprisingly small amount of migration is sufficient to swamp out the differentiation that arises from random drift of unselected genes. If two populations, each of size N, exchange a proportion m of their genes in each generation, the absolute value of the difference in gene frequency between them will be, on the average

$$d = 2\sqrt{\frac{\bar{p}(1-\bar{p})}{(1+4Nm)}} \tag{11}$$

which is small for even a moderate value of Nm. For example, suppose $N = 10^4$, $m = 0.001$, and the average allelic frequency in both populations is $\bar{p} = 0.5$. Then

$$d = 2\sqrt{\frac{0.5 \times 0.5}{1+40}} = 0.156$$

which is about comparable to gene-frequency differences observed between populations. Thus even a migration rate as small as one individual in a thousand per generation is sufficient to prevent differentiation between populations of moderate size.

As for the heterozygosity arising from neutral mutations, the critical parameters for the effect of migration on random differentiation appear only as their product, Nm. But since the migration rate m is expressed as "migrant individuals per unit of population," Nm is an absolute number of individuals. Thus from equation (11) we can calculate that if there are 10 migrant individuals exchanged between populations in each generation, *irrespective of population size*, the average value of d will be 0.15. If there are 100 migrant individuals, irrespective of population size ($Nm = 100$), d is 0.05. A small absolute number of migrants would thus be sufficient to explain any observed degree of similarity between populations.

The same result arises for more realistic models although the formulas are not so transparent. The differentiation of populations can be measured by the ratio of the variance in allele frequency between them to the variance that would be observed if there were complete differentiation, with p of the populations fixed for one allele and $1-p$ for an alternate allele. Suppose that the populations of a species are of size N and form a network exchanging individuals only with neighboring populations at a rate m, but very occasionally receiving a long-distance migrant from a remote and random part of the species range at a rate b. Then, if there is no selection, the ratio of observed to maximum variance will be

$$f=\frac{\sigma_p^2}{p(1-p)} \cong \frac{1}{1-12mN\ [1/(\ln 2b)]} \tag{12}$$

Equation (12) is virtually insensitive to the amount of long-distance migration b, over many orders of magnitude, provided it is small; we are at liberty to suppose that long-distance dispersal is as unlikely as we please. If b is of the order of a tenth of an individual or less per generation, equation (12) comes very close to being

$$f=\frac{1}{1+mN}$$

The variation in gene frequencies among populations of *D. pseudoobscura* in the United States, shown in table 26, gives an average f over loci, not including the third chromosome, of 0.03. This corresponds to only about 30 migrant individuals received from neighboring populations in each generation.

If the variation in allelic frequencies among populations is entirely the result of the breeding structure of the species, including the history of population size fluctuations and migration, then all loci should show the same degree of interpopulation variation, as measured by the inbreeding coefficient f. If we calculate $\hat{f}=\sigma_p^2/\bar{p}(1-\bar{p})$ for each locus separately, each $\hat{f}$ will be an estimate of the true f for the species, since nothing has gone into determining f but the recent history of the breeding structure, which is the same for all loci. The $\hat{f}$ values should be homogeneous, varrying only because of observational sampling error.

Conversely, if only some loci are neutral, if others are being selected in all populations toward the same balanced equilibrium, and if still others are being selected for different locally adapted types, the $\hat{f}$ values from different loci will not be homogeneous. For loci that are being identically selected in all populations, $\hat{f}$ will be low, whereas it will be very high for loci involved in local adaptation, and intermediate for unselected loci.

Table 43 shows the range of $\hat{f}$ values estimated by Cavalli-Sforza (1966) for a number of human genes. The values of $\hat{f}$ vary from 0.029 for the *Kell* locus, which has virtually no population differentiation, up to 0.382 for the R_0 allele of the *Rh* locus, which varies tremendously from race to race and population to population. Lewontin and Krakauer (1973) have shown that the theoretical variance of $\hat{f}$ values if all loci are subjected to some forces should be

$$\sigma^2_{\hat{f}} = \frac{2\bar{f}^2}{n-1}$$

where n is the number of populations characterized. In the case of table 43, $\hat{\sigma}^2_{\hat{f}} = 0.0007301$, which is ten times the observed $s^2_{\hat{f}}$, so it is not possible to explain all the variation in allele frequencies as the result of a uniform process of random differentiation with migration. There must be selection. From the data alone we cannot tell whether the high $\hat{f}$ values of the *Rh, Gm* and *Duffy* loci are the result of diversifying selection, with the other loci being representative of the breeding structure of the species. or whether the low $\hat{f}$ values represent selection toward an equilibrium that is the same in all pop-

TABLE 43

Variation of several gene frequencies among human populations expressed as $\hat{f}$, the ratio of the variance in gene frequency σ^2 to the maximum possible variance $\bar{p}(1-\bar{p})$

Locus	*Allele*	*No. of populations*	$\hat{f} = \sigma^2/\bar{p}\,(1-\bar{p})$
ABO	I^A	125	0.070
	I^B	125	0.055
	i	125	0.081
MNS	*MS*	45	0.071
	Ns	45	0.094
Rh	R_0	75	0.382
	R_1	75	0.297
	R_2	75	0.141
	r	75	0.172
Duffy	*Fy*	62	0.358
Diego	*Di*	64	0.093
Kell	*k*	64	0.029
Haptoglobin	Hp^1	60	0.096
Gm	Gm^a	25	0.226
Gc	Gc^1	42	0.051

Note: From Cavalli-Sforza (1966).
$\bar{f} = 0.148$ and $s^2_{\hat{f}} = 0.007410$.

ulations, with the high $\hat{f}$ loci being a randomly differentiating group. Ancillary data strongly suggest the latter alternative, however, because the high $\hat{f}$ values are in large part the result of the deviant gene frequencies of small, isolated human groups like the Eskimos, Australian Aborigines, Papuans, and American Indians. There is no reason why selection should so strongly differentiate these people from each other and the rest of the world, but their small population size and isolation would cause considerable random divergence.

The $\hat{f}$ values for *Drosophila pseudoobscura*, if we exclude the third chromosome genes and ignore rare alleles at each locus, are three times more variable than they should be if all the alleles were neutral. This excess is significant ($P < .01$), but most of the variance is contributed by two loci, *xanthine dehydrogenase* and *esterase-5*. Unfortunately, the statistical test for heterogeneity of $\hat{f}$ values is not very powerful unless a large number of loci and populations have been examined, so it cannot be applied to the data so far gathered for other species or for the electrophoretic variation in man. Nevertheless, the heterogeneity of $\hat{f}$ values is potentially a very useful general criterion for selection.

EVIDENCE FROM CLOSELY RELATED SPECIES AND INCIPIENT SPECIES

Especially for animals, it is ordinarily impossible to rule out a small amount of migration between neighboring populations or some occasional long-distance dispersal. Because the differentiation of populations is so sensitive to small amounts of migration, we can never argue convincingly from the similarity of gene frequencies. However, *species* are completely isolated from each other genetically, as are some subspecies and incipient species, so that a comparison of allelic frequency distributions between these reproductively isolated groups can be very illuminating.

The data of chapter 4 on the allelic similarities and differences between closely related groups show cases of close similarity and of wide divergence. Nothing is gained by choosing a particular locus in one pair of species and exhibiting it triumphantly as proving the balance or classical case. Under any hypothesis, there will be some loci that are identically monomorphic in two closely related forms, some that are fixed for different alleles, and some that are polymorphic with varying degrees of similarity of distribution. There is, however, a difference in the proportions of the different patterns predicted by alternative views. Under the neoclassical theory, most loci should either be fixed for different alleles in the different isolated groups or, if the loci are polymorphic, there should be no apparent relationship between the patterns of frequency distribution. Since electrophoretic variation is nearly all neutral, there will be no force holding species identically monomorphic. Polymorphic loci will have frequency distributions that are purely the result of the particular neutral mutations that happen to have risen to high frequency in each isolated group. On the balance theory, however, most loci should be identically monomorphic between groups or, when polymorphic, should show a fairly obvious similarity between species or subspecies in the pattern of allelic frequencies.

The strongest evidence for the neoclassical-neutralist view comes from the two subspecies of *Mus musculus* (p. 180). Out of 41 loci, 2 are fixed at different alleles in the subspecies, and 8 have different alleles predominating in polymorphic distributions. Although 25 loci are still identically monomorphic in the two subspecies, it can be argued plausibly that the subspecies have been separated for too short a period for the majority of loci to have drifted apart by chance.

The other comparisons in chapter 4 are strongly against a hypothesis of random drift of allelic frequencies. *Drosophila pseudoobscura* and *D. persimilis*, which have been isolated for more than a million generations, have only 4 "diagnostic" loci out of 39 loci examined. The great majority of loci are either identically monomorphic or show remarkably similar patterns of gene frequency (table 38). Multiple allelic loci like *amylase* and *leucine aminopeptidase* have the same predominant allele. *Esterase-5*, with 6 alleles segregating in *D. persimilis* and 8 in *D. pseudoobscura*, has a particularly interesting pattern. Each species is segregating for a series of alleles that are in adjacent mobility classes, but the entire distribution for *D. persimilis* is displaced toward the higher mobilities. The two distributions are certainly not haphazardly related.

The extensive data on the four sibling species of the *willistoni* group, given in table 39, form a compelling body of evidence for selection. Despite the fact that the loci in table 39 are "diagnostic" in the sense that at least one species comparison is definitive, virtually every case shows remarkable similarity of allelic distribution for pairs or triplets of species. *Drosophila equinoxialis* and *D. paulistorum*, for example, are virtually identical at the multiple allelic loci *lap-5, est-5, aph-1*, and *mdh-2*. *Drosophila willistoni* and *D. tropicalis* are identical in distribution for the six alleles at the *est-7* locus and nearly as similar at the *lap-5*, *adh-2*, and *hk-1* loci, which all have more than two alleles.

The data in table 39 dispose completely of one alternative "neutralist" explanation that underlies all comparisons of similar frequency distributions. What we have been calling "alleles" are really electrophoretic mobility classes. From a given amino acid sequence there are many single amino acid replacements along the polypeptide chain that could produce essentially the same change in isoelectric point. Thus it might be, and in some cases must be, that an electrophoretic mobility class is heterogeneous and really consists of a collection of allelic forms If there were a large number of allelic forms in each mobility class, then the frequency of a class would simply reflect the number of possible different mutations in that class; we would expect, by the law of large numbers, that independently evolving populations would nevertheless have the same frequency distribution of classes. The data on the *willistoni* group rule out this explanation. If *D. equinoxialis* and *D. paulistorum* have identical frequency distribution because we are seeing the frequencies of mutational classes, why are *D. willistoni* and *D. tropicalis* not also identical to them? The fact that pairs and triplets of species have very similar allelic frequencies but that not all four species are identical means that we must look elsewhere for the explanation of the similarities. It is hard to see how the hypothesis that the observed electrophoretic variation is neutral can explain the patterns of similarity and difference among the four species of Drosophila in table 39.

EVIDENCE FROM AVERAGE EVOLUTIONARY RATES

An important argument in support of the neoclassic theory is that the rate of evolution of amino acid sequences has been too rapid for a selective explanation. It may seem strange to claim that evolution has been *too fast* for natural selection which is, after all, a kind of motive force, but the claim rests again on the genetic load argument. The replacement of alleles in evolution by natural selection demands differential survival and reproduction, and there clearly must be a limit to how rapidly genes are replaced, a limit imposed by the total reproductive excess of a species. On a logarithmic scale, the total genetic load accumulated in replacing one allele by another is approximately

$$L = -2 \ln p_0 \qquad (13)$$

where p_0 is the initial frequency of the allele that is eventually fixed. It is interesting that the total load is independent of the selection intensity, since the greater the selection, the greater the load per generation, but the fewer the generations required to complete the gene replacement.

If an evolutionary line is in a steady state with K substitutions per generation, and if the initial frequency p_0 of a new allele is simply the mutation rate μ, then the load from replacement per generation is

$$L = -2K \ln \mu$$

measured on a logarithmic scale. The number of amino acid substitutions per generation, K, can be estimated from the number of substitutions per generation for a selected group of proteins and an estimate of the number of genes in the genome. King and Jukes (1969) surveyed the amino acid substitution in seven polypeptides whose amino acid sequences had been studied in a variety of vertebrates. Using estimates of the time since divergence from a common ancestor, they obtained estimated numbers of amino acid substitutions per year, giving an average of 1.6×10^{-9} per year. From molecular weight studies, Vogel (1964) estimates 4×10^{9} nucleotides per genome for man. With three nucleotides per codon, there are then 1.3×10^{9} amino acid codes per genome, the equivalent of about 10^{7} normal-sized genes. Then

$$K = (1.3 \times 10^{9})\,(1.6 \times 10^{-9}) = 2 \text{ per year} = 6 \text{ per generation}$$

on the very rough assumption of three generations per year for most of the history of the mammals. Finally, if we take the mutation rate to be of the order of 10^{-5} we have

$$L = (-2)\,(6)\,(\ln 10^{-5}) = 68$$

with the patently absurd result that the average fitness of the population is e^{-68} compared with the most fit individual possible in the population, that is, one who is homozygous for the genes that are in process of substitution.

This comparison is spurious, however, since such an ideal genotype does not exist in a population. The actual load depends in part on how important competition is within a population, so that fitnesses must be scaled according to the individuals actually present. Not all selection is competition, however, and genetic load has an absolute component that derives from the relative probability that different genotypes will survive rigors of the physical environment (Felsenstein, 1971). If all of the selection to substitute new alleles was in response to a constantly decaying physical environment that was now inhospitable to old genotypes, then the load calculated as e^{68} is correct and impossible. However, if all the selection is "soft" selection (Wallace, 1968), so that competition for resources in short supply is the sole determinant of fitness, then we must consider only those genotypes actually present, and the load becomes considerably less. Ohta and Kimua (1971a) calculate the load in the latter case by using the concept of "most probable value." If a variable x is normally distributed, then in a sample of size n the most probable extreme deviation from the mean is

$$d = \sqrt{2\sigma_x^2 \ \ln (0.4n)} \tag{14}$$

In the case we are considering, x is the logarithm of fitness, and Ohta and Kimura show that

$$\sigma^2 \ln (W) = K\bar{s}$$

where $\bar{s}$ is the average selection in favor of the allele that is spreading. Even if n is moderately large, perhaps 10^5, and $\bar{s}$ is 0.1, we see that

$$L = \sqrt{2 \times 6 \times 0.1 \times \ln (4000)} = 3.15$$

Relative to the mean, the most fit individual only needs to leave $e^{3.15} = 23$ offspring.

A second reservation about the calculation of the cost of gene replacement enters into all calculations of evolutionary rates: How many genes are there? For purposes of displaying the argument, I have accepted the estimate of 1.3×10^9 codons, based on the total weight of DNA in a human sperm. But do we really believe there are 10 million structural genes specifying enzymes and proteins? To keep things in perspective, we should note that there are just 432 vertebrate enzymes in Dixon and Webb's exhaustive list (1964), although this figure includes only those enzymes for which kinetics have been measured and so excludes those whose activities have been detected but not characterized quantitatively. Nevertheless, the number of vertebrate enzymes definitely known to exist is only on the order of 10^3. If we suppose that the number of structural genes is, for example, 100,000 rather than 10 million, the rate of amino acid replacement in evolution is only 0.06, and the load from substitution is a mere $e^{0.68} = 1.97$ even on the most unfavorable model of absolute selection. This leaves the perennial problem of how to account for the remaining 99 percent of the DNA, only a small proportion of which is highly redundant. At any rate, we cannot accept the argument that the substitution load contradicts the selective hypothesis, since 100,000 genes could be candidates for adaptive evolution without creating a noticeable cost.

Another, similar argument concerns the rate of fixation of favorable alleles in finite populations. If species are in a steady state of gene replacement, then the rate of substitution must be equal to the rate of origin of new mutations multiplied by the probability that a new mutation will eventually be fixed. This latter probability is only $2s$ for alleles of small selective advantage, a surprisingly low rate. Then the rate of mutant substitution will be

$$k = (2NV)(2s) \qquad (15)$$

where N is the breeding size of the populations that have made up the phyletic line and V is the total rate of mutation to favorable alleles. We can then estimate from equation (15) the rate of mutation to favorable alleles as

$$V = k/4Ns \qquad (16)$$

Kimura and Ohta suggest $N = 5 \times 10^4$ and $s = 0.001$ as reasonable values; in this case, for $k = 6$ we get $V = 0.03$ per gamete, which they regard as an absurdly high rate of favorable mutation, since it is about the same magnitude as the total lethal and semilethal mutation rate per gamete. But, again, if there are only 10^5 genes, then $k = 0.06$ and the rate of mutation to genes of very small advantage is only 1 percent as great as for lethals and semilethals. If the reader has the feeling by now that there is nothing in the arguments but arbitrary number-juggling that can be made to support any preconceptions, he has rightly understood my message.

Although the rate of substitution of amino acids in evolution is suspiciously fast under an adaptive theory, especially if the total number of genes is very high, the rate is perfectly consonant with random, nonadaptive substitution. At a steady state, the rate of substitution of amino acids must be equal to the rate of introduction of new mutants multiplied by the probability that a new mutant will be eventually fixed. For any sort of mutation the total rate of introduction per gene must be the mutation rate per gamete, μ, multiplied by the total number of gametes, $2N$. Moreover, for nonselected mutants, the probability that a newly arisen mutant will eventually become fixed is $1/2N$. Therefore the rate at which new alleles destined to be fixed arise must be

$$k = \frac{1}{2N} 2N\mu = \mu$$

and this must also be equal to the rate of substitution at equilibrium.

King and Jukes estimated a substitution rate per codon of 1.6×10^{-9} per year and, again supposing three generations per year and about 130 codons for a typical polypeptide, we get a substitution rate per locus per generation (and therefore a mutation rate per locus to nonselected alleles) of 6.4×10^{-7}. This is certainly a possible and even reasonable value for the mutation rate, indeed somewhat on the low side, so that the rate of substitution can be quite adequately accounted for by the occasional fixation of neutral mutations.

Mutation rates to electrophoretic variants are difficult to measure directly because there is no selective screen for them. However, in Drosophila it is possible to carry a chromosome through hundreds of generations in a balanced state and so observe the accumulation of hundreds of replications when a single genome is finally tested. In this way Mukai (1970) found a rate of allozyme mutation of 4×10^{-6} pooled over three loci for a total of 10^5 gene replications. In 669,904 gene replications pooled over ten loci, Tobari and Kojima (1972) estimated a rate of 4.5×10^{-6}. In each case we are dealing with a handful of mutations (3, in the Tobari and Kojima study, involving only two of the ten loci) so the estimates can only be correct to an order of magnitude.

If we accept the theory that the vast majority of amino acid substitutions in evolution have been the result of the random fixation of neutral alleles, we must be prepared to accept some extreme consequences with respect to the kinds of mutation that can occur. Even the mildest form of the neutral substitution hypothesis leads to extreme conclusions about mutations. This contrast arises from the difference between the *a priori* and *a posteriori* distributions of allelic changes.

One result of the steady-state stochastic theory of gene substitution is that the probability of eventual fixation of a newly arisen neutral mutant, P_o, is

$$P_o = \frac{1}{2N}$$

whereas the probability of evenutal fixation of a mildly adaptive mutant with fitness advantage s is very close to

$$P_s = 2s \quad \text{if} \quad Ns >> 1$$

P_s/P_o expresses the probability that an advantageous mutant will be fixed relative to the chance of fixation of a neutral mutant. The relative frequency of adaptive gene substitutions actually observed at the *end* of the substitution process will be this ratio multiplied by the relative frequency of new mutations of the two types at the *beginning* of the process. That is,

$$\frac{n_s}{n_o} = \left(\frac{P_s}{P_o}\right)\left(\frac{\mu_s}{\mu_o}\right) = 4Ns\left(\frac{\mu_s}{\mu_o}\right) \tag{17}$$

or

$$\frac{\mu_s}{\mu_o} = \left(\frac{n_s}{n_0}\right)\left(\frac{1}{4Ns}\right) \tag{18}$$

where n_s and n_o are the number of selective and neutral substitutions that have taken place in evolution.

The neoclassical theory states that the bulk of completed substitutions in evolution is neutral because adaptive mutations are rare. But how rare must they be? Let us consider mutants with a very small selective advantage, for example $s = 0.001$, and suppose that the population size is moderately large, 10^5. Let us further take the neoclassical theory in its mildest possible *a posteriori* form, that 10 percent of all substitutions in evolution are adaptive. To admit any substantially greater proportion of adaptive mutations would be to assert an adaptive theory, and all the claimed objections to such a theory would apply. Ohta and Kimura (1971b) have suggested 10 percent as an upper limit for adaptive substitutions, based on observed variation in substitution rate. So if we take n_s/n_o as 0.1 we find from equation (18) that

$$\frac{\mu_s}{\mu_o} = \frac{1}{4000}$$

Thus even if the neoclassical theory allows as much as 10 percent of substitution in evolution to be adaptive, it must assert that neutral mutations are 4000 times more frequent than mutations with a very slight advantage. If a stronger version of the neutral theory is advanced, one that accords better with the general tenor of neoclassical arguments, n_s/n_o would be more like 10^{-2} or 10^{-3}. For example, Haldane's estimate that the standard rate of adaptive gene substitution should not exceed about 1/300 per generation has been repeatedly contrasted by the neoclassicists with their estimated observed rate of 2–3 substitutions per generation in protein evolution (see, for example, Kimura and Ohta, 1971a, p. 25). But if n_s/n_o is as small as 1/100, neutral mutation rates must be 40,000 times greater than mutations with a 0.1 percent average advantage! If population sizes are larger than 10^5, the problem becomes proportionally greater. We must bear in mind that natural selection is a sieve that retains adaptive mutations and vastly enriches the frequency of these events among the end products of the evolutionary process. To assert that very few of the final results of evolution are adaptive is necessarily to assert that vastly fewer were adaptive *ab initio*.

This argument, like so many others, depends strongly on assumptions about population size, *N*. The models of gene substitution involve population size but there is a blurring of the distinction between the breeding size of a population and the breeding size of a species. The models are of fixation of genes in a population, but the results are applied to gene substitutions in the evolution of species, so it follows that the correct value of *N* in all calculations must be the effective breeding size of the stream of germ plasm that continuously connects one extant species with another through their common ancestor. This almost certainly means that the correct values for *N* are much larger than the 10^4 or 10^5 that are so often used as illustrative examples.

Total gene substitution may occur in two ways. First, there may be pure phyletic evolution without speciation so that a species originally homozygous for an allele a comes to be entirely characterized by an allele a' either by random fixation of a mutant allele or by selection. If the species is broken up into virtually isolated populations, such a replacement process cannot occur in all populations, and the extinction of all but the substitutional population would be required, an unlikely event. Alternatively, if there is sufficient migration to allow the mutant to spread to all populations, then the effective number is the total species number during the process of spread. As shown by Moran (1962), migration of even one individual per generation into each population from the species pool at large makes the entire species a single effective breeding unit.

Second, gene substitution may occur during the formation of a new species, which frequently will involve a small founding population in isolation from its parental group. The founding group is likely to increase in size quickly after its original colonization, and the history of gene fixation will reflect both the small initial size and the large subsequent population. To see the effect of such a history on the relative probabilities of fixation of selected and unselected alleles we must revert to the more general formulation for fixation probability. In a population whose effective size is N over many generations, the probability of fixation of an allele with selective advantage s is

$$P_s = \frac{1 - e^{-4Nsp}}{1 - e^{-4Ns}} \tag{19}$$

where p is the frequency of the allele at the beginning of the process. For a neutral allele, however,

$$P_o = p \tag{20}$$

independent of population size. In the usual model a newly arisen allele has the frequency $1/2N$, and substitution of this value for p in equations (19) and (20) yields the standard expressions $P_s = 2s$ and $P_o = 1/2N$, given before. Let us suppose, however, that a population is founded by a small number of individuals, M, budded off from a large species population. Most of the rare alleles will not be represented in this small sample but a few will by chance be included as single heterozygotes, and their frequency will then be $1/2M$. Because of the drastic sampling process, there will be no difference in this respect between neutral genes and those with a slight selective advantage. The new colony will expand rapidly to fill its new territory, reaching a much larger population size N. During this process it will more or less faithfully reproduce the initial gene frequencies of the first small colony. Presumably this is the history of the Bogotá population of *Drosophila pseudoobscura*, whose population size has grown tremendously in the last ten years and whose gene frequencies are typical of a small sample from the main species distribution but include abnormally high frequencies of an occasional rare allele like $pt\text{-}8^{0.80}$, which is 87 percent in Bogotá but about 1 percent elsewhere. The subsequent probability of fixation of alleles will follow equations (19) and (20), with $p = 1/2M$. If $N_s > 1$ then

$$P_s \cong 2s\left(\frac{N}{M}\right)$$

and

$$P_o = \frac{1}{2M}$$

The ratio P_s/P_o is still $4Ns$, as it was without considering colonization. We are forced to conclude, then, that N is likely to be a very large number in relation to the fixation of genes in an entire species, unless we are prepared to assume that most species have undergone long periods (of the order of $2N$ generations) at very small total species population sizes. In considering gene substition in evolution it is extremely important not to confuse single population sizes with the size of the total species pool.

EVIDENCE FROM COMPARATIVE EVOLUTION

Three observations about the detailed pattern of protein evolution are often advanced as agruments in favor of a neoclassical hypothesis. The first observation is that the amino acid composition over many proteins and species is in fairly good agreement with what would be expected if the purine and pyrimidine bases were arrayed at random along the DNA string, except that there is a striking deficiency of arginine (King and Jukes, 1969). But such an agreement is really irrelevant to the issue. There is nothing in an adaptive theory of evolution of proteins that suggests that some amino acids should *on the average* be more favored than others, although in a particular position in a particular molecule it may make a great difference which amino acid is substituted. Therefore the general agreement of amino acid composition with random expectation is not informative.

This line of argument is very revealing, however, of the logic and sociology of the dispute about selection. Recent analysis of amino acid substitution data by Clarke (1970) has shown that substitution of amino acids by those with similar physicochemical characteristics is more common than expected by chance. Moreover, Subak-Sharpe (1969) has shown by nearest neighbor analysis in viruses that DNA doublet frequencies are nonrandom, and King and Jukes (1969) and Kimura and Ohta (1972) reach the same conclusions by statistical estimations. Yet *these* observations are regarded by Kimura and Ohta and by King and Jukes as evidence for the neoclassical hypothesis, because they can be interpreted as showing that natural selection discriminates against some kinds of substitutions and places constraints on the arrangement of bases.

It must be remembered that selective *constraints* are an important part of the neoclassical theory and are perfectly in accord with purifying and normalizing selection. In the words of Kimura and Ohta (1972); "Through these analyses we have been led to the view that amino acid composition of proteins is determined largely by the existing genetic code and the random nature of base changes in evolution. Small but significant deviations from such expectation can be accounted for satisfactorily by assuming selective constraint of amino acid substitution."

The way in which these constraints are included in the neoclassical view can be seen in the development of the argument on substitution probabilities at different sites in a protein. Originally King and Jukes proposed as evidence for "non-Darwinian" evolution that the number of substitutions at different sites in the evolution of three proteins followed a Poisson distribution, provided a certain number of invariant sites were removed from the calculation. Thus it appeared that except for a small number of conserved sites, all amino acid positions were equally likely to be substituted.

Later Fitch and Markowitz (1970) radically revised this conclusion for cytochrome *c*, showing that only 10 percent of amino acid positions were free to vary in any phyletic line but that the identity of the 10 percent changed from phyletic line to phyletic line. Moreover, the position of these "concomitantly varying codons" was such as to bring the variable amino acid sites close to each other on the surface of the folded molecule. Apparently there is strong interaction between sites in determining where substitutions will occur. Whereas an adaptive theory would argue that the first substitution *predisposed* the neighboring sites to be substituted, the neoclassical theory would argue that the first substitution *prohibited* all but the neighboring sites from accepting substitutions.

Finally, the observations by Clarke (1970) and Epstein (1967) that substitutions of amino acids of like properties are more probable than substitutions at random are now claimed by Kimura and Ohta (1972) to be most plausibly explained by *neutral* substitutions.

Thus the neoclassicists have the best of both worlds. Both randomness and nonrandomness are interpreted as evidence in their favor. They do not tell us what observations might not confirm the theory.

Second, strong emphasis is placed on the uniformity of substitution rates over geological time in different phyletic lines. Indeed the "remarkable constancy of the rate of amino substitution in each protein over a vast period of geological time constitutes so far the strongest evidence for the theory that the major cause of molecular evolution is random fixation of neutral or nearly neutral mutations," according to Ohta and Kimura (1971b). Table 44, taken from Ohta and Kimura's paper, illustrates this constancy for three different proteins of quite different average substitution rates. Although there is some variation in the rates, which may not be significant since they depend heavily on rough estimates of the number of years since divergence from a common ancestor, the uniformity within proteins is indeed striking.

The relevance of this uniformity to our problem is completely destroyed, however, when we look at the first column of figures in table 44, the estimated divergence times. At the least, 80 million years of evolution separate the gorilla and the monkey, and the time is 750 million years for several comparisons. Typically we are dealing with times of 200 million years, putting the common ancestor 100 million years ago, in the middle of the Cretaceous. Since that time every phyletic line has undergone numerous episodes of rapid and slow evolution, so that the substitution rates shown in the table are averages over vast periods of time and many cycles of

TABLE 44
Estimates of the number of years of evolution separating two forms, expressed as two times the number of years since a common ancester, $2T$, and the average rate of amino acid substitution per codon, k, for a number of phyletic lines

Comparison	$2T \times 10^{-8}$	$k \times 10^{9}$
Hemoglobin β		
Spider monkey–mouse	1.6	1.225
Human–rabbit	1.6	0.631
Horse–bovine fetal	1.0	2.319
Llama–bovine	1.0	1.806
Human δ–sheep (A)	1.6	1.288
Rhesus monkey–goat	1.6	1.184
Pig–sheep (c)	1.0	2.231
Average		1.526
Hemoglobin α		
Human–bovine	1.6	0.769
Gorilla–monkey	0.8	0.450
Rabbit–mouse	1.6	1.326
Horse–sheep	1.0	1.442
Pig–carp	7.5	0.877
Average		0.973
Cytochrome *c*		
Human–dog	1.6	0.699
Kangaroo–horse	2.4	0.290
Chicken–rabbit	6.0	0.136
Pig–gray whale	1.6	0.121
Snapping turtle–pigeon	6.0	0.136
Bullfrog–tuna	7.5	0.207
Rattlesnake–dogfish	7.5	0.384
Average		0.281

Note: From Ohta and Kimura (1971b).

speciation, extinction, and phyletic evolution. The claimed "constancy" is simply a confusion between an average and a constant. It is like claiming that the temperature never varies in Chicago because the total number of degree days measured there was the same in the last fifty years of the nineteenth century as it was in the first fifty years of the twentieth. The immense variations in evolutionary rate documented in Simpson's *Tempo and Mode in Evolution* (1944) show how erroneous it is to average rates over 200 million years.

Finally, the neoclassical theory makes a prediction about the rates of evolution of different proteins that is at variance with the prediction from an adaptive theory. Since the neoclassical theory regards most natural selection as purifying, as putting constraints on evolution by cleaning out deleterious mutants, ascribing most successful substitutions to the fact that they are neutral and thus unconstrained, it follows that the most rapid evolution should be in those polypeptides with the fewest constraints. Thus the more "useless" a protein is physiologically, the more rapidly it should evolve. In contrast, a theory which holds that most of the gene substitution in evolution has been adaptive would predict that "useless" proteins would be among the slowest evolving.

Table 45 shows present estimates of evolutionary rates for 11 proteins that have been sequenced over a phylogenetic range. Of these, the most rapidly evolved and the third most rapidly evolved are good candidates for the class of "useless" proteins. Fibrinopeptide A (20 amino acids) is a residue that is clipped off fibrinogen to convert it to its active form, fibrin, in a blood clot. Similarly, proinsulin polypeptide C (35 amino acids) is removed from proinsulin to convert it to the normally active insulin. The active moiety has evolved ten times more slowly than the discarded safety catch. Dickerson (1971) argues, on the basis of three-dimensional configuration and function, that cytochrome *c* should be the slowest evolving of the group in table 45 because it has the most constraints on it and that, in general, the order of proteins in the list is predictable from such constraints. The same reasoning applied to the different regions of

TABLE 45
Rates of amino acid substitution per site per year on the average in vertebrates

Protein	*Substitutions per amino acid per year* ($\times 10^{10}$)
Cytochrome *c*	2.5
Glyceraldehyde-3-phosphate dehydrogenase	2.7
Insulin A and B	3.6
Trypsinogen	4.6
Lysozyme	9.4
Hemoglobin α	9.9
Hemoglobin β	13
Ribonuclease	25
Proinsulin polypeptide C	27
Immune light chain	33
Fibrinopeptide A	45

Note: Data are from King and Jukes (1969) and Dickerson (1971)

the cytochrome *c* molecule explains why some positions are invariant in evolution (positions 70–80, for example) and why the hydrophobic residues are never substituted by charged amino acids. Yet this kind of argument can also be applied to the so-called useless proteins with similar results. Stebbins (in Stebbins and Lewontin, 1972) has shown that there is a marked nonrandomness in the distribution of amino-acid substitutions in the fibrinopeptides. Most changes are concentrated near the amino terminal end; these are nearly all radical changes, substituting hydrophobic for hydrophilic, or positively for negatively changed, amino acids. Moreover, the region of greatest change in fibrinopeptide A, around residue 13, is apparently important in governing the rate at which the molecule is split off from fibrinogen.

All such functional arguments are somewhat weakened by being *post facto* rationalizations rather than true predictions, but it is a problem of functional explanation in evolution that is not restricted to molecular events. If similar functional arguments were applied *a priori* to a set of proteins before they were sequenced, a very powerful support for the neoclassical view would result. Certainly the data of table 45 should cause disquiet to those who believe that most amino acid substitutions in evolution are adaptive.

Direct evidence on whether most gene substitutions have been adaptive would be extremely difficult to obtain, but not impossible. It would involve the demonstration that a particular molecular form performs best in the internal and external environment of its own species, and more poorly in other species. *Drosophila pseudoobscura* and *D. persimilis* overlap in many of their polymorphisms. Table 38 (p. 172) shows that for *D. pseudoobscura* the allelic range of *esterase-5* is from *0.95* to *1.16* electrophoretic mobility classes, whereas the alleles of *D. persimilis* range from *1.07* to *1.33*, with the most frequent allele, *1.20,* just outside the *D. pseudoobscura* range. Since occasional mutations to this class must occur in *D. pseudoobscura*, we could ask whether the different ranges of the two species' allelic repertoire are adaptive or not, by studying the physiology of *1.20* mutants in *D. pseudoobscura*.

Many opportunities for this sort of comparison exist in the sibling species of *D. willistoni*, for example at the *mdh* locus, where the rare variants already exist in the different species (see table 39, p. 176). To make the comparison valid, it would be necessary to show that the molecular form, rare in one species and common in the other, was indeed identical in sequence in both organisms.

When such a comparison has been made *in vitro*, it has not revealed any functional difference between specific variants. If cytochrome *c* from any species is combined with bovine cytochrome oxidase there appears to be no average adverse effect of the heterologous combination on the kinetics. But such an *in vitro* comparison leaves out nearly every variable of internal and external environment except the fit of the molecule into the organelle, so that a negative result is rather uninformative.

OBSERVATIONS OF SELECTION

Attempts to establish whether natural selection is maintaining genetic variation in populations, by means of fitting a few observed statistics to theoretical predictions, are bound to be unsatisfactory. Population genetic theory is well developed but it is still only a first-order theory. It still depends upon vast numbers of simplifying assumptions and primitive models, so that contradiction between observations and theoretical predictions may only reveal the shallowness of the predictions. Why do we not put aside this interminable number juggling, this genetical *pilpul*, and go straight to the heart of the matter? Let us go into nature and measure natural selection directly.

Any program to judge the importance of natural selection by detecting its operation in nature must have a form similar to our methodological program for detecting variation itself. Just as we could not know what proportion of the genome of an organism was polymorphic by collecting cases of polymorphisms, so we cannot know the over-all importance of balancing selection by demonstrating that it exists. Of course it exists. The problem is, What proportion of observed genic variation is maintained by selection? The problem cannot be solved by compiling a list of a dozen or even a hundred cases from as many species in which selection is in some way implicated in the control of a polymorphism. It can only be solved by taking some arbitrary set of genetic polymorphisms and attempting to establish, for each case, the selective forces involved, through an exhaustive study of natural history and demography.

This is the strategy that has been adopted by the school of

"ecological genetics," which owes its inspiration largely to E. B. Ford (Creed, 1971; Ford, 1971). It is their hope that by studying polymorphisms in large, countable, nonsecretive and genetically amenable organisms like snails, moths, butterflies, and man, they will be able to establish case after case of selective polymorphism. They are frankly partisan in their belief that polymorphism is in general balanced, but such bias is necessary for the success of this research strategy, for it is a strategy of confirmation rather than exclusion (Lewontin, 1972b).

Suppose that 100 polymorphisms are subjected to a search for balancing selection and that the search is successful in 98. Then no reasonable person could doubt, if the polymorphisms were chosen without prior knowledge or bias about the forces of selection, that balancing selection is the chief cause of polymorphism. Suppose, however, that only 2 cases were proved. Then it might be that balancing selection is unimportant, but it also might be that the investigators did not try hard enough to find the selective forces, or were not ingenious enough. That is why they must want very much to find selection in the first place.

Consider the case of the snail *Cepaea nemoralis*. *Cepaea nemoralis* is highly polymorphic for a variety of shell characteristics, especially the pattern of banding and the color. The presence or absence of bands results from a pair of alternative alleles at a single locus, and another locus determines whether shell color is pink or yellow. The frequency of banding and of shell color alternatives varies widely over the distribution of the species in northern and central Europe.

In an extensive study of French populations of *C. nemoralis*, Lamotte (1951) came to the conclusion that the polymorphism for the presence or absence of bands was due essentially to random drift. He looked for, but failed to find, consistent evidence of differential predation by thrushes on the banded and bandless forms, and he was unable to find any correlation between allele frequency and climatic variables. On the other hand he did find that the variance of allele frequency among large colonies was less than among small colonies, and this suggested to him that random drift was important.

On the assumption that the allele frequencies in different populations represented one of Wright's (1937) stationary distributions under the joint pressures of mutation, migration, selection, and drift,

Lamotte fitted the data of one geographical region to the theoretical distribution and estimated from the best-fit curve the parameters of migration, mutation, and selection, all confounded with population size since they always appear in the form Nm, $N\mu$, and so on. He concluded that there was about a 7 percent selection against bandless homozygotes, with heterozygotes nearly as unfit, and that the polymorphism was maintained despite this directional selection by an extremely high mutation rate to the bandless allele. Although the fit of the observed data to the theoretical curve was excellent, it would be a strange curve indeed that could not be fit with four parameters!

Working in parallel on English snails, Cain and Sheppard (1950, 1954) showed a correlation between vegetational background and the allelic frequencies for both banding and shell color. This correlation was most striking when banding and shell color were considered together, a finding that illustrates the difficulty of the strategy of exhaustive confirmation: If no selection has been detected with respect to a single polymorphism, the possibility can always be held out that consideration of two loci simultaneously will reveal selection. When main effects fail, there may always be interactions. But such an escape clause puts the selectionist in an unassailable position since the number of interactions between polymorphic loci is effectively infinite, even if we only consider pairs, and we can never exhaust our search. Later, Lamotte too (1959) found a correlation of both banding and color with environment, banding with mean summer temperature and color with mean winter temperature. Whether these variations explain the polymorphism is another issue, but there seems no doubt at present that the various morphs are not indifferent to the action of natural selection.

The case of *Cepaea* is regarded as a paradigm by selectionists, but other polymorphisms have not so far yielded to a persistent attack. The most glaring failure of selectionist explanation is the variety of human blood-group polymorphisms, none of which has yet been explained by natural selection although the data are more copious and reliable than for any other case and the natural history and reproductive schedules are better known than for any other species. Several attempts have been made to invoke interactions between polymorphisms as an explanation for the blood group variation, notably ABO and Rh incompatibility (Levine, 1958), but how important this interaction is in maintaining the polymorphisms is uncertain.

The case of *Cepaea* raises another issue that is somewhat glossed over in ecological genetics but that lies at the core of our theoretical and experimental problem. A completely satisfactory demonstration of the balance hypothesis demands not only that genotypes be differentially selected but also that the selection be of such a nature as to preserve genetic variation. Only special forms of selection will do that. In particular, heterozygotes must be more fit than homozygotes, or some form of frequency-dependent selection favoring rare genotypes must operate, or a pattern of temporal and spatial heterogeneity of a particular kind must exist, in order to stabilize variation.

In the end it is not enough to show that hot summers favor unbanded snails and that yellow snails are more frequent where winters are cold. It must also be shown that heterozygotes are superior in all environments or that some other form of balancing selection holds. The balance theory is a very strong assertion, for it maintains not only that there is selection, but that there are very special kinds of selection. Much of the work on human blood-group polymorphism ignores this point. Even if the association between ABO blood groups and various diseases like ulcers, cancer, and anemia were firmly established (Vogel, 1970), and there is some doubt of this (Weiner, 1970), there is nothing to suggest any form of balancing selection. The situation is far worse for ABO incompatibility and Rh hemolytic disease, since both select against heterozygotes and are thus completely unstable.

DIRECT MEASUREMENT OF NET FITNESS

The most direct approach to assessing selection is to measure the reproduction of the various genotypes at a locus and to calculate fitness values. For species with synchronized populations, like annual plants or univoltine insects, a straightforward measurement of the number of fertilized zygotes produced per fertilized zygote in the previous generation, for each genotype, will give direct estimates of the fitnesses. For populations with overlapping generations, the age-specific mortality and fecundity schedules for each genotype need to be measured, and these must be combined in a special way to give fitness estimates (Bodmer, 1968; Charlesworth, 1970; Charlesworth and Giesel, 1972). Although there is no difficulty in theory in estimating fitnesses, in practice the difficulties are virtually insuperable. *To the present moment no one has succeeded in measuring with any accuracy the net fitnesses of genotypes for any locus in any species in any environment in nature.* Let us see why.

The direct measurement of fecundity of a genotype in females requires the identification of individuals and the capturing of their total zygotic output. For plants this can in fact be done because seed set can be counted, but for most animals it is much more difficult, being possible in practice only for those species that produce well-defined clutches in accessible places. Assessment of the contribution of males of different genotypes depends either on the formation of stable monogamous pairs or else on the possibility of analyzing the offspring of females of known genotype to determine the population fertility of males. Plants are again ideally suited for the latter method, but any species will serve in which the zygote can be carried without significant mortality to the age at which it can be classified genetically. Then, knowing the maternal genotype, one can estimate the relative contribution of different male genotypes to the pool of zygotes.

Having estimated the fecundity of genotypes, one must now measure the probability that a newly formed zygote of each type will survive to reproduce and will be included at the stages when fecundity is assayed. There is virtually no organism in which a direct measurement is possible because the genotypes of newly formed zygotes cannot usually be identified nor the individuals followed until they reach sexual maturity. A substitute for actually measuring the probability of survival of genotypes is to estimate this probability from the proportions of adults of various genotypes compared with the proportions among the zygotes that gave rise to them, assuming one could reconstruct the zygotic proportions from the fecundities and frequencies in the previous generation. Unfortunately, in a real population, adult frequencies reflect not only the survival of zygotes but also immigration and emigration, both as propagules and as adults.

Let us take the most favorable case, an annual plant that can be raised in a greenhouse. In year 1 we classify genotypes in a field of plants just as seed capsules are ripening. We record seed set per plant and germinate a sample of seed to reconstruct the effective pollen fertility of various genotypes. We now have an estimate of the actual genotypic composition of the seed crop. In year 2 we again assay adults and compare the frequencies of different genotypes with the seed-crop estimate from year 1. But we cannot estimate the survival of genotypes because we do not know what proportion of the adult plants comes from the local seed crop and what proportion was blown in, carried in, or washed in. Of course we may choose a well-isolated population, or a species in which seed movement is essentially nonexistent or can be measured accurately, but these restrictions only underline the point that a species must satisfy a restrictive set of requirements before components of fitness can be measured.

The practical problem is nearly beyond solution in the case of species with overlapping generations because we need to have the complete age-specific mortality and fecundity schedules. The only species for which there is any hope of following cohorts of identified genotypes through their reproductive lines is man, but the expense of follow-up studies has been discouraging.

Christiansen and Frydenberg (1973) have completely analyzed the statistical problems of estimation of fitness in an ideal system like that of our hypothetical plants. They were able to divide the total fitness into mating, fertility, and viability components and to show how each could be separately estimated and tested for significance. They then applied their theory to an esterase polymorphism in a live-bearing fish, *Zoarces viviparus*, in which it is possible to score mother-fetus combinations. The total numbers of fertile mothers, sterile females, and males in the sample were 782, 69, and 431, respectively. Table 46 shows the outcome of the tests for all the components of selection except female fecundity, which had previously been shown to be nonsignificant. The last column of the table gives the percentage deviation from the expectation under the hypothesis of no selection, that would need to obtain before the investigators could have been 50 percent sure of detecting it in their samples. Given the sample sizes that were necessary to achieve even this much statistical power, it would obviously be no small matter to detect selective differences of the order of 1 percent, an order that is surely closer to realistic values for the vast majority of polymorphic loci.

TABLE 46
Tests of the components of fitness and net fitness differences between genotypes, and the smallest selection differences that could have been detected with a 50 percent probability, for an esterase polymorphism in the eelpout, *Zoarces viviparus*

Component	Degrees of freedom	χ^2	*P*	% Difference detectable
Gametic selection in females	1	.34	$>.50$	10
Random mating of breeding fish	2	1.37	$>.50$	11
Differential male mating success	1	1.07	$>.30$	14
Differential female mating success	2	.37	$>.50$	33
Zygotic selection	3	.33	$>.95$	7
Net fitness (assuming equal female fecundity)	9	3.48	$>.90$	

Note: From Christiansen and Frydenberg (1972).

It was at one time supposed that net fitness could be measured directly by comparing the frequencies of genotypes at any point in the reproductive cycle with the Hardy-Weinberg frequencies predicted from the genotypes in the previous generation at the same point in the cycle. For example, suppose a population is segregating for a balanced lethal in the egg stage. The only adults in generation 1 will be heterozygotes, the estimate of gene frequency from these adults will be $p = q = 0.5$, and the predicted frequency of adults in the next generation would be 0.25 AA:0.50 Aa:0.25 aa if there were no selection. But because of the balanced lethal only heterozygotes will be actually observed in generation 2, and so the fitnesses of the homozygotes will be estimated as zero, correctly.

In a fundamental paper on fitness estimation, Prout (1965) showed that the method just described is erroneous and does not take proper account of differential fertility. As an example, suppose that instead of a balanced egg lethal, a population is segregating for a balanced *sterile* condition. Then all three genotypes will be present in adults in every generation in a 1:2:1 ratio and no selection will be detected, yet there is very strong heterosis for fertility.

Another consequence of Prout's finding is that the component of fitness due to fertility differentials will cause an erroneous appearance of frequency-dependent selection, with the rarer genotypes apparently having the highest fitness, and all fitness differences apparently disappearing at equilibrium. Unfortunately, Prout's discovery was slow in reaching the consciousness of many geneticists, and at least one report of frequency-dependent selection is based on this erroneous method of estimation (Kojima and Yarbrough, 1967).

PARTIAL FITNESS

Even though net fitness cannot be measured in nature except in special circumstances, so far not fulfilled, individual components of fitness can be estimated even though they give only a partial picture. Thus it is relatively easy to estimate seed set in plants, or completed family size in man, or the genotypic frequencies in different age classes of fish. But the retreat to the measurement of partial fitness represents a major change in objective. Originally we wished to measure fitness in order to explain polymorphism, and this meant showing not only that selection was operating but also that the fitness relations among the genotypes were such as to produce a stable equilibrium of allele frequencies.

If we are willing to settle for the measurement of partial fitness, then we cannot hope to demonstrate stability, for there is no necessary correlation between any component of fitness and total fitness, except insofar as parts are correlated with wholes. Net heterosis may arise because of opposite fitness relations in different parts of the life cycle, one homozygote being favored by survival, the other by fecundity. In the case of continuously breeding organisms, the relationship between any component of fitness and total fitness is even more uncertain that for synchronized species. Total life-time fecundity, for example, may be very poorly correlated with net fitness, especially when there are marked differences in the way fecundity is distributed over age classes (Dobzhansky, Lewontin, and Pavlovsky, 1964). The measurement of partial fitness can only serve to demonstrate that there is *some* differential fitness associated with the alleles segregating. Yet this demonstration is probably sufficient. If alleles are segregating at intermediate frequencies and any selection at all can be demonstrated to operate on them, it would be difficult to avoid the conclusion that they are held by some form of balancing selection.

Occasional alleles may be in the process of replacement, but these must be very rare. Since the length of time spent by an allele in any frequency interval dq is approximately proportional to $dq/q(1-q)$, transient alleles will almost always be close to fixation. Sometimes a disadvantageous allele may be associated with a meiotic drive mechanism, like the t alleles in mice, maintained by the balance between the loss of the allele in unfit homozygotes and the gain of the allele in the distorted gametic pool of heterozygotes (Dunn, 1956; Lewontin and Dunn, 1960). But this phenomenon, too, is apparently not common. Certainly a demonstration that some component of fitness is affected by the segregation of most allozyme variants in populations would rule out any neoclassical-neutralist explanation of heterozygosity.

The measurement of even one component of fitness in nature is not trivial. A much larger sample size is required than investigators usually realize, and some methods of fitness measurement have hidden snares. In order to be x percent sure that an observed difference between two populations will be significant at the y percent level, if the true difference between the populations is k percent of their mean, we need a sample size n, given by

$$n = \frac{2(t_x + t_y)^2 V^2}{k^2}$$

where V is the coefficient of variation of the character and t_x and t_y are, for even moderate samples, the standardized normal deviates corresponding to the x and y percent probability levels. For example, suppose we wish to be 95 percent sure that an experiment will reveal a 1 percent difference in fecundity, if it exists, with the conventional .05 level of significance. Then $t_x = 1.65$, $t_y = 1.96$, and $k = 1$. The coefficient of variation for fecundity is generally large [an average of 70 over a disparate group of human populations, for example (Cavalli-Sforza and Bodmer, 1971)], so that $V^2 = 5000$. Then the number of females of each type whose fertility must be measured is

$$n = \frac{2(1.65 + 1.96)^2\,(5000)}{1} = 130{,}321$$

This tremendously high number can be reduced to only 38,000 if the investigator is willing to take a 50 percent chance that the true difference will not be detected and can be further reduced to an actual, practical experiment ($n = 380$ for each of the genotypes) only if he is also willing to give up detecting differences in fecundity smaller than 10 percent. But selective forces on single-locus polymorphisms are unlikely, in general, to be as great as 10 percent although they may be for an occasional locus of drastic effect. One percent selection differences are undoubtedly much nearer the mark if balancing selection is operating on thousands of loci, and we see that almost impossibly large samples are needed to detect them.

A similar calculation made by Cavalli-Sforza and Bodmer for detecting viability differences in the progeny of different crosses gives the same order of values for sample size, and Hiorns and Harrison (1970) find large sample sizes necessary to detect differential death rates from a difference in frequency in two age classes. It seems unlikely that the volume of data required will be obtainable from any species except *Homo sapiens*, and so far no one has thought it worthwhile to measure the completed family sizes or genotypic distributions in hundreds of thousands of human families categorized by genotype.

In addition to problems of statistical power, estimates of net fitnesses can be quite erroneous. The most common, because the simplest, method of estimating partial fitness in natural populations is to compare the distribution of genotypes at some advanced age with the Hardy-Weinberg frequencies expected from the allele frequencies. It is reasoned that zygotes will be formed in Hardy-Weinberg frequencies if there is no assortative mating for the genotypes involved, so that deviations from these theoretical proportions among adults will measure relative survival probabilities. The appeal of the method is that only a single measurement is required—the frequencies of the different genotypes among the adults. This procedure is misleading, however, and has no statistical power for realistic selection values (Lewontin and Cockerham, 1959).

Because the allele frequencies on which the Hardy-Weinberg expectations are based are estimated from the genotypes *after* selection, the procedure does not test whether fitnesses of the three genotypes, W_{AA}, W_{Aa}, and W_{aa}, are equal, but whether

$$(W_{AA})\,(W_{aa}) = W_{Aa}^2$$

Although this includes equality of the Ws, it also includes any multiplicative fitness relations, for example, $W_{AA} = 1.00$, $W_{Aa} = 0.9$, $W_{aa} = 0.81$, and so will have little power against most cases of intermediate dominance and weak selection. Second, an excess of heterozygotes over expected numbers does not necessarily mean heterosis, but only that

$$(W_{AA})\,(W_{aa}) < W_{Aa}^2$$

so that simple intermediacy of the heterozygote will appear as heterozygote excess. Only if gene frequencies are not changing does the excess of heterozygotes give strong evidence of heterosis in viability. This ambiguity needs to be taken account of in interpreting the numerous reports of heterozygous excess in natural populations (for example: Dobzhansky and Levene, 1948; Lewontin and White, 1960; Richmond and Powell, 1970).

The demonstration of a significant excess of heterozygotes does demonstrate that selection of some kind is operating, even if it is not proven heterosis. This selection must be rather strong, however, to be detectable, especially if heterozygotes are intermediate. The viability differences most easily detected by this method are those that arise from true heterosis, but 10 percent heterosis requires a sample size of 4000 to be even 90 percent sure of detection, and a 1 percent heterosis would need a sample of about 400,000. It should not surprise us that selection for single-locus polymorphisms has not often been found by direct measurement in nature. Only when there are very large viability differences, on the order of 10 percent or more, as for the inversions in *Drosophila pseudoobscura* (Dobzhansky and Levene, 1948) or in the grasshopper *Moraba scurra* (Lewontin and White, 1960), will we expect to find significant deviations from Hardy-Weinberg expectations with reasonable sample sizes.

CORRELATIONS WITH ENVIRONMENT

The alternative in natural populations to direct measurement of fitness or fitness components is to show at least that selection must be operating, even though it cannot be measured, by correlating the frequencies of alternative alleles with temporal or spatial differences in environment.

The information obtainable from temporal variation is much

greater than for spatial. If allele frequencies undergo repeatable cyclic fluctuation during the course of repeated seasonal cycles, as do the third chromosome inversions of *Drosophila pseudoobscura* in southern California (Dobzhansky, 1943), it is possible to estimate the magnitude of selective differences from the speed of the change. No serious attempt at accuracy is worthwhile since the cyclic nature of the changes shows that the selection coefficients must be changing, and even if they remained constant over a few generations the standard errors of fitness estimation are much too great. Nevertheless, limits can be put on selection intensities, especially if at one season of the year there is a repeatable, rapid change. In this way Dobzhansky was able to estimate selection coefficients of 0.1 to 0.4 at some times of year for the *D. pseudoobscura* inversions.

In principle the same kind of information is obtainable if frequency changes are not cyclic but are nevertheless *well* correlated with some aperiodic feature of the environment. I emphasize well correlated because there is a considerable danger here. Suppose the frequency of some genotype is observed to change steadily over a number of generations or years and then levels out at a new equilibrium. If one then institutes a *post facto* search through environmental and biotic records of the period, one is likely to find some aspect of the environment that will show either a distinct alteration at the same time that the allele frequency began to change, or a parallel progressive alteration over the same period, given the medium-term instability of temperate environments especially, and the very rapid alteration that man is producing in the environment. Not to find any environmental element that is roughly in phase with the genetic change would be extraordinary. But nothing is proved thereby.

As an example, Dobzhansky (1963) found a marked increase in the frequency of the Pikes Peak gene arrangement on the third chromosome of *Drosophila pseudoobscura* from Mather, California, from 0.3 percent in 1946 to 8.7 percent in 1962, at the same time that the Chiricahua arrangement dropped from 17.1 to 2.2 percent. Other inversion types also changed but not so dramatically. A plausible hypothesis could be built up to explain the first five years of the change, from relation with the amount of winter precipitation and the intensity of the summer drought. This explanation was given further credence by the fact that those inversions that usually increase during summer months in the regular within-year cycle also generally increased over years when the average precipitation was lower. But the later years of the frequency changes showed no positive correlation with precipitation and even showed some very sharp contrary correlation.

Dobzhansky demonstrated that although such diverse factors as climate, competing species density, insecticide use, and smog seemed well correlated with some sequences of changes, none held up over the whole span of 16 years. Had a much shorter span of years been recorded, compelling evidence could have been presented for one or another cause of the genetic change, especially in view of the tremendous range of environmental variables that was considered.

Spatial variation in allelic frequencies, unlike temporal changes, cannot provide any information on the intensity of selection, because, in the absence of historical evidence, it must be assumed that the spatial pattern is an equilibrium state, and equilibrium frequencies depend only on the ratio of fitnesses and not on their absolute value. The equilibrium frequency of an allele A at a heterotic locus is given by

$$p = \frac{1}{1 + t/s} \tag{20}$$

where $1 - s$ is the fitness of the homozygote AA and $1 - t$ is the fitness of aa. The same equilibrium, for example $p = 0.67$, will result if these fitnesses are 0.9 and 0.8, 0.99 and 0.98, or 0.999 and 0.998. Any variation in equilibrium over different populations, then, tells nothing about the magnitude of selection differences between populations.

Some spatial correlations are with discrete qualitative variables, like the correlation of shell patterns in *Cepaea* with vegetational type. Others are clinal in the sense that allele frequency shows some more or less regular trend with the increase or decrease of an environmental variable like temperature, or density, or some combination of environmental variables that is used in constructing an environmental index. Still others may be topographical clines that clearly must be related to another environmental variable, even though the important factors have not been definitely pinned down. Examples are altitudinal clines for terrestrial species and depth clines for aquatic organisms, or distance clines along river systems and distances from shore lines. All such clines, provided they are not spurious correlations, show the working of selection.

Geographical clines are dangerous because we may not be able to distinguish an equilibrium pattern, which would imply a pattern of environment, from a historical pattern resulting from migrations or even the spread of genes from a point of original mutation. The situation is even more uncertain when there is a mosaic distribution of gene frequencies rather than a true cline. If populations near each other resemble each other in gene frequency so that a coarse-grained pattern of gene-frequency variation is observed, it may indicate the relic of historical processes, but even if it is an equilibrium pattern, it can arise entirely from processes of random drift and local migration.

Both Malécot (1959) and Kimura and Weiss (1964) have shown that there will be a patchy distribution of allelic frequencies in the absence of selection, with the correlation between colonies x distance apart given approximately by

$$r(x) = \frac{e^{-x\sqrt{4b/m}}}{\sqrt{x}}$$

where b is the long-distance migration and m the migration between adjacent colonies. A superb example of this effect is given by Selander (1970), who found clear patchiness in gene frequency distribution of an allozyme polymorphism in the house mouse within single barns occupied by many semi-isolated mouse families. Figure 20 shows one of these intra-barn distributions. Although this looks for all the world like a patchy distribution of allele frequencies over a species range, it is all taking place in an area 192 feet × 48 feet. It would be absurd to imagine we are observing selection gradients. For this reason we cannot deduce selection from patchy and mosaic gene-frequency distributions. For example, the geographical patchiness observed by Selander, Yang, and Hunt (1969) in several loci in the house mouse, both between regions within Texas and between larger regions of the Central States, as a whole may be nothing but the barn pattern writ large.

The same is true for the differentiation between Pacific Island groups and the similarity within the groups of the polymorphisms of *Drosophila ananassae* (F. M. Johnson, 1971). The island study, especially, is in perfect accord with expectations from the theory of

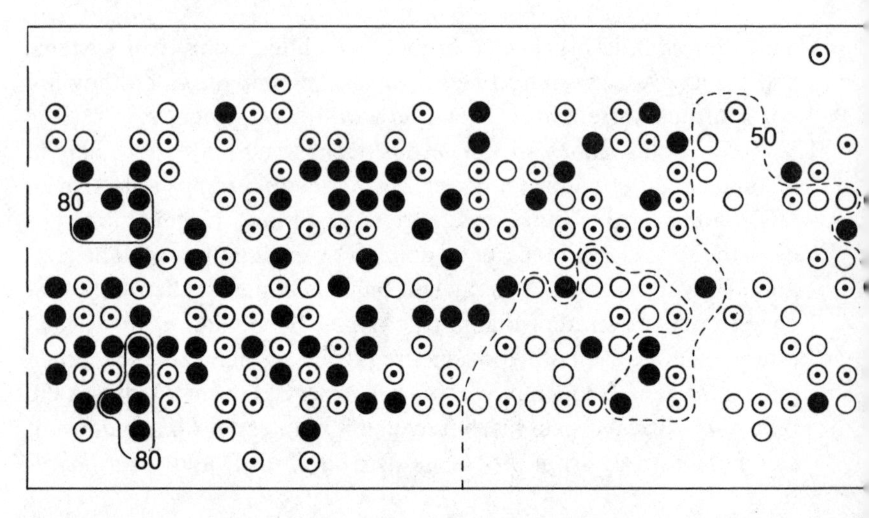

FIGURE 20

Allelic "geography" of an esterase polymorphism in the house mouse in half of a single barn in Texas. Symbols show the *esterase-3* genotypes (*MM, MS,* and *SS*) of 379 mice trapped in a grid pattern. Contours are 50 percent and 80 percent isofrequency lines for allele *s*. From Selander (1970).

random drift of neutral mutations, with some migration between neighboring islands and occasional long-distance dispersal.

At present the evidence for selection of allozyme or serological variation from studies of environmental correlations is sketchy and contradictory. Negative evidence is hard to come by because workers seldom report the fact that they have searched for an environmental correlate of allelic variation but have failed to find it. Only occasionally will such negative evidence be of general enough nature to seem positively interesting.

The lack of any appreciable reduction in heterozygosity at the ecological margins of the distribution of *Drosophila robusta* (Prakash, 1973) or *D. willistoni* (Ayala, Powell, and Dobzhansky, 1971), despite the marked loss of inversion polymorphisms in marginal populations, might be regarded as evidence against the selective importance of the genic variation although, as I have argued on page 151, it need not. In our own work we have looked for but failed to find seasonal variation in allozyme loci on the X, second, and

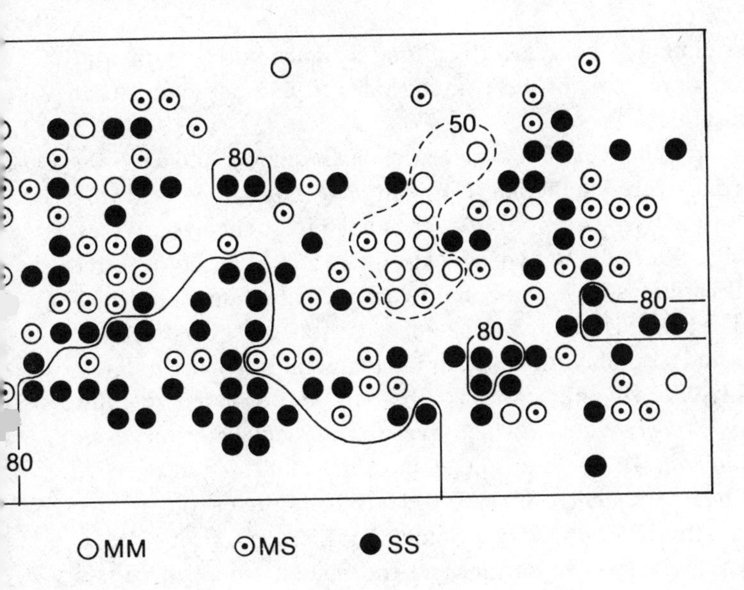

fourth chromosomes of *D. pseudoobscura* from Strawberry Canyon, although inversions show seasonal variations in such low-altitude populations, so that significant environmental variation is certainly taking place.

There have been, however, a few reports of clear-cut correlations. Berger (1971) found that the frequency of the "slow" allele at the *α-gdh* locus in *D. melanogaster* doubled in four populations during the 20 weeks from mid-June to mid-November. Since this phenomenon occurred in two distant localities in two different years for an allele in the frequency range of 10 to 25 percent, there is little doubt of its significance. The speed of change would demand selection differences of the order of 10 percent, a very high value for a single locus.

In the encrusting bryozoan, *Schizoporella unicornis*, Schopf and Gooch (1971) found a significant cline in the frequency of an allele at a *leucine aminopeptidase* locus over a distance of only 20 miles of shoreline, associated with a 6 C change in August water temperature. This cline was paralleled by a long-distance north-south cline along the eastern coast of the United States from Massachusetts to North Carolina.

A more complex relation between gene frequency and environment was discovered in the harvester ant, *Pogonomyrmex barbatus*, by Johnson and associates (1969). The allelic frequencies at two esterase loci and a malic dehydrogenase locus and a variety of environmental variables were subjected to a principal-components analysis, and a significant relationship was found between allele frequency and a combination of rainfall, temperature, and elevation.

A highly convincing case is that of the crested blenny, *Anoplorchus purpurescens*, which lives in the rocky intertidal zone (M. S. Johnson, 1971). An allele A^1 of a lactate dehydrogenase locus changes frequency from 0.021 to 0.305 over only three degrees of latitude around Puget Sound, but there is a strong deviation from the cline at the entrance to the Sound, where there is a great deal of turbulence and nutrient mixing. Allele frequency has a correlation of 0.69 with August surface temperature, 0.71 with summer oxygen concentration, and 0.96, almost total, with the relative frequency of another species, *A. insignis*. The temperature relation was confirmed in laboratory studies of death rates at different temperatures.

Sometimes the relation between allelic frequency and environment can be misleading if not all the genetic facts are taken into account. At first sight, there appears to be strong evidence from studies of the wild oat, *Avena barbata*, that the allelic composition at many loci is correlated with climate (Clegg and Allard, 1972; Hamrick and Allard, 1972). In that part of its range in California that is characterized by dry, hot summers. *A. barbata* is nearly monomorphic at five enzyme loci studied, whereas in the cooler, moister regions it is highly polymorphic for these loci. In a population lying on a vegetational transition, changes in microhabitat from mesic to xeric over a distance of less than 200 meters are accompanied by dramatic changes in allele frequencies. The oats in the most xeric microhabitat are monomorphic for the "xeric" alleles, but the oats in the mesic microhabitats are highly polymorphic. The similarity between the broad geographical pattern and the pattern over different habitats in the same locality is hard to explain on any but a direct selective hypothesis until one remembers that *Avena barbata* is almost completely self-fertilized. As a result there are very few heterozygotes in the population and thus there is virtually no recombination. Because of the absence of recombination, alleles at different loci are not randomly associated with each other (see chapter 6); a single locus that is undergoing selection will carry along with it all other loci on the chromosome. It is highly unlikely that five randomly chosen loci in *A. barbata* will all be related to adaptation for xeric and mesic environments. The obvious explanation is that in this highly selfed species the geographical distribution of alleles is a linkage effect with one or more unknown loci that are under selection.

When we turn from direct correlations with environmental gradients to geographical clines, there are a few pertinent studies. Grossman, Koreneva, and Ulitskaya (1969) found that in southern Asia, the Caucasus, and European Russia there is a correlation between altitude and the frequency of an allele at an alcohol dehydrogenase locus in *D. melanogaster*. The frequency of the allele A_1 was between 0.15 and 0.42 in mountainous localities, between 0.07 and 0.14 in foothills, and from 0.00 to 0.08 in lowlands.

Koehn and Rasmussen (1967) found a dramatic cline in the frequency of an esterase allele, *est-I*^{a}, with latitude in the fish *Catastomus clarkii* from the Colorado River drainage basin. The allele increases linearly in frequency from 0.18 to 1.00 over a latitude range of 7.6 degrees. On the basis of laboratory studies, Koehn and Rasmussen concluded that the critical variable is temperature. Since the drainage runs east-west, the more northerly latitudes are no higher in the drainage system, and thus temperature seems the obvious variable. It is not clear, in this case, how exhaustively the workers tested other environmental factors in laboratory conditions. A latitudinal cline in a hemoglobin allele of the marine bivalve *Anadara trapezia* was found by O'Gower and Nicol (1968) along the east coast of Australia, and Frydenberg and co-workers (1965) found a hemoglobin cline in the cod populations off Norway.

Although the studies I have enumerated in the last few paragraphs do not exhaust the literature, they come close to doing so. What such a collection of cases can do is to show that polymorphism for electrophoretic variants is indeed under the influence of selection in some cases and that those cases cannot be *too* hard to find, since the community of workers in the field is not large and the time available for work has not been very great, about five years at the most, since the first demonstrations of high heterozygosity for allozymes in natural populations. Nevertheless, such a miscellaneous collection of studies cannot, in itself, answer the question, How much selection goes on in nature?

SELECTION IN THE LABORATORY

If selection cannot be measured or even demonstrated in nature, the advocate of selection can take yet another step back and make a still weaker demonstration. If it could be demonstrated that in laboratory conditions there was selection for one or another allele at a polymorphic locus, then it would be established that the substitution of such an allele does, in fact, make a significant physiological difference to the organism.

Although laboratory conditions are certainly far from nature, this fact does not matter for the purposes of our problem. We must remember that the neoclassical proposition is that the allozyme variants for enzymes and other proteins are all really wild-type because the developmental, physiological, and behavioral apparatus of organisms is insensitive to the differences that we can detect by electrophoresis or serology. A demonstration that these gene substitutions matter to the organism in any environment in which the organism can be husbanded, or even in any stress environment that the species is likely to encounter with fair frequency, would contradict the fundamental premise of any neutralist theory. The advantage of working in the laboratory, of course, is that the sensitivity of the analysis can be so much greater, environments can be manipulated and optimal experimental designs concocted.

There is one serious danger of laboratory experiments that seems not to be appreciated by some workers. If a single chromosome bearing an electrophoretic allele, say E^F, and one bearing an allele E^S are made homozygous and these homozygous lines are crossed to produce a segregating progeny, any measures of fitness on homozygotes and heterozygotes are really measures of the effect of homozygosity and heterozygosity of an entire chromosome, not of alleles at the marker locus. Any experiment that begins with a single E^F and a single E^S allele is simply a form of the chromosome replication scheme discussed in chapter 3, slightly disguised, and the result can be predicted. In 99 cases out of 100, both homozygotes E^F/E^F and E^S/E^S will be lower in fitness than the heterozygotes E^F/E^S because the inbreeding depression of an entire chromosome is being manifested, a depression that might as easily arise from a single partly recessive deleterious gene at a different locus on each chromosome as from overdominance at allozyme loci.

This confusion between the effects of an allozyme marker and of the whole chromosome is not much relieved by using two or three chromosomes of each type as founding material. Nor is it much help to take two lines and allow them to recombine or backcross them to a common line for a number of generations. The time taken for the initial association of loci to be broken down is much greater than ordinarily realized. Consider a gene r centimorgans away from a marker locus. The association ρ between the two loci will decay exponentially, and in *Drosophila*, with no crossing over in males,

$$\rho_t = \rho_0 e^{-rt/200} \tag{22}$$

After 10 generations of random mating, a gene five centimorgans away from the marker will retain 78 percent of its original association and will require 35 generations to lose just 50 percent of the association. For a locus one centimorgan removed from the allozyme marker, there will still be a 95 percent association after 10 generations, and 140 generations would be needed to reduce this correlation to 50 percent. Since there are about 100 cistrons per centimorgan in *Drosophila,* the hope of randomizing the background of a given marker locus by a few or even a few dozen generations of recombination between two chromosomes is a false one.

Experiments to measure selection must be carried out with, at the very least, a dozen and preferably two dozen or more independently derived E^F and E^S alleles from nature. The danger of starting with even two or three strains for each allele is shown dramatically in figure 21, taken from Berger's work (1971). Four replicate laboratory populations were founded by two or three strains each of the mdh^F and mdh^S alleles from a natural population. Three replicates show the erratic behavior with some trend that is expected if there is either no selection or a weak directional force. The fourth replicate, however, shows the smooth and rapid drop expected of a drastic allele with considerable selection against the heterozygote. Clearly the allele mdh^F is by chance associated with a severely deleterious

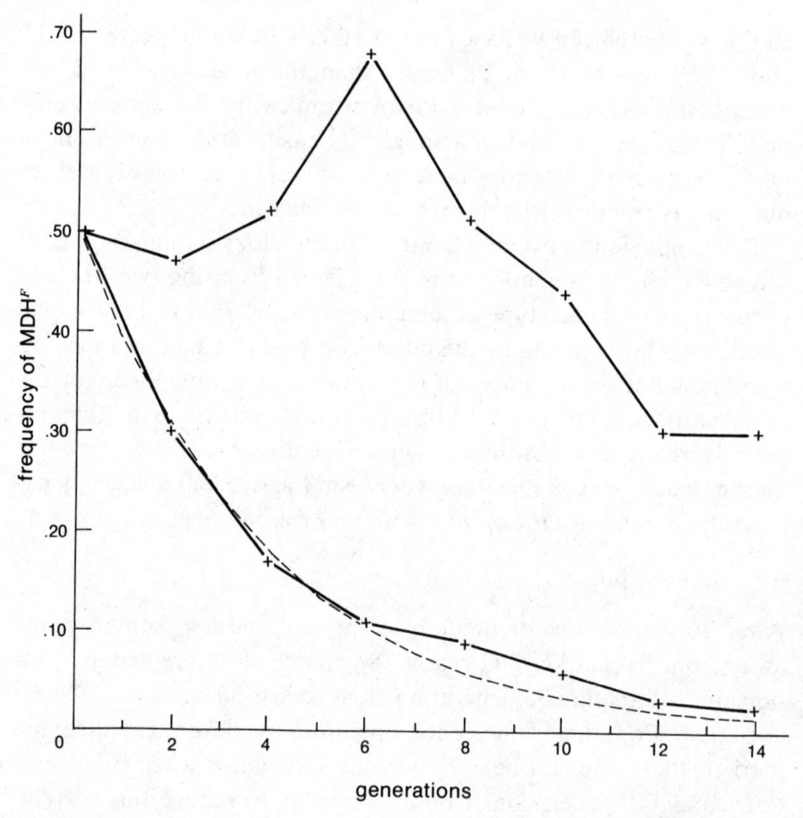

FIGURE 21

Frequency through time of chromosomes marked with an MDH polymorphism in *Drosophila melanogaster* in two replicate laboratory populations. The dashed line is the theoretical course of selection if $W_{AA} = 1$, $W_{AA} = 0.75$, and $W_{aa} = 0.4$. Data are from Berger (1971).

mutant in the founding chromosomes and this association is responsible for the rapid and smooth selection.

The most direct way to observe fitness differences in the laboratory is to create laboratory populations under different conditions and to observe the changes in frequency of the alleles over long periods. From the direction and speed of change in replicated populations it is possible to esimate fitnesses and test whether there has been any significant trend. Unfortunately several such experiments have begun with single inbred lines and so are useless for detecting selection at the enzyme locus that is being followed.

Berger's experiments (1971) seem to be on the edge in this respect, so I am uncertain of their meaning. Three loci. *mdh, adh,* and *α-gdh,* all went to intermediate equilibria in replicate populations (except the one replicate shown in figure 21), but several lines of internal evidence suggest the importance of unique linkages.

Yarbrough and Kojima (1967) found moderately fast, significant secular changes toward intermediate equilibrium in the frequency of *esterase* alleles in *D. melanogaster* population cages, but these cages were founded by strains that came from a laboratory population that itself had been founded by two inbred lines some 50 or so generations before. It is difficult to say whether or not the observed selection was a consequence of residual associations of the *esterase* locus with blocks of genes that Huang, Singh, and Kojima (1971) estimate to have been about five centimorgans long.

I know of only one convincing case of evolution of allozyme frequencies in laboratory populations. Gibson (1970) found a change in the frequency of the fast allele of an alcohol dehydrogenase locus in *D. melanogaster*, from 0.5 to 0.73 and 0.82 in two replicates after 18 generations on medium to which ethanol had been added. No change took place in control replicates.

Two other experiments, designed especially to find evidence of evolution of allele frequencies in laboratory populations, found none. Yamazaki (1971), studying the sex-linked esterase polymorphism of *Drosophila pseudoobscura*, set up laboratory populations segregating for a fast and a slow electrophoretic allele. All populations were founded with 22 independently derived strains of each allele. Replicate populations with different initial allele frequencies were husbanded at two temperatures and two levels of nutrition. In no case was there any significant trend in allele frequencies over a three-year period (25 to 35 generations, depending upon temperature).

In parallel with Yamazaki, the Chicago laboratory has followed the *pt-8* polymorphism in *D. pseudoobscura* for nearly five years. Two replicates at each of two initial frequencies at each of two temperatures have been sampled for more than 54 months (40 to 60 generations) with the results given in table 47. With the exception of cage V at 18C, which has changed markedly in the past year, the populations are essentially at the same frequency after 5 years as at the start of the experiments. That is not to say that nothing occurred. There was a decrease in the frequency of *pt-8*^{81} in all four populations at 25 C toward the middle of the experiment, but there

TABLE 47
Frequency of the *pt-8*[81] allele in population cages of *Drosophila pseudoobscura*

		25 C				18 C			
Months	*Cage:*	*I*	*II*	*III*	*IV*	*V*	*VI*	*VII*	*VIII*
0		.69	.80	.20	.30	—	—	—	—
6		.75	.68	.23	.27	.74	.67	.24	.28
12		.69	.72	.29	.27	.70	.69	.31	.27
18		.62	.51	.11	.10	.60	.70	.31	.22
24		.50	.62	.09	.09	.59	.60	.32	.20
30		.41	.59	.13	.08	.61	.64	.36	.21
36		.41	.64	.17	.12	.61	.68	.40	.21
42		.37	.64	.14	.09	.59	.66	.39	.18
48		.58	.66	.28	.14	.45	.68	.36	.28
54		.60	.77	.30	.20	.38	.67	.34	.32

was no synchrony in the high-frequency cages. The low-frequency populations, III and IV, did indeed go down together, reaching a minimum after two years, but the subsequent rise of population III is much greater than of IV.

Some of the fluctuation is only apparent, since the standard error of the values is between 0.02 and 0.04. Nevertheless, the difference between replicates is consistent after the second year, and no doubt is real. Many interesting fantasies can be constructed from a close look at the variations in frequency over time, but any long, random, autocorrelated sequence suggests trends, cycles, and patterns if it is viewed optimistically. No convincing case for selection can be made from the data.

Following the changes in allele frequency in a laboratory population is an easy way to look for selection, but it is by no means the most sensitive. For example, if the homozygotes are 99 percent as fit as the heterozygotes, a population started at 0.8 of one allele and 0.2 of the other will progress toward its eventual equilibrium at an initial rate of 0.00097 per generation, and this rate will decrease as equilibrium is approached. In the 50 generations of the Chicago populations, such a selection would produce a total change of only 0.05 in gene frequency, which would go quite unnoticed.

The alternative is to attempt to measure one or more components of fitness by viability and fecundity tests. For many years the methodological theory of fitness estimation has been undeveloped. As a result, the power of estimation procedures has been grossly overestimated by workers who never had a remote chance of finding any realistic difference in the experiments they designed, and serious errors have been made in the biological interpretation of the outcome of experiments. Recently, in a series of fundamental papers that must be the cornerstone for future experimental work on fitness estimation, Prout (1965, 1969, 1971a) has proved what can and cannot be deduced from various standard procedures and has designed an optimal method for fitness estimation. In particular he has shown how the fertility component of selection has been lost from many experiments that supposedly measured net fitness, and he has given optimal methods for separately estimating the viability and fertility fractions.

These methods have been applied to model experiments in Drosophila by Wilson (1968, 1972), Prout (1971a, b), and Bundgaard and Christiansen (1972); all have shown that indeed the fertility component, in both males and females, is of much greater importance than viability. Moreover, the standard errors of the estimates from these optimal methods are small enough to give confidence in the results. Table 48 shows Wilson's (1972) estimates of

TABLE 48
Relative net fitness and the fertility and viability components of fitness estimated by Wilson (1972) for nine mutant genotypes on chromosome IV of *Drosophila melanogaster*

Genotype	*Fertility* ± S.E.	*Viability* ± S.E.	*Net fitness* ± S.E.
$\frac{ss\ in}{ss\ in}$	0.49 ± 0.13	0.64 ± 0.05	0.31 ± 0.04
$\frac{ss\ in}{ss\ ri}$	0.21 ± 0.03	0.75 ± 0.05	0.16 ± 0.05
$\frac{ss\ ri}{ss\ ri}$	0.31 ± 0.26	0.37 ± 0.04	0.12 ± 0.03
$\frac{ss^a\ in}{ss\ in}$	0.24 ± 0.04	0.87 ± 0.05	0.21 ± 0.05
$\frac{ss^a\ in}{ss\ ri}$	1.00 —	1.00 —	1.00 —
$\frac{ss^a\ ri}{ss\ ri}$	0.07 ± 0.02	0.82 ± 0.04	0.06 ± 0.05
$\frac{ss^a\ in}{ss^a\ in}$	0.22 ± 0.07	0.81 ± 0.05	0.18 ± 0.05
$\frac{ss^a\ in}{ss^a\ ri}$	1.32 ± 0.15	0.94 ± 0.05	1.23 ± 0.05
$\frac{ss^a\ ri}{ss^a\ ri}$	0.14 ± 0.06	0.91 ± 0.06	0.13 ± 0.05

the fertility and viability components of net fitness for nine genotypes segregating at two mutant marker loci in *Drosophila melanogaster*.

Thus far the only application of Prout's techniques to laboratory measurement of allozyme fitnesses has been in Yamazaki's (1971) study of the *esterase-5* locus in *D. pseudoobscura*. Optimal designs showed no significant differences between genotypes in either viability or fertility in either sex (even though observed differences were as large as 17 percent), despite an immense amount of labor on the investigator's part (see table 49). If selection is as weak as 1 percent, it is going to prove very difficult to find.

Wills and Nichols (1971, 1972) found rather subtle differences at the *octanol dehydrogenase* locus in *Drosophila pseudoobscura*, but the interesting aspect of their finding is that the viability differences occurred only if octanol was present in the medium and did not show up at a physiologically unrelated locus, *esterase-5*. Moreover, the differences could not be demonstrated when the genetic background was heterozygous.

Tsakas and Krimbas (1970) found a greater resistance to dimethoate insecticides in *Dacus oleae* carrying an active allele of an esterase gene, and a greater adult movement in heterozygotes for allozymes at another esterase locus.

All in all, the record of detected selection of allozyme loci in laboratory conditions is not a very large or convincing one. The most carefully designed and controlled work, that of Yamazaki, revealed no selection. Yet the statistical problems are formidable and prevent us from seeing selection at the level of a few percent.

FREQUENCY-DEPENDENT SELECTION

The classical models of selection assign some fixed fitness W_{ij} to a genotype A_iA_j and proceed to build theoretical expectations from that assumption. But there is no reason to suppose that fitness is fixed, even when the physical factors of the environment are constant. Indeed the W_{ij} are not fixed in a truncation model of selection because, as the population changes composition, the relative probability that two genotypes are past the truncation point is a function of the population distribution and the location of the truncation point. These fitnesses are then *frequency dependent* and must be

TABLE 49
Estimates of viability, fertility, and net fitness for esterase allozymes in *Drosophila pseudoobscura*

	Females			*Males*	
	S/S	*S/F*	*F/F*	*S*	*F*
Viability	1.0245 ± 0.1682	1.00	1.0364 ± 0.1371	1.00	1.0551 ± 0.2565
Fecundity	1.1048 ± 0.2345	1.00	1.1308 ± 0.2392	1.00	0.9655 ± 0.1855
Net fitness	1.1319 ± 0.1520	1.00	1.1720 ± 0.1934		

Note: Data are from Yamazaki (1971).

written as $W_{ij}(g)$, where g is some description of the genotypic composition of the population.

The general theory of frequency-dependent selection has been developed by Wright (1949), Lewontin (1958), and Fisher (1958), among others. The most important theoretical conclusion for us is that a stable equilibrium of gene frequencies is possible without heterosis, indeed even with an inferior heterozygote. A particularly interesting case from our standpoint occurs when the rarest genotype is the most fit. Then a rare allele will increase in frequency but will not become fixed in the population, because as it gets commoner the fitness of its carriers decreases, and the other alleles are now favored. It is clear that this scheme leads to a stable equilibrium.

Rare-genotype advantage is an attractive hypothesis because it solves at one blow the dilemmas of genetic load and of inbreeding depression. If the fitness of homozygotes is a decreasing function of frequency, there may be no fitness differences at all at equilibrium, and thus no genetic load, but strong fitness differences away from equilibrium so that the stability is maintained. For example, let us take the fitnesses

$$\begin{array}{ccc} AA & Aa & aa \\ 2(1-p) & 1 & 2(1-q) \end{array}$$

When the population is virtually pure AA, $p=1$ and the fitnesses are 0, 1, and 2, while at the opposite end of the frequency scale, when aa is nearly fixed, the fitnesses are 2, 1, and 0. But at equilibrium, $p=q=0.5$ and the fitnesses are 1, 1, and 1.

In general such a model of selection allows an arbitrarily small fitness variance at equilibrium yet an arbitrarily large selective force driving the population toward the equilibrium. The problem of inbreeding depression is solved because in a competitive experiment to measure inbreeding depression, like that of Sved and Ayala (1970), the multiple homozygote is in an intermediate frequency in the competing mixture and so will have an intermediate relative fitness, whereas in a normal segregating population the occasional rare multiple homozygote will have a very high fitness.

It might seem that such a specialized model of fitness is unlikely to apply to vast numbers of polymorphisms, but it can be justified from the simplest ecological consideration. If resources are in short supply and if each genotype exploits the resources in a slightly different way, then an individual of a given genotype is in more intense competition with others of its own type than with individuals of other genotypes that are slightly ecologically displaced from it (Gustafsson, 1953). But if a genotype is its own worst enemy, its fitness will decrease as it becomes more common. The difficulty with this explanation is that it supposes an extraordinary complexity of the resource space, such that each genotype does in fact utilize a slightly different niche and that competition for this complex resource is critical to the species It is really a rather extreme hypothesis, but perhaps no more extreme than the postulate of heterosis at most loci.

Evidence of frequency-dependent selection, especially viability of larvae competing for resources, is abundant. Levene, Pavlovsky, and Dobzhansky (1954) found that the relative fitnesses of inversion karyotypes in laboratory populations of *Drosophila pseudoobscura* were strongly dependent on the kind and proportion of other karyotypes competing with them, and Spiess (1957) found the same phenomenon in *D. persimilis*. Lewontin (1955) and Lewontin and Matsuo (1963) found that larval viability in *D. melanogaster* varied with mixture proportions and that in some cases the less frequent a genotype the greater is relative viability. A similar facilitative phenomenon is known in plants (de Wit, 1960; Harper, 1968). There are also many cases known in insects (see Dobzhansky, 1970, for a review).

The most striking phenomenon of all is the minority mating effect in *Drosophila* (Petit, 1958; Ehrman, 1967; Spiess, 1968). If two sorts of males are competing for females in a test apparatus, the rarer type has an advantage. This advantage appears whether the rarer type is genetically different or has been raised in a different environment from the common competitor. Since the effect depends on olfactory cues, it may be of particular significance in enhancing the effect of migration or the prevention of inbreeding since it favors males that have been raised at a different food site. A generalized intrapopulation advantage for rare males with respect to almost any genotype could not exist, if for no other reason than that *every* male is genotypically rare in a population with 10 percent heterozygosity and 40 percent polymorphism!

Recently K. Kojima and his colleagues have presented several lines of evidence that frequency-dependent selection is operating for allozyme polymorphisms, maintaining the polymorphisms by higher fitness of minority genotypes. The original report concerned the relative viability of homozygotes and heterozygotes at the *esterase-6* locus in *Drosophila melanogaster* (Kojima and Yarbrough, 1967). However, the method of estimating the fitnesses by comparison of adult frequencies with those expected from the gene frequencies in the previous generation was exactly the one warned against by Prout (1965), who predicted the appearance of just this form of frequency-dependent selection as an artifact of the method. A second experiment testing the same hypotheses at the *alcohol dehydrogenase* locus (Kojima and Tobari, 1969), but correcting for the fertility component by using pre-mated females whose fecundities had been estimated, still showed evidence of decreasing viability with increasing frequency.

The result at the *esterase-6* locus was pushed further by an experiment conducted with preconditioned medium (Huang, Singh, and Kojima, 1971). Larvae of one of the three genotypes *FF*, *FS*, and *SS* were allowed to grow in fresh culture medium and were then killed in the medium by freezing. Subsequently a counted number of larvae of the same or another genotype were cultured on this preconditioned medium, and the proportion surviving was observed. The results are shown in table 50; viabilities are expressed relative to heterozygotes conditioned by heterozygotes. The effect, which Kojima (1971) claims to be a very general one, is certainly apparent. However, Yamazaki (1971) looked for, but failed to find, the effect in *Drosophila pseudoobscura*, and such a facilitation was not the general rule in Lewontin and Matsuo's experiments.

At the moment, the importance of this potentially powerful stabilizing phenomenon is in doubt. It suffers from the same paradox as the minority mating advantage. Although *est-6*^{F}/*est-6*^{F}

TABLE 50
The viability of the three genotypes at the *esterase-6* locus in *Drosophila melanogaster*, when grown on "preconditioned" medium

Tested	*Preconditioning genotypes*		
genotypes	*FF*	*FS*	*SS*
FF	0.923 ± 0.053	1.068 ± 0.056	1.130 ± 0.058
FS	1.090 ± 0.057	1.000 ± 0.057	1.087 ± 0.057
SS	1.146 ± 0.059	1.078 ± 0.057	0.928 ± 0.053

Note: Data are from Huang, Singh, and Kojima (1971).

individuals may be common with respect to that locus, they are rare with respect to other loci, and it is a little difficult to see how the environment, especially of a Drosophila food vial, can be so finely subdivided along so many dimensions as to allow for ecological displacement with respect to most polymorphic loci. The problem is essentially one of information, and all that the hypothesis of frequency-dependent selection has done is to transfer this problem from the sorting of genotypes (genetic load) to the sorting of environmental niches.

IN VITRO ACTIVITY

The selectionist has still one more step he can take in his retreat from the direct measurement of fitness in nature. The neoclassicist-neutralist theory is based ultimately on *a priori* arguments about the physicochemical properties of protein molecules. When the three-dimensional structure of proteins has been reconstructed, most of the charged amino acids have been found on the surface, while the interior consists of tightly packed hydrophobic groups. It is in the interior of the molecule that metallic coenzymes are bound and the active site for operation on the substrate is located. Moreover, in evolution, most of the substitutions have been on the exterior surface of the molecule, where there is little stereospecificity (Dickerson, 1971). In addition, different molecules like the proteases α-chymotrypsin, trypsin, elastone, and thrombin, which apparently have evolved from a common ancestral gene by duplication and which may differ from each other at 50 to 70 percent of their amino acid sites, have rigidly preserved their three-dimensional structure and the conformation of their active site (Hartley, 1970).

Therefore the neoclassical-neutralist theory predicts that the vast majority of substitutions observed to be segregating in populations will be substitutions on the surface of the molecules, where they have no effect on the physicochemical properties of the enzymes. It follows that the kinetics of such variant enzymes will be indistinguishable from each other, which is why they are selectively neutral. A valid attack on the neutralist theory would then be a demonstration that the kinetics of different allozymic forms are indeed different. If a large proportion of allozyme variants were detectably different from each other in their *in vitro* kinetics, it would be difficult, although not impossible, to maintain that nevertheless the *organism* could not detect the difference. This would certainly put the neoclassical theory in a shaky position. Conversely, the failure to find a kinetic difference would not mean much, since the demands on a molecule *in vivo* are certainly much more complex than *in vitro*.

What is the evidence? A number of studies of the activity of allozyme alleles have shown significant differences, sometimes correlated with clines in nature. For example, Koehn (1969), following up the cline of esterase polymorphism in *Catostomus clarkii*, measured total enzyme activity at low and high temperature. Homozygotes for the "northern" allele I^b were ten times more active in the cold than either the homozygote for the "southern" allele, I^a, or the heterozygote. At 37 C, however, the homozygotes were reversed in their activities but the heterozygote was intermediate. Koehn also reports, although he gives no data, that at intermediate temperature the heterozygotes were superior to both homozygotes. This last, if true, is a startling result, since the enzyme does not form hybrid dimers and so heterozygotes are only a mechanical mixture of the two alternative molecular forms.

In like manner, Merritt (1972) showed that a sharp north-south cline in a lactic dehydrogenase polymorphism in the minnow, *Pimephales promelas*, is in accord with *in vitro* activity. Both homozygotes and the heterozygote have an identical K_m that increases linearly with temperature until, at 25 C, the K_m of the F/F homozygote begins to rise much more rapidly, signifying a loss of affinity for the substrate. This corresponds to a frequency cline from $p_F = 1.00$ in North Dakota to $p_F = 0$ in Oklahoma.

In man a number of enzymes show differential activities for polymorphic allozymes. For example, in *acid phosphatase* the activities of three alleles are in the ratio 1:1.5:1.0. In *6-phosphoglycerate dehyrogenase* the two forms are in the ratio 1:1.15, in *glucose-6-phosphate dehydrogenase A* and *B* forms, 1:1.25, and in *adenylate kinase,* 1:1.5 (see Modiano et al., 1970, for a review).

In addition to normal kinetic differences, some electrophoretic variants alter the stability or solubility of the molecule with respect to heat or ionic concentration of the milieu. For example, sickle cell anemia arises from the insolubility of the deoxygenated form of hemoglobin S molecule. Both hemoglobin S and the mildly deleterious hemoglobin C are variants at position 6 on the surface of the protein. The three alleles of human red-cell acid phosphatase produce proteins that differ in their heat stability in the order $P^A < P^B < P^C$, which is the same order as their activities at normal temperature (Luffman and Harris, 1967). Placental alkaline phosphatase allozymes also differ in heat stability, the most stable being three to five times more so than the least (Thomas and Harris, 1972).

In *Drosophila melanogaster* the fast and slow alleles of the *alcohol dehydrogenase* locus differ not only in activity but also in inducibility and thermostability (Gibson, 1970), with a remarkable result that is shown in table 51. Not only is the *F/F* homozygote most active, but it increases relatively more when induced by growing the larvae on ethanol-containing medium. The heterozygote is more thermostable than either homozygote. Such heterosis directly at the molecular level is conceivable for multimeric enzymes, but it is unlikely that monomeric enzymes will show this kind of heterosis since heterozygotes are a simple mixture of two molecular species. However, one should be cautious about supposing that even dimeric

TABLE 51
Specific activities of alcohol dehydrogenase in normal larval extract, in extract heat-treated for ten minutes at 40 C, and in extract from larvae raised in a medium with 6 percent ethanol

Genotype	*Normal extract*	*Heat-treated*	*Alcohol medium*
F/F	8.5 ± 0.33	1.2 ± 0.092	12.4 ± 0.57
F/S	5.4 ± 0.32	2.0 ± 0.108	6.8 ± 0.23
S/S	2.9 ± 0.13	1.5 ± 0.188	3.5 ± 0.23

Note: From Gibson (1970).

enzymes will often show heterosis at the molecular level. Latter (1972) points out that the average heterozygosity in *Drosophila pseudoobscura* and man is almost identical for polymorphic enzymes that do not form hybrid enzyme (0.23) and for those that do (0.22).

Bulmer (1971) has made the interesting observation that in multiple allelic loci in *D. pseudoobscura*, if there are common and rare alleles, the rare ones always are at the extremes of the electrophoretic mobility distribution, whereas the common alleles have intermediate mobility. The lack of randomness in the ordering is highly significant ($P < 10^{-4}$). Although Bulmer interprets this ordering as indicative of selection, it can as easily be explained by the simplest classical hypothesis. If we suppose that one-step mutational transitions occur between adjacent mobility classes, and that all the observed alleles are selectively neutral, then the steady-state distribution will have a mode corresponding to an intermediate allele. Slow, medium, and fast alleles with equal mutation rates between S and M and between M and F will settle down at equilibrium to the frequencies 0.29:0.42:0.29. By adjustment of the relative forward and backward mutation rates, any observed ratio can be produced.

Kinetic studies and stability measures are showing more and more that the amino acid substitutions in polymorphism make a functional difference. Table 52 presents the current state of knowledge about human electrophoretic polymorphisms. Of 23 enzymes, 15 are known to have a differential activity between allozymes, either because of altered kinetics or different stabilities. Of the remaining 8, most have not yet been investigated. The evidence is overwhelming that the substitutions that are being detected as allozyme polymorphisms do make a difference to enzyme function. Even those amino acid substitutions in evolution that were once thought to be neutral, like those of cytochrome *c*, may make a functional difference since they are known to change the ion-binding capacities of the molecule (Barlow and Margoliash, 1966).

It might be argued that the differences in activity between allozymes *in vitro* cannot be detected by the physiological apparatus of the organism, or that it is so buffered that the magnitude of the variation seen is immaterial to fitness. Perhaps our laboratory techniques are much more sensitive than nature. If there is ordinarily an

TABLE 52
Quantitative effects of polymorphic allozymes in man

Polymorphisms for which quantitative differences between the common phenotypes have been found	*Polymorphisms for which no quantitative differences have been reported*
Glucose-6-phosphate dehydrogenase	Adenosine deaminase
Red cell acid phosphatase	Peptidase D (prolidase)
Phosphogluconate dehydrogenase	Pancreatic amylase
Adenylate kinase	Pepsinogen
Placental alkaline phosphatase	Glutamate-oxalate transaminase
Peptidase A	Phosphoglucomutase
Peptidase C	locus pgm_1
Galactose-1-phosphate uridyl transferase	locus pgm_3
Glutathione reductase	
Liver acetyl transferase	
Red cell NADase	
Salivary amylase	
Serum cholinesterase	
locus E_1	
locus E_2	
Alcohol dehydrogenase	
locus adh_2	
locus adh_3	

Note: From Harris (1971).

excess of enzyme so that reactions are substrate-limited, a variation in enzyme activity would not be reflected in the flux along biochemical pathways. This is the usual model for dominance, and if a drastic allele like a lethal or semilethal is recessive, it is because the organism can carry on perfectly with only half the normal amount of enzyme. By this argument one allozyme homozygote would have to be about half as active as another before the difference in activity would begin to tell in fitness even between homozygotes.

Neoclassicists can use the argument only with considerable embarrassment, however, since it has been a cornerstone of their theory up to the present time that the recessivity of deleterious mutation is incomplete, and that slightly deleterious alleles are even less recessive than severely deleterious ones. They have been at some pains to demonstrate the partial dominance of deleterious genes experimentally and seem to have done so in several cases, although proponents of the balance theory have attempted to throw the evidence into doubt (see p. 72). The irony of the present situation is that the two parties will now want to reverse their attitudes toward these experiments.

We must beware of making easy assumptions about the relationship between gene dose, enzyme concentration, and physiological sensitivity. It is not true that dominance is simply the result of there being an excess of enzyme produced by wild type, providing a buffer against genotypic or environmental disturbances in function.

An example of the complex relation that exists among environmental stress, dominance, and the effects of allelic substitutions is given in Sang and McDonald's (1954) experiments on eye-facet number in *Drosophila.* The mutant *ey* is normally a completely recessive reducer of facet number. The number of facets is also reduced by increasing concentrations of sodium metaborate. Figure 22 shows the response of eye size to these two factors jointly. First we see that the dominance of wild type disappears at higher concentrations of metaborate and that the heterozygote becomes almost exactly intermediate between mutant and wild-type homozygotes. The usual explanation would be that increasing concentrations of metaborate somehow use up, inactivate, or otherwise interfere with the normal excess of gene product, so that the loss of gene product in the heterozygote becomes detectable. The metaborate has, so to speak, saturated the buffering capacity of the system, and the genetic defect in the heterozygote is now effective. But this explanation will not suffice because, despite its low facet number at all salt concentrations, the homozygous mutant is much *less* sensitive than heterozygotes are to intermediate ranges of metaborate concentration, whereas the simple buffering hypothesis predicts that it will be more so. Moreover, there is a suggestion that metaborate actually increases eye size of *ey/ey* at low concentrations.

No general *a priori* arguments about enzyme concentrations and activities will stand up under experimental test. Whether a 20 percent decrease in enzyme activity for an allozymic variant will or will not be detected by the physiological apparatus of the organism depends in a nonadditive way on the other factors, environmental and genetic, that are acting on the developmental and metabolic system.

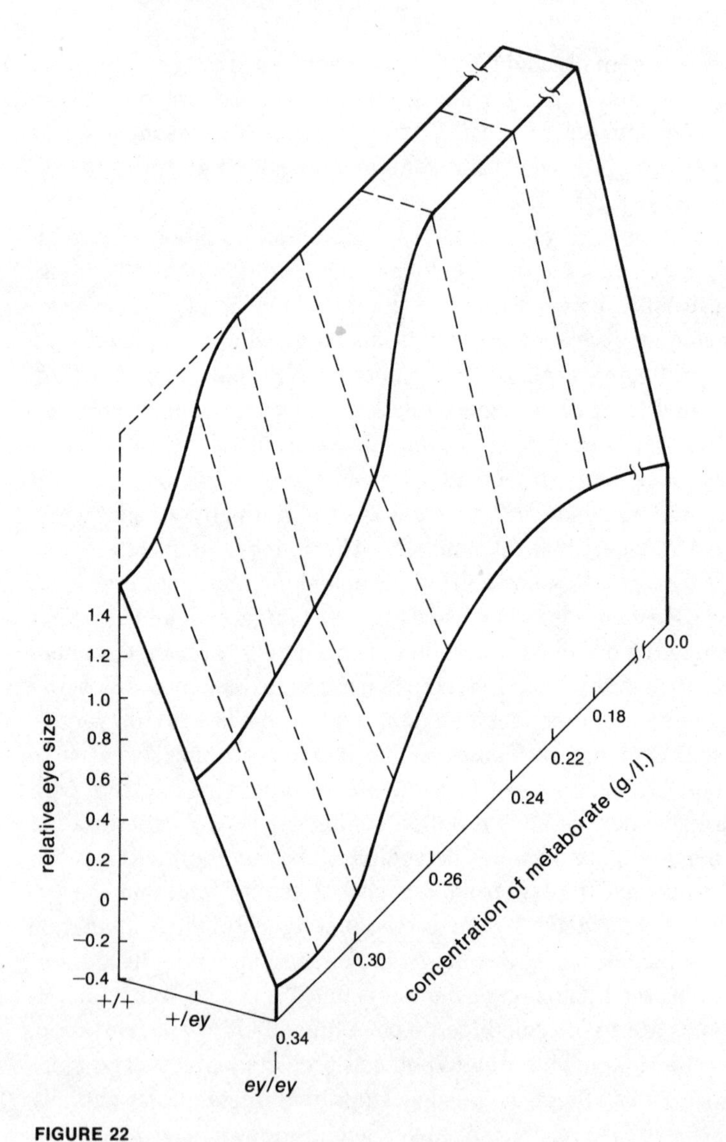

FIGURE 22

The relative eye size of three genotypes, +/+, +/*ey*, and *ey*/*ey*, over a range of concentrations of metaborate treatment. From Sang and McDonald (1954).

PROBLEMS IN THE THEORETICAL STRUCTURE OF POPULATION GENETICS

Our review of the theory of polymorphism and the evidence for and against competing theories has left us in a state of unresolved tension. On the one hand, there are strong reasons for rejecting a balance theory because it predicts tremendous inbreeding depressions that are not observed, because the rates of evolution of different molecules strongly suggest that the least functional evolve fastest, because heterozygosity does not seem to be sensitive to ecological stringency, and because selection has proved extremely difficult to find in operation. On the other hand, the theory that standing variation and most substitutions of amino acids have been neutral also strains our credulity to the limit, because it requires us to believe that population sizes for all species are effectively the same, because it requires adaptive mutation to be several orders of magnitude less frequent that neutral changes, because there is too much variation from locus to locus in the amount of divergence between populations, because of striking similarities in allelic frequency distributions in closely related species, and because the majority of polymorphic substitutions do alter the functional properties of enzymes.

How can such a rich theoretical structure as population genetics fail so completely to cope with the body of fact? Are we simply missing some critical revolutionary insight that in a flash will make it all come right, as the Principle of Relativity did for the contradictory evidence on the propagation of light? Or is the problem more pervading, more deeply built into the structure of our science? I believe it is the latter.

First, population genetics is not an *empirically sufficient* theory in the sense that I defined such a theory in chapter 1. Built into both deterministic and stochastic theory are parameters and combinations of parameters that are not measurable to the degree of accuracy required.

Deterministic theory depends critically on the parameters of selection. There is virtually no qualitative or gross quantitative conclusion about the genetic structure of populations in deterministic theory that is sensitive to small values of migration, or any that depends on mutation rates. But selection is absolutely critical, and small perturbations in parameters can make big differences. Even though selection in favor of an allele may be very weak, that weak bias over long time periods will have a cumulative effect sufficient to replace the original gene completely. That is, the dynamic forces may be extremely weak but, because time is so long, the initial and end states of the historical process may be very different.

At the same time the statistical power of our estimation procedures is very low, and unavoidably so, because we are counting and not measuring. No technological breakthrough can alter the fact that under ideal conditions the standard error of a proportion is never smaller than $\sqrt{(pq/N)}$, so that if we want to be 95 percent sure that we know some proportion within 2 percent of its true value, we must count 10,000 cases. In a real world with experimental error we must count more. A theory whose parameters and variables are frequencies or the ratios of frequencies will always be bound to the error limits of binomial variance. Its variables and parameters cannot be very accurately measured without immense effort, one that could certainly not be expended over and over again. We must give up the idea of directly measuring selection differences of the order of 1 percent, and 1 percent is probably large for most loci if we average over all other genes.

Stochastic theory is much worse. By its nature a stochastic theory makes no predictions about individual events but only about distributions and averages. In population genetic theory, much of which is formally similar to the theory of diffusion processes, the distributions and their moments turn out to depend critically on the ratio of the mean deterministic force to the variance arising from random processes. Since the variance that is usually considered arises from the finiteness of population size, it is proportional to $1/N$. Thus quantities like $N\mu$, Nm, and Ns, the product of population size and a deterministic force of mutation, migration, or selection, appear over and over again in stochastic theory. The deterministic parameters μ, m, and s are very small but we do not know how small. We do not even know what their order of magnitude is although we know mutation rates for certain classes of alleles more accurately than either migration or selection. This follows from our ability to measure mutation in the laboratory under arbitrary conditions, whereas migration and selection are parameters of natural populations, especially migration, which has no meaning out of its context. Equally, N is very large and also impossible to measure except tautologically by using the very genetic variables it is meant to predict and some assumptions about other quantities. Even such tautological estimates of N involve the reciprocal of the difference between very small numbers that are themselves subject to great experimental error (Dobzhansky and Wright, 1941; Wright, Dobzhansky, and Hovanitz,

1942). Yet the predictions of stochastic theory depend critically on the sizes of Nm, Ns, and $N\mu$. The most that such a theory can do is to tell us not to be surprised at anything we see and to caution us against supposing that only one hypothesis will explain our observations.

A second problem is that population genetics is an *equilibrium theory*. Although we do have the dynamical laws of change in deterministic theory and are able to describe the time dependence of distribution functions in a few cases of stochastic theory, in *application* only the equilibrium states and the steady-state distributions are appealed to. Equilibria annihilate history. It is the nature of an equilibrium point that all paths in the dynamical space lead to it (at least locally), so that the particular history of change is irrelevant and, once the system is at equilibrium, there is no trace left of historical information.

Evolutionary geneticists are anxious to purge historical elements from their explanations. In part this desire stems from sociopolitical convictions about stability that are deeply held and are characteristic of the present stage of social and political development of the West. This preoccupation with stability was quite uncharacteristic of the middle of the nineteenth century, when the successful bourgeois revolution led to the prominence of notions of progress and directional movement in scientific theories. As the bourgeois revolution became consolidated, however, notions of progress and secular change gave way to themes of stability and equilibrium in a variety of natural and social sciences.

The banishing of history is almost a necessity because there is no historical record. Some selectionists wish to explain human blood group polymorphisms as a relic of past selection. "Blood groups used to be adaptive," they say, "but the conditions of human existence have changed so much that selection has ceased." The difficulty of that explanation is that although it might be true there is no way to find out. However, that is not a sufficient reason to reject the explanation. Many of the apparently paradoxical features of our problem may stem entirely from the application of equilibrium explanation to historical phenomena.

As an example, consider the ABO blood group frequencies in Britain and Ireland. Group A is over 50 percent in East Anglia and the south but declines in frequency in the north and west, reaching frequencies as low as 25 percent in western Ireland. The blood groups B and AB have the opposite trend, rising from 8 percent in the east and south to as high as 18 percent in Scotland, Wales, and Ireland. Were we dealing with Drosophila we would search for an equilibrium explanation involving temperature, rainfall, and other factors. But we know, in fact, that these distributions are the relics of four major invasions and displacements within the last 100 generations, beginning with the Beaker Folk and ending with the Vikings and their parvenu grandchildren.

An example of how a historical explanation can be an alternative to a steady-state theory is Haigh and Smith's (1972) examination of the frequencies of hemoglobin variants in man. To explain the lack of variants in a frequency range between 10^{-2} and 10^{-3} and contributing about 5 percent of all current alleles, whose existence is predicted under the theory of neutral mutation and drift, they propose two alternatives. One is that evolution of hemoglobins has been adaptive. The other is that the human population passed through a severe bottleneck in numbers in the recent past, one that lasted for $2N$ generations. For example, if N were 10,000, the bottleneck would have had to last for 20,000 generations or 400,000 years and would have had to end about 600 generations or 12,000 years ago.

The dynamics of gene-frequency change has a built-in historicity of a complex sort. If environment fluctuates so that the selective forces are changing, the gene frequency will take a complicated, wandering path. This path, however, only poorly reflects fluctuations in the selective forces because the change in gene frequency is not only a function of the selection coefficient but also of the present gene frequency.

For even the simplest case of no dominance, the change in p at time t is

$$\Delta p = pqs(t)$$

Then Δp will be quite small if p or q is close to zero no matter what the value of $s(t)$, and in general $p(t)$ will be a strongly autocorrelated process. The average value of p during some short time, perhaps 50 generations, will be independent of the distant past but it may also be independent of the recent history of selection if pq is close to zero. Conversely, if pq is near 0.5, the recent history of selection will be the most important factor determining the average allele frequency because the system is in a highly responsive state.

One consequence of the serial autocorrelation of $p(t)$ is that the average allele frequency over any period is a function not only of the distribution of environments but of their exact temporal sequence (Lewontin, 1967b). Thus two populations that have undergone a long series of fluctuating environments may have radically different average gene frequencies over their entire history, even though the mean, variance, and every other moment of the distribution of environments is the same in the two cases, *because the environments occurred in a different order.* The geneticist will search in vain for the missing environmental factor that explains why one population had a high average gene frequency and the other a low average. The answer lies in the historical sequences.

A third problem of theory is that the usual treatment of the genome as a collection of single loci ignores both the physiological interaction between loci and the fact that the genes are arranged on chromosomes and therefore do not segregate independently. If different loci are correlated in their allelic distributions, then the dimensionality of the dynamic system is much greater than the number of loci and is an exponential of the number of loci. That is, the usual theory is not *dynamically sufficient.* The attempt to build a dynamically sufficient theory of a genome with many genes is the most pressing problem of theory, because only in that way will it be possible to understand the evolution of the vast genetic variety in populations.

CHAPTER 6 / THE GENOME AS THE UNIT OF SELECTION

One of the best ways to test the adequacy of a theory is to ask how the theory could be used to explain some simple observations whose causes are known in advance. Could we have inferred those causes from the observations or would we have been led to some erroneous result or to a contradiction that could not be resolved within the framework of the theory. The appearance of such contradictions demands the formulation of new theories for their resolution and is a precondition for an advance in our understanding of the real world.

Suppose we wish to study the dynamics of an inversion polymorphism in an insect. In nature the frequency of the inversion *IN* is, for example, 0.65, while the standard arrangement *ST* is 0.35. We will use the standard perturbation technique for a laboratory study of selection. A large number of fertilized females are sampled from nature to establish a large number of laboratory strains. By choosing among these strains we can create laboratory populations in which the inversion is at any arbitrary initial frequency. We start one set of replicated populations at a ratio of 0.05 *IN* : 0.95 *ST*, and another set at the opposite extreme, 0.95 *IN* : 0.05 *ST*. A pair of populations, one from each set, is founded every ten generations so that if there are secular changes in laboratory conditions they will average out over replications.

The results of this *gedanken-experiment* are shown in figure 23. Populations started with a low frequency of the inversion progress regularly to an intermediate equilibrium not far from that observed

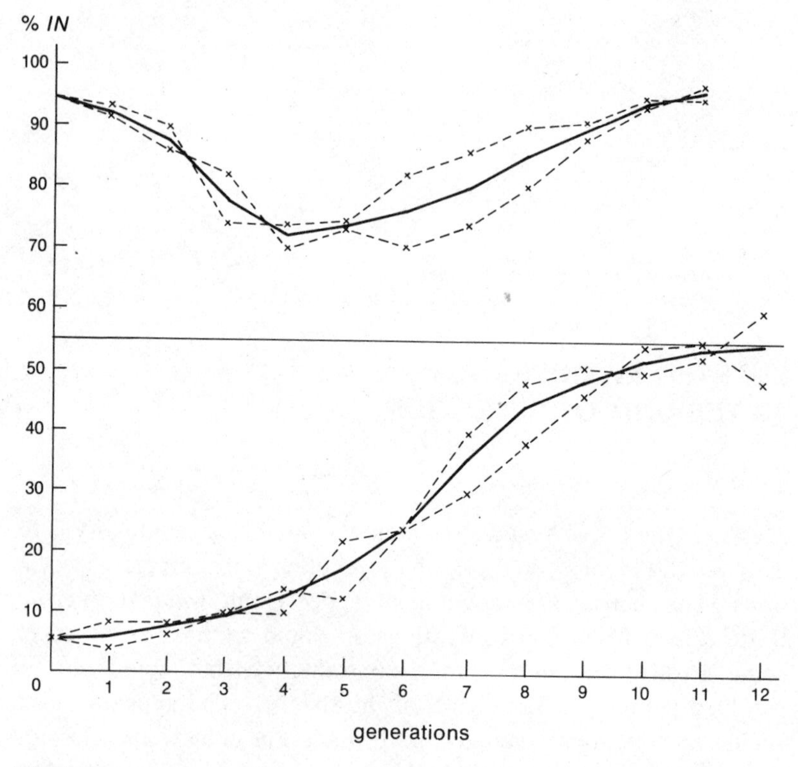

FIGURE 23

The frequency of an inversion *IN*, in hypothetical laboratory populations. The heavy lines represent the average behavior of replicates while the x's represent individual data points.

in nature, following the S-shaped trajectory that we expect from simple selection theory for a gene that is initially in low frequency. The populations starting at a high frequency of the inversion begin to move downward toward the same equilibrium but then, mysteriously, start back up again and appear, at the end of the experiment, to be on their way to fixation of the inversion. How can we explain this extraordinary result? Clearly we are not dealing with a change in selection coefficients resulting from a change in external conditions, since the experiment was performed in pairs, and all replicates behaved alike despite their not being simultaneous in absolute time.

The obvious explanation seems to be frequency-dependent selection. Even with fairly simple forms of frequency dependence it is possible to have many stable and unstable points of gene frequency.

For example, there might be a stable point at $p = 0.55$ and an unstable point at $p = 0.65$, so that populations beginning at lower inversion frequencies would evolve to an equilibrium at $p = 0.55$, whereas those at initial frequencies above the unstable point would go to fixation at $p = 1.0$. The behavior of the upper curve in figure 23 cannot be explained, however, by frequency-dependent selection, no matter how complicated, *because these populations have passed through the same set of gene frequencies, going in opposite directions*. The frequency of the inversion has both decreased and increased at every value of p, depending upon the stage of population evolution. If the inversion is known to be at a frequency $p = 0.85$, for instance, it is impossible to know whether the frequency will be higher or lower in the next generation unless we already know whether the frequency was increasing or decreasing in the previous generation.

In the terms of chapter 1, the gene frequency alone is not a *sufficient dimension*, because we cannot describe the evolution of the system unambiguously in terms of that dimension and a set of parameters whose values are determined externally to the system. We are in the same situation as the pilot of a space capsule who knows only his position and velocity. He cannot plot his course unless he also knows whether he is speeding up or slowing down. Some dimensions are missing from our description of the evolution of the inversion system, without which our theory is dynamically insufficient. By isolating the frequency of the inversion as a variable, we have destroyed the possibility of predicting its historical progression.

As a matter of fact, the hypothetical changes in figure 23 are not the product of idle fancy projected to show how complicated life might be. They are the changes that would occur in the frequency of the *Blundell* (*BL*) inversion of the *CD* chromosome element in the Australian grasshopper *Moraba scurra*, based on actual fitness estimates from nature (Lewontin and White, 1960). There is only a single dimension missing from the dynamically sufficient description of this polymorphism, and that is the frequency of an inversion, *Tidbinbilla* (*TD*), on a different chromosomal element, *EF*, that is also segregating in populations of *Moraba scurra*.

When both polymorphisms are taken into account, there are nine genotypes; Lewontin and White estimated the fitnesses of these genotypes in nature by comparing the frequencies of adults in one

TABLE 53
Estimated fitnesses for the nine genotypes with respect to two polymorphic inversion systems, Blundell/Standard on chromosome *CD* and Tidbinbilla/Standard on chromosome *EF* in *Moraba scurra*

	Chromosome CD		
Chromosome EF	*ST/ST*	*ST/BL*	*BL/BL*
ST/ST	0.791	1.000	0.834
ST/TD	0.670	1.006	0.901
TD/TD	0.657	0.657	1.067

Note: Data are from Lewontin and White (1960).

generation with the predicted Hardy-Weinberg projection from the gene frequency in the previous generation.* The results for the population of Royalla, near Canberra, in 1956 are given in table 53. Without any calculations at all it is clear from table 53 that there are strong interactions between the two inversion systems in determining fitness. For example, there is heterosis for the *CD* polymorphism, provided the *EF* chromosome pair is either homozygous *ST/ST* or heterozygous *ST/TD*, but the heterosis is changed into dominance when the *EF* chromosome pair is *TD/TD*. Moreover, *TD/TD* homozygotes are superior to either *TD/ST* heterozygotes or *ST/ST* homozygotes when the *CD* chromosome pair is homozygous *BL/BL*, but the reverse for other genotypes of the *CD* pair.

In general, the relative fitnesses of the three genotypes *ST/ST*, *ST/BL*, and *BL/BL* are functions of the frequencies of the other inversion polymorphism. Therefore the evolutionary changes that take place on the *CD* chromosome pairs cannot be predicted without knowing the state of the *EF* polymorphism, and vice versa.

This phenomenon should not be confused with "frequency-dependent selection." The fitnesses of the nine genotypes are assumed to be fixed constants. The dependence of the evolutionary history of one polymorphism on the history of the other polymorphism arises entirely from the epistatic interaction of the two in their determination of fitness. For there to be no dependence of the polymorphisms on each other, the fitnesses of the nine genotypes would have to be

*The shortcomings of this method (see p. 241), although severe, are not relevant to the point I am making here.

exactly the products of separate fitnesses for the two separate polymorphisms Table 54 gives the general scheme of fitnesses for two loci (or inversions). The average or *marginal fitnesses*, $\bar{W}_{AA}$, $\bar{W}_{Aa}$, $\bar{W}_{aa}$, $\bar{W}_{BB}$, and so on, are the fitnesses of the genotypes averaged over all combinations in which they appear, weighted by the frequency of those combinations. Thus

$$\begin{aligned} \bar{W}_{AA} &= P^2 W_{AABB} + 2PQ W_{AABb} + Q^2 W_{AAbb} \\ \bar{W}_{Aa} &= P^2 W_{AaBB} + 2PQ W_{AaBb} + Q^2 W_{Aabb} \\ &\cdot \quad \cdot \quad \cdot \quad \cdot \\ &\cdot \quad \cdot \quad \cdot \quad \cdot \\ &\cdot \quad \cdot \quad \cdot \quad \cdot \\ \bar{W}_{bb} &= p^2 W_{AAbb} + 2pq W_{Aabb} + q^2 W_{aabb} \end{aligned} \tag{23}$$

and the mean fitness of the entire population $\bar{W}$ is

$$\begin{aligned} \bar{W} &= p^2 \bar{W}_{AA} + 2pq \bar{W}_{Aa} + q^2 \bar{W}_{aa} \\ &= P^2 \bar{W}_{BB} + 2PQ \bar{W}_{Bb} + Q^2 \bar{W}_{bb} \end{aligned} \tag{24}$$

The relation between the frequency p of A in two successive generations depends upon the marginal fitness of the genotypes at the A–a locus at any instant.

$$p' = \frac{p(p\bar{W}_{AA} + q\bar{W}_{Aa})}{\bar{W}} \tag{25}$$

If no account is taken of the fact that these are marginal fitnesses, because it is assumed that other loci do not affect them, equation (25) appears to be a relation involving only the allelic frequencies at the A–a locus and a set of constants. But in general the marginal fit-

TABLE 54
Fitnesses of the nine genotypes for the general two-locus model

	Locus A				
Locus B	AA	Aa	aa	Mean fitness	Frequency
BB	W_{AABB}	W_{AaBB}	W_{aaBB}	$\bar{W}_{BB}$	P^2
Bb	W_{AABb}	W_{AaBb}	W_{aaBb}	$\bar{W}_{Bb}$	$2PQ$
bb	W_{AAbb}	W_{Aabb}	W_{aabb}	$\bar{W}_{bb}$	Q^2
Mean fitness	$\bar{W}_{AA}$	$\bar{W}_{Aa}$	$\bar{W}_{aa}$	$\bar{W}$	
Frequency	p^2	$2pq$	q^2		

TABLE 55
Fitness relations when two loci are multiplicative in their action

Locus B	Locus A: AA	Aa	aa	Mean fitness	Frequency
BB	$W_{AA} \cdot W_{BB}$	$W_{Aa} \cdot W_{BB}$	$W_{aa} \cdot W_{BB}$	$\bar{W}_{AB}$	P^2
Bb	$W_{AA} \cdot W_{Bb}$	$W_{Aa} \cdot W_{Bb}$	$W_{aa} \cdot W_{Bb}$	$\bar{W}_{Bb}$	$2PQ$
bb	$W_{AA} \cdot W_{bb}$	$W_{Aa} \cdot W_{bb}$	$W_{aa} \cdot W_{bb}$	$\bar{W}_{bb}$	Q^2
Mean fitness	$\bar{W}_{AA}$	$\bar{W}_{Aa}$	$\bar{W}_{aa}$	$\bar{W}$	
Frequency	p^2	$2pq$	q^2		

nesses are functions of the frequencies at the B–b locus, as shown in equation (24), and if the B–b locus is evolving, these marginal fitnesses change. In general, equation (25) is a function of the frequencies at both loci, and there will be an equivalent expression for changes at the B–b locus, also a function of the frequencies at both loci. If there are n loci segregating, there are n equations in n independent frequencies that need to be evaluated simultaneously.

In the special case that the fitnesses of combination genotypes are the products of fitness constants applied to each locus separately, the dependence of the loci on each other disappears. This is easy to see. Table 55 shows the fitnesses as products of six "basic" single-locus values, W_{AA}, W_{Aa}, W_{aa}, W_{BB}, W_{Bb}, and W_{bb}. It is easily verified that in this case the relations (23) and (24) become

$$
\begin{aligned}
\bar{W}_{AA} &= W_{AA}\bar{W}_B \\
\bar{W}_{Aa} &= W_{Aa}\bar{W}_B \\
\bar{W}_{aa} &= W_{aa}\bar{W}_B \\
&\vdots \\
\bar{W}_{bb} &= W_{bb}\bar{W}_A
\end{aligned}
\tag{26}
$$

and

$$\bar{W} = p^2W_{AA}\bar{W}_B + 2pqW_{Aa}\bar{W}_B + q^2W_{aa}\bar{W}_B = \bar{W}_A\bar{W}_B \tag{27}$$

where

$$\bar{W}_A = p^2W_{AA} + 2pqW_{Aa} + q^2W_{aa}$$

$$\bar{W}_B = P^2W_{BB} + 2PQW_{Bb} + Q^2W_{bb}$$

On substituting the special values of equations (26) and (27) into the equation for the gène frequency in two successive generations, equation (25), we obtain by cancellation of $\bar{W}_B$

$$p' = \frac{p(pW_{AA} + qW_{Aa})}{\bar{W}_A} \tag{28}$$

which is indeed independent of the *B–b* locus and is solely a function of the fitnesses at the *A–a* locus, taken in abstraction. It is *only* when the fitnesses at the two loci have a multiplicative relationship that they can be treated independently.

The treatment of the *Moraba scurra* problem as a two-dimensional evolution, with both polymorphisms taken into account simultaneously, is shown in figure 24. Starting from different initial conditions, the trajectories are calculated from the continuous analogue of equation (24) for both loci simultaneously (see Lewontin and Kojima, 1960), and there is no ambiguity about the paths of evolution of the total system. The trajectories are shown superimposed on contour lines of mean fitness, $\bar{W}$. To a first approximation, gene frequencies move along the surface of $\bar{W}$ toward higher values and reach a stable equilibrium at maxima of the surface, *P*. An unstable equilibrium on the surface is at the saddle point *S*, and the directions of evolution of the system depend upon the relation of the initial composition to the saddle point and the maxima. There is a kind of "continental divide" that runs through the saddle point from the top right of the field to the center left. Initial compositions above this line evolve to the upper stable point at 0 percent *Tidbinbilla* and 55 percent *Blundell*, whereas initial conditions below this line evolve to the fixation of both *Blundell* and *Tidbinbilla.*

Trajectories 1 and 4 correspond to the two experimental outcomes shown in figure 23, when only the *CD* polymorphism is observed. In trajectory 1 the frequency of *Blundell* increases steadily from 0.05 to 0.55, where it reaches a stable equilibrium. At the same time *Tidbinbilla* has been lost from the *EF* chromosome pair. In trajectory 4, Tidbinbilla becomes fixed, having risen steadily from an initial frequency of 0.10, but during this process the *CD* inversion, *Blundell*, has first fallen in frequency from 0.95 to 0.72 and then risen again to eventual fixation.

A number of further mysterious complexities would appear if each of the polymorphisms were followed in isolation. For example, trajectory 5 would show a decrease to nearly 0.50 of *Tidbinbilla*

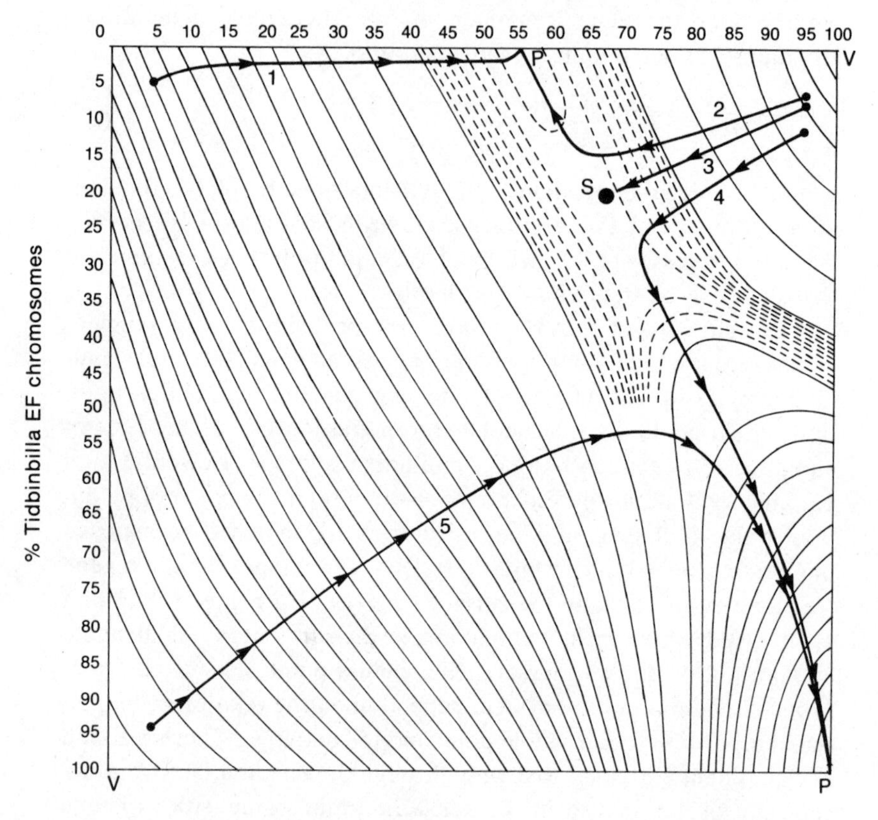

FIGURE 24

Projected changes in the frequency of two polymorphic inversion systems in *Moraba scurra* from different initial compositions, based on fitness estimates from nature. The trajectories, shown by arrow-marked lines, are calculated by the solution of differential equations of gene frequency change. Lines crossing the trajectories are contours of equal mean population fitness, $\bar{W}$.

before it rose and eventually became fixed. Also, very small variations in the initial frequency of *Tidbinbilla* in the upper right hand corner of the field would result in quite different outcomes for the *Blundell* polymorphism. When *Tidbinbilla* was 0.10 initially, *Blundell* would decrease and then increase again to fixation, but if *Tidbinbilla* were initially 0.05, then Blundell would evolve to the same equilibrium, $p = 0.55$, as for trajectory 1. If one were not aware of the *EF* polymorphism, it would appear that the evolution of the *CD* system was indeterminate, sometimes going to equilibrium and sometimes to fixation, even though the initial frequency was identical ($p = 0.95$).

Indeterminate outcomes, including reversal in the direction of frequency changes during the course of the evolution of laboratory populations, were observed by Dobzhansky and Pavlovsky (1953, 1957) when they followed selective changes in the third-chromosome inversion polymorphism of *Drosophila pseudoobscura* and by Spiess (1959) for inversions of *D. persimilis*. The indeterminacy was deliberately induced by Dobzhansky and Pavlovsky, as reported in their later paper, by initiation of replicated populations with only 20 founders, thus promoting variation in the initial frequencies of genes not included in the inversions.

The indeterminancy and the self-contradictory paths of inversion-frequency change in these populations are reflections of the dimensional insufficiency of a single-locus treatment. Even though we may be interested in following only one segregating entity, say a third chromosome inversion in *D. persimilis*, an understanding of evolution along that one dimension requires *first* a synthetic treatment of the genotype and *then* an abstraction of the single system of interest from the complex mass. We cannot reverse the process, in general, building a theory of a complex system by the addition or aggregation of simple ones.

ASSUMPTIONS ABOUT THE DIMENSIONALITY OF EVOLUTIONARY PROCESSES

The argument of the preceding section shows that a sufficient dimensionality for describing changes in the frequency of alleles at one locus must involve at least the frequencies of alleles at other loci that interact with it in determining fitness. This description in n dimensions is not really complete, however. A total genetical description of a population is an enumeration of all its diploid genotypes. If there are a alleles at each of n segregating loci there will be a^n different gametic combinations that can be produced, and when these gametes come together at fertilization they can form $a^n(a^n + 1)/2$ distinguishable zygotic combinations. The space of complete description of a population would have $a^n(a^n + 1)/2 - 1$ independent dimensions.

This number of dimensions is successively reduced in practice by making a series of assumptions about the way in which allelic frequencies, zygotic frequencies, and gametic frequencies are related. For example, if the gametes can be assumed to come together completely at random at fertilization, the frequencies of the $a^n(a^n + 1)/2$ distinguishable zygotic types will be completely determined from the $a^n - 1$ independent gametic frequencies; an immense reduction in dimensionality, by a factor of $(a^n + 1)/2$, will have been achieved. The reduction is compounded of the assumptions that genotypes mate at random and that gametes within mating pairs do not show preferential fertilization. On these assumptions all loci are in Hardy-Weinberg equilibrium at the moment of fertilization.

For loci that are involved in sexual attractiveness or timing of sexual maturity, or morphological factors that cause assortment of mates (for example, size), assortative mating by genotype must occur. Moreover, in species that are partially self-fertilizing there will be an excess of homozygotes from partial inbreeding. For annual plants in which both flowering time and self-fertilization are important elements in determining the pattern of fertilization, loci will deviate far from Hardy-Weinberg proportions at fertilization.

The second major reduction in dimensionality can be made if one assumes that the alleles at different loci are distributed at random with respect to each other in gametes. Then the frequency of each of the $a^n - 1$ independent gametic types is completely predictable from the $n(a - 1)$ independent allelic frequencies. This means that the description of the entire population is simply the collection of descriptions of the individual loci, without any concern for the joint distribution of frequencies. The genotypic descriptions of allozyme polymorphisms like those in tables 26 and 32 (pp. 131, 153) are proper genetic descriptions of Drosophila and man only if the enumeration of the allelic frequencies at each locus separately gives complete information about the frequencies of total genotypes. For genes on the third chromosome of *Drosophila pseudoobscura*, alleles at different loci are not associated at random (Prakash and Lewontin, 1968, 1971), and a correct description of the genotypic composition of populations requires the specification of gametic frequencies.

Finally, the dimensionality for the evolutionary dynamics of a single locus can be reduced to $a - 1$ if, as the preceding section showed, the relative fitnesses of the genotypes at the locus are independent of all other loci. Nearly all the theoretical formulations of population genetics and nearly all its experiments and natural historical observations are at this last, lowest dimensionality. A single locus is usually observed in isolation and its evolution explained in isolation. Moreover, that evolution is nearly always framed in terms of the frequency of alternative alleles at the locus. The presumption that an adequate picture of evolutionary dynamics is contained in a space whose sole dimensions are the frequencies of alternative alleles at single loci is justified only if the various assumptions about independence of fitness relations and independence of genic assortment are correct or nearly so. Table 56 summarizes the reductions in dimensionality that are possible as more and more assumptions are made about the genetic structure of the population.

THE COMPLICATION OF LINKAGE

The assumption that gametic frequencies are predictable from individual allele frequencies is based upon the notion that recombina-

TABLE 56
Sufficient dimensionality required for the prediction of evolution of a single locus with a alleles, when there are n segregating loci in the system

			Example with:	
			$n = 2$	$n = 3$
Dimensionality	*Level of description*	*Assumptions implicit*	$a = 2$	$a = 3$
$a^n(a^n + 1)/2 - 1$	zygotic classes	none	9	336
$a^n - 1$	gametic classes	random union of gametes	3	26
$n(a - 1)$	allele frequencies	random union of gametes random association of genes at different loci (linkage equilibrium)	2	6
$a - 1$	allele frequencies	random union of gametes random association of genes at different loci no epistatic interaction	1	2

tion, no matter how infrequent, would eventually randomize genes with respect to each other, even if at some time in the history of a population there was an excess of certain gene combinations. In the absence of natural selection or mutation, the dynamics of the randomization process is easily derived (Geiringer, 1944).

In all that follows I will use a binary notation for alleles at a locus rather than the more conventional upper- and lower-case letters. Thus an individual whose genotype is usually given as *Abc/aBC* will be *100/011*. Suppose there are two loci with two alleles each. There are four gametic types, *00, 01, 10*, and *11*, and on the assumption of random mating, the population composition can be completely specified by their frequencies, g_{00}, g_{01}, g_{10}, and g_{11}. Any three are sufficient since their sum is one.

If the alleles at the two loci are distributed at random, then

$$\begin{aligned} \tilde{g}_{00} &= qQ \\ \tilde{g}_{01} &= qP \\ \tilde{g}_{10} &= pQ \\ \tilde{g}_{11} &= pP \end{aligned} \tag{29}$$

where p and q $(= 1 - p)$ are the frequencies of the *1* and *0* alleles at the first locus, P and Q $(= 1 - P)$ the frequencies at the second locus, and $\tilde{g}_{ij}$ the gametic frequencies when the loci are at random with respect to each other. Geiringer showed that at any moment in the history of these loci, the actual gametic frequencies could be represented by

$$\begin{aligned} g_{00} &= qQ + D \\ g_{01} &= qP - D \\ g_{10} &= pQ - D \\ g_{11} &= pP + D \end{aligned} \tag{30}$$

where D, called by Lewontin and Kojima (1960) the *coefficient of linkage disequilibrium*, is

$$D = g_{11}g_{00} - g_{10}g_{01} \tag{31}$$

From equations (30) and (31) it follows that

$$\tilde{D} = \tilde{g}_{11}\tilde{g}_{00} - \tilde{g}_{10}\tilde{g}_{01} = pPqQ - pQqP = 0$$

so that D measures the deviation of the actual gametic frequencies from random combination. If D is positive there is an excess of "coupling" gametes *11* and *00*, whereas if it is negative there are too many *10* and *01*.

The set of relations given by equations (29) and (30) makes clear how the dimensionality of the problem is changed by the dependence of the loci or each other. If the loci are independent, the gametic frequencies are entirely specified by the two independent variables, P and p. When the loci are not independent there are three dimensions, given by any three of the gametic frequencies, g_{ij}. These can equally well be represented by the three variables p, P, and D, the relations between the two parametric representations of the dynamical space being given by equations (30).

By analogy with the analysis of variance, the three degrees of freedom of the gametic frequencies can be assigned to two *main effects* (the allele frequencies at each locus) and an *interaction* (the linkage disequilibrium measure D). If there are more than two loci, the same partition can be carried out. For three loci, there are eight gametic frequencies, $g_{000}, g_{001}, \ldots, g_{111}$, but since their sum must be one, there are only seven independent variables. These can be parametrically represented by three main effects (the allele frequency at each of the three loci), three first-order interactions (the D values for the three pairwise combinations of the loci), and a second-order D (measuring the deviation of the gametic frequencies from independence not accounted for by the pairwise D values). For example, if

$$g_{000} = g_{011} = g_{101} = g_{110} = 0$$

and

$$g_{111} = g_{100} = g_{010} = g_{001} = 1/4$$

all the pairwise Ds are exactly zero so that each pair of loci is in perfect linkage equilibrium, yet the gametes as a whole are obviously not since their random frequencies would be

$$g_{000} = g_{001} = \ldots = g_{111} = 1/8$$

Geiringer showed that over time D approaches zero because of recombination. More precisely, if R is the recombination factor between the loci, then after t generations of recombination

$$D_t = (1 - R)^t D_0 \tag{32}$$

or for small R

$$D_t = D_0 e^{-Rt} \tag{33}$$

D goes to zero asymptotically, but even if genes are closely linked, for instance $R = 0.005$, after 1000 generations there would be less than 0.5 percent of the original correlation between the loci.

The evolution of the population can be represented by the trajectory of a point in a unit tetrahedron whose three edges are three independent gametic frequencies, g_{00}, g_{01}, and g_{11} (figure 25a). Inside the tetrahedral space is a hyperbolic paraboloid that is the surface of independence, $D = g_{11}g_{00} - g_{01}g_{10} = 0$. The points above the surface have a positive association (an excess of "coupling" types),

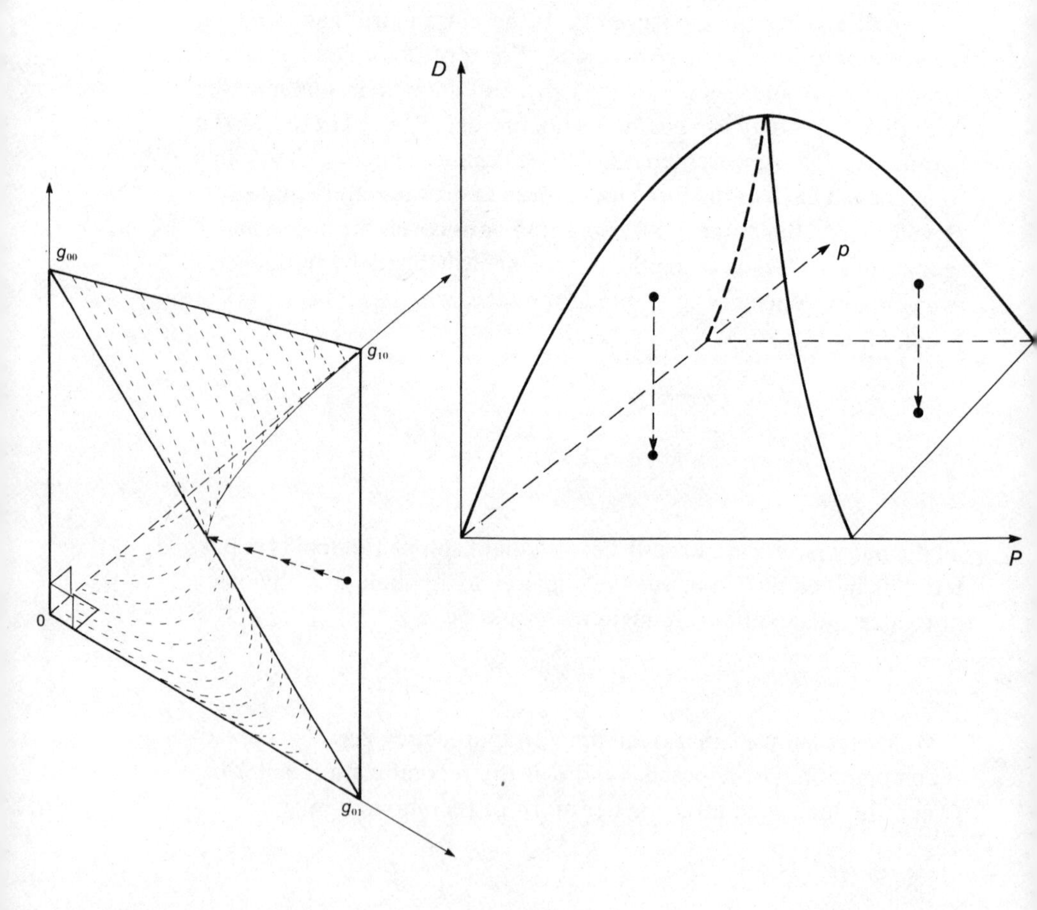

those below, a negative association. A population beginning at some arbitrary point inside the space will evolve along a line determined by the restriction that the allele frequencies at the two loci do not change, arriving finally at the surface of independence. These two restrictions determine the planes $g_{00} + g_{01} = q$ and $g_{00} + g_{10} = Q$; the trajectory will be the straight line that is their intersection.

Alternatively, the population may be seen as evolving in the space of p, P, and D, shown in figure 25b. Again the trajectories are straight lines leading to the plane $D = 0$.

Since most selective changes must require thousands of generations to complete, it would appear intuitively obvious that linkage cannot play an important role. Especially for balanced polymor-

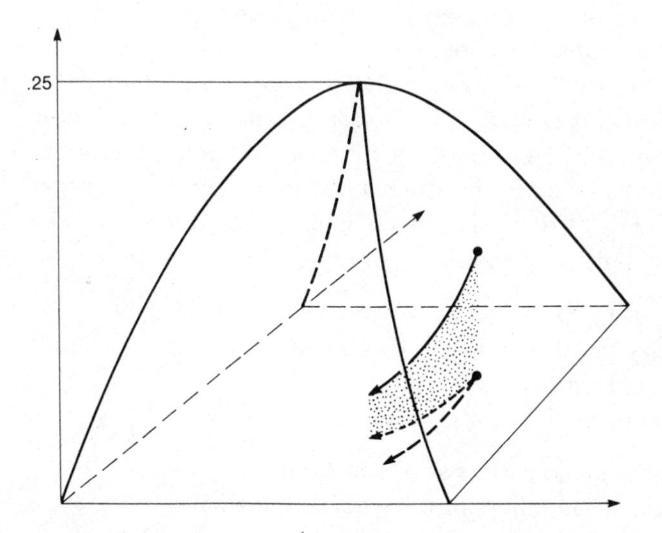

FIGURE 25
Far left, unit tetrahedron representing the dynamic space for a system with four gametic types. The three axes are the frequencies g_{00}, g_{01}, and g_{10}. Inside the tetrahedron is a surface D=0. The dashed line shows the evolution of a population under recombination without selection. Center, the equivalent space with the dimensions *p, P,* and *D*. Again, trajectories without selection are straight lines leading to the surface *D*=0. Above, trajectories of genetic evolution under natural selection in the *p, P, D* space. The solid line is the true trajectory; the dashed line is the trajectory in the *p, P* plane that would be calculated on the assumption of constant linkage equilibrium; the dotted line is the projection of the true trajectory onto the *p, P* plane.

phisms in which allele frequencies are held at an intermediate equilibrium for extremely long periods, it would seem that any association between loci would certainly be broken up by recombination.

SELECTION AND RECOMBINATION

In *The Genetical Theory of Natural Selection* (1930), Fisher proposed that natural selection would favor closer linkage of two genes influencing the same polymorphism. Unfortunately he offered no proof of the assertion, which contradicts the intuitive conclusion that recombination is not a critical issue in selection. If he had offered proof, theoretical population genetics would be considerably advanced over its present status. The first exact treatment of Fisher's suggestion was made only in 1956, when Kimura investigated the dynamics of Batesian mimicry polymorphism dependent upon two epistatically interacting loci.

In the process of investigating the mimicry problem, Kimura derived general equations for the change of gametic frequencies in continuous time; the analogous equations for discrete generations were derived by Lewontin and Kojima (1960) in their general paper on the evolutionary dynamics of complex polymorphisms. The changes in gametic frequencies, g_{ij}, are given by

$$\begin{aligned} \Delta g_{11} &= [g_{11}(\bar{W}_{11} - \bar{W}) - W_H RD]/\bar{W} \\ \Delta g_{10} &= [g_{10}(\bar{W}_{10} - \bar{W}) + W_H RD]/\bar{W} \\ \Delta g_{01} &= [g_{01}(\bar{W}_{01} - \bar{W}) + W_H RD]/\bar{W} \\ \Delta g_{00} &= [g_{00}(\bar{W}_{00} - \bar{W}) - W_H RD]/\bar{W} \end{aligned} \tag{34}$$

where $\bar{W}_{ij}$ is the marginal fitness of gamete ij
$\bar{W}$ is the mean fitness of the whole population
W_H is the fitness of the double heterozygote $AaBb$
R is the recombination function between the loci

and $D = g_{11}g_{00} - g_{10}g_{01}$ as before.

These equations for two loci are special cases of a general form (Lewontin, 1964a) for any number of gametic types

$$\Delta g_i = \frac{g_i(\bar{W}_i - \bar{W})}{\bar{W}} + \rho_i(g, R) \tag{35}$$

where $\rho(g,R)$ is a complicated function of all the gametic frequencies and all the recombination functions among the loci involved. For example, the functions $\rho(g, R)$ in the three-locus case consist of 18 terms, each the product of two reciprocal gametic frequencies—a recombination fraction and a fitness of a double or triple heterozygote (Feldman, Franklin, and Thomson, 1973).

Equations (34) are not independent, since there are only three independent variables and the sum of the Δg_i must be zero. They can be reduced to three by eliminating one of the gametic frequencies, and these three equations can be used to trace the trajectory of gametic frequency change in the tetrahedral space of figure 24a. The trajectories will no longer be straight lines, in general, because the allele frequencies will also be changing under the influence of natural selection. Alternatively three parametric equations Δp, ΔP, and ΔD can be written and the evolution of the populations followed in the space of figure 25b.

The claim that most of the evolution of loci occurs in a state of effective linkage equilibrium amounts to the claim that the evolutionary trajectory of the population lies on, or very near to, the plane $D = 0$ on figure 25b, because the force of recombination exerts a constant pressure toward the surface of independence. Another possibility is that, even though the population trajectory may lie well off the $D = 0$ plane, its projection onto the plane may be only trivially different from the path that would be calculated assuming constant linkage equilibrium. This amounts to supposing that *allele* frequencies at the loci undergo changes or come to equilibria that are essentially the same, irrespective of how they are assorted into gametes. Figure 25c shows these relations graphically.

If gene frequencies were unaffected by linkage, irrespective of what might happen to gametic frequencies, it might be supposed that linkage could be discounted, but even that is not true. The *zygotic* distribution depends very much on the gametic frequencies, rather than the allele frequencies, and thus the distribution of phenotypes and of fitnesses may be greatly affected. To take an extreme case, suppose there were two loci at which balanced lethals were segregating. At both loci the allele frequencies would be held at a stable equilibrium of $p = q = 0.5$. If the genes were in linkage equilibrium, all four gametic types would be in equal frequency; the frequency of double heterozygotes, the only viable class, would be 1/4 and the mean fitness of the population would then be $\bar{W} = 0.25$. However, if the only gametes were *00* and *11*, half the population of zygotes formed would be double heterozygotes and the fitness would be $\bar{W} = 0.5$. While the case just described is extreme, the same principle applies for any intensity of selection. The mean fitness of the population is enhanced if deleterious alleles are in strong coupling association, because the death of one individual will carry away the alleles at both loci simultaneoulsy, one of the deleterious genes being given a "free ride" by the elimination of the other. It is this principle that led Fisher to postulate a selective advantage for closer linkage of balanced polymorphisms.

A second interesting effect of the linkage is that it increases the variance in fitness (from 0.1875 at linkage equilibrium to 0.2500 for complete association, in the example of the balanced lethals) at the same time that it decreases genetic load. Thus a considerable genetic variance in fitness may exist without necessarily reducing the mean fitness excessively, an important point in understanding the paradox of so much heterozygosity. This effect of nonrandom association of genes is similar to the nonindependence of gene effects in determining fitness, as postulated by the threshold models of selection. In one case it is the nonindependence of physiological effects of genes that reduces the calculated genetic load, in the other the nonindependence of assortment of genes. Both reduce the average genetic load for a given variance in fitness. We shall see that the two phenomena are, in fact, interrelated in a causal way.

INTERACTION OF LINKAGE AND SELECTION

When the equations (34) for genetic frequency change are applied to various selection models, the trajectory of the population does not lie in the $D = 0$ plane during the course of selection but departs from it more or less, as determined by the amount of recombination and the form of selection. Moreover, the departure from linkage equilibrium may grow greater, as well as smaller, during the selection, and *always* grows greater at first if the initial gametic frequencies are close enough to $D = 0$.

Even if the genes are in perfect linkage equilibrium to start with, selection will force them out of equilibrium immediately. The only exception to this rule occurs when the fitnesses at the two loci are perfectly multiplicative. Then an initial $D = 0$ will remain, but it is unstable and if even the smallest linkage disequilibrium appears by chance deviation of gametic frequencies, D will grow under the force of selection. Several numerical examples are given by Lewontin and Kojima (1960), and the general case was proved by Felsenstein (1965).

Figure 26 shows the changes that take place in allele frequencies for a simple, symmetrical fitness model (table 57) and the changes that are occurring at the same time in the linkage disequilibrium, as measured by D', the ratio of D to its maximum possible value, given the allele frequencies. Although there is little effect in the gene-frequency plane, considerable linkage disequilibrium is generated during the course of selection *despite an initial linkage equilibrium.* For $R < 0.0625$, the disequilibrium continues to rise throughout the evolution of the system, while for $R > 0.0625$, D goes through a maximum. These results suggest that even at gene frequency equilibrium there may be permanent linkage disequilibrium if recombination is small enough.

The general equilibrium solution to equations (34) has not proved to be possible explicitly (although numerical solutions are easy to obtain), but several symmetric models of fitness have been solved. The general, totally symmetric model studied by Kojima and Lewontin has fitnesses

	Locus A		
Locus B	*11*	*10*	*00*
11	a	b	a
10	c	d	c
00	a	b	a

The equilibria for this model are

$$\hat{g}_{00} = \hat{g}_{01} = \hat{g}_{10} = \hat{g}_{11} = {}^1/_4, \quad \hat{D} = 0 \tag{36}$$

$$\hat{g}_{00} = \hat{g}_{11} = {}^1/_4 \pm {}^1/_4 \sqrt{1 - \frac{4Rd}{a+d-b-c}},$$

$$\hat{g}_{01} = \hat{g}_{10} = {}^1/_2 - \hat{g}_{00},$$

$$\hat{D} = \pm\, {}^1/_4 \sqrt{1 - \frac{4Rd}{a+d-b-c}} \tag{37}$$

Provided that there is some overdominance, with $d > a$ and $|a - d| > |b - c|$, the gene frequencies will come to an intermedi-

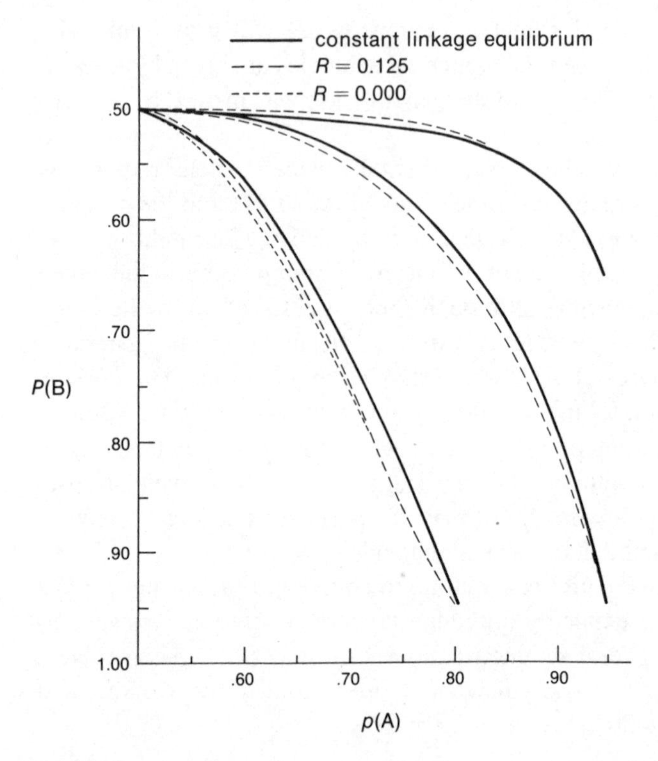

FIGURE 26

Paths of evolution for the two-locus case whose fitnesses are given in table 55. Above, trajectories of gene frequency change. At right, linkage disequilibrium relative to the maximum, D', along the trajectories shown above.

ate stable equilibrium at $p = P = 0.5$, and one or the other of the gametic equilibria (36) or (37) will apply. Otherwise the system will go to fixation of one or both loci. Considering only those cases in which a stable heterotic equilibrium exists at both loci, we see that solution (36) corresponds to the expectation when linkage is ignored and the loci are treated independently. The paired solutions of (37), however, although they still are at $p = P = 0.5$, may have considerable linkage disequilibrium.

The pair of solutions in (37) is stable under two conditions. First, it is required that

$$\epsilon = a + d - b - c > 0 \tag{38}$$

The quantity ϵ measures the deviation of the scheme of fitnesses

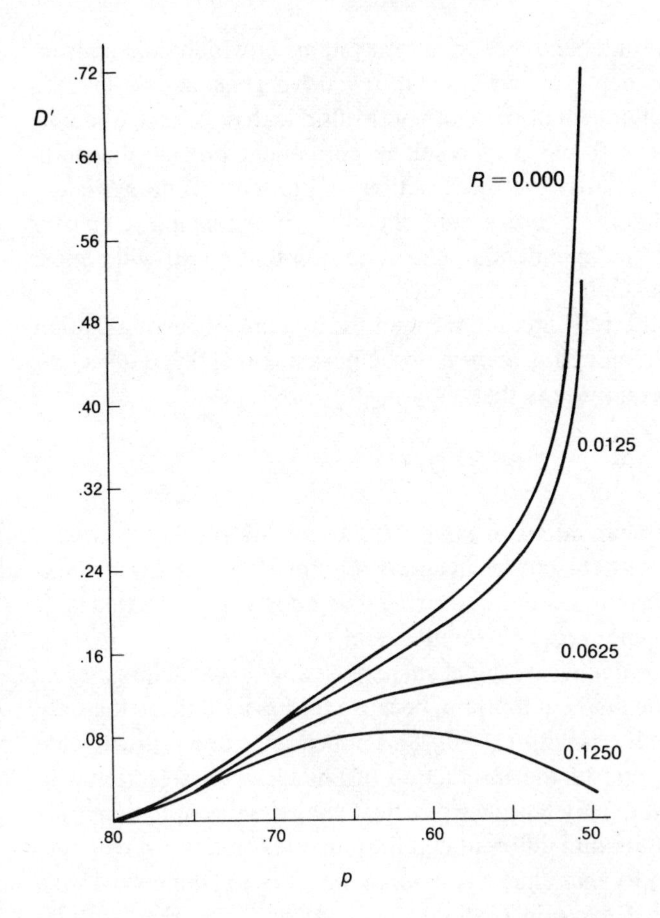

from additivity. If the fitnesses are perfectly additive, then only the linkage-equilibrium solution (36) is the stable one, but if there is any nonadditivity in fitness, there is the possibility that the linkage-disequilibrium solution will be the stable one. This is a rather dif-

TABLE 57
Fitnesses for the two-locus model whose evolution is shown in figure 26

	Locus A		
Locus B	*1/1*	*1/0*	*0/0*
1/1	0.250	0.375	0.250
1/0	0.550	1.000	0.550
0/0	0.250	0.375	0.250

ferent form of independence than for systems in which linkage is ignored, where departures from *multiplicative* gene action lead to dependence of one locus on another. Multiplicative gene action corresponds to $\epsilon \neq 0$ and may result in permanent linkage disequilibrium. Since additivity of gene action on the scale of fitness is exceedingly unlikely, whereas multiplicativity is the simplest kind of physiological "independence," the condition that $\epsilon \neq 0$ will almost always be satisfied.

Second, if there is interaction, then the amount of recombination must be small enough if there is to be permanent linkage disequilibrium. The condition is that

$$R < \frac{\epsilon}{4d} \tag{39}$$

For the numerical values of table 57, the condition is $R < 0.0625$, which agrees with the curves in figure 26. For $R > 0.0625$, the linkage disequilibrium generated during the course of selection will disappear at gene-frequency equilibrium.

Relation (39) defines a critical value of R, above which linkage has no effect on the final equilibrium, because for looser linkage the only stable equilibrium is with $\hat{D} = 0$. This appearance of a critical value is a characteristic of the interaction of linkage with selection. In some cases there may be several critical values, separating intervals of R that lead to quite different equilibrium conditions. An example of the complexity that can arise even in the simple symmetrical case was found by Lewontin (1964a) and completely worked out by Ewens (1968), in which there is (1) strong linkage disequilibrium for $R < 0.10$; (2) no stable intermediate equilibrium of gene frequencies of any kind for $0.10 < R < 0.373$, so the genes go to fixation; (3) weak linkage disequilibrium for $0.373 < R < 0.375$; and (4) stable equilibrium of gene frequencies with $D = 0$ for $0.375 < R \leq 0.50$.

The two general requirements, nonadditivity of fitness and recombination that is sufficiently small, can be summed up in what we may call, by analogy with physical systems, a *coupling coefficient*.

$$C \propto \frac{\epsilon}{R}$$

If C is large either because there is strong interaction or because recombination is low, the two loci can be said to be strongly coupled systems and there will be large departures from independence, even when gene frequencies are at equilibrium. When the coupling coefficient is small, the loci are weakly coupled systems because their fitness interaction is small or their recombination large, and their final equilibrium states will be weakly dependent. Moreover, there is a critical value of C, below which there is complete independence of the loci at equilibrium.

Explicit solutions for the equilibrium gametic frequencies have so far been found only for generally symmetrical models (Karlin and Feldman, 1970), of which the models of Lewontin and Kojima and of Kimura are special cases. Numerical solutions of any model can be obtained, however, and the full range of effects of linkage and selection is manifest in the asymmetrical cases. Table 58 gives the numerical results for an asymmetrical model with conditional heterosis (Lewontin, 1964a). For smaller recombination values there are two alternative stable equilibria possible at each value of recombination, a "coupling" and a "repulsion" equilibrium, and these have different intensities of linkage disequilibrium, different gene frequencies, and different mean fitnesses. The critical value of R is above 0.50, so that these fitness relations will always result in some linkage disequilibrium, no matter how much recombination occurs between the genes. *They may even be on different chromosomes.*

It is sometimes supposed that genes on different chromosomes ($R = 0.50$) are necessarily always in linkage equilibrium with each other. They are not, as the table shows. "Independent assortment" is a misleading phrase and there can be gametic correlations for genes that assort "independently." Table 58 demonstrates that every aspect of the genetic composition of an equilibrium population is affected by linkage. Not only are gamete frequencies and $\hat{D}$ radically different for the loosest and tightest linkages, but equilibrium gene frequencies are changed, especially at the A locus, and tightening of the linkage results in a 22 percent increase in mean fitness.

THE EFFECT OF LINKAGE ON FITNESS

Every case of linkage and selection that has been numerically investigated has shown a higher mean fitness at equilibrium with tight linkage. In the special symmetrical case of Lewontin and Kojima it is an analytical result that fitness is increased at equilibrium by $4\epsilon\hat{D}^2$.

TABLE 58

Equilibrium compositions of populations with different amounts of recombination between two loci having the fitness relations described below

Locus B	*Locus A*		
	11	*10*	*00*
11	0.5000	0.5000	0.3750
10	0.5625	1.0000	0.3125
00	0.3750	0.4375	0.3750

R	g_{00}	g_{01}	g_{10}	$\hat{g}_{11}$	p	r	D	D'	$\overline{W}$
.00	.55556	.00000	.00000	.44444	.55556	.55556	+.24691	+1.00000	.72223
	.00000	.50000	.50000	.00000	.50000	.50000	−.25000	−1.00000	.68750
.01									
	.01664	.48928	.48593	.00815	.50592	.50257	−.23762	− .96684	.67849
.02	.54063	.02385	.01668	.41884	.56448	.55731	+.22604	+ .93128	.70255
	.03563	.47750	.47063	.01624	.51313	.50626	−.22415	− .90940	.66738
.03	.53282	.03652	.02543	.40523	.56934	.55825	+.21499	+ .89423	.68779
	.05457	.46552	.45443	.02548	.52009	.50900	−.21016	− .89190	.65730
.05	.51637	.06352	.04396	.37615	.57989	.56033	+.19144	+ .81325	.67350
	.10201	.43688	.41605	.04506	.53889	.51860	−.17717	− .79727	.63669
.07									
	.16945	.39738	.36821	.07036	.56683	.53226	−.13225	− .65273	.61463
.075									
	.19509	.38244	.34280	.07967	.57753	.53789	−.11556	− .59187	.60815
.10	.46805	.14242	.09854	.29099	.61047	.56659	+.12216	+ .55351	.62830
.15	.41262	.21957	.15828	.20953	.63219	.57090	+.05170	+ .24621	.59970
.20	.38645	.24803	.18406	.18146	.63448	.57051	+.02447	+ .11734	.59356
.35	.36977	.26391	.19969	.16663	.63368	.56946	+.00891	+ .04271	.59138
.50	.36582	.26743	.20328	.16347	.63325	.56910	+.00544	+ .02606	.59101

Note: From Lewontin (1964a).

The most general results that have been proved are that (1) $R = 0$ is always a local maximum of fitness so that a slight loosening of linkage from $R = 0$ always results in a decrease of fitness, and (2) $R = 0$ is a global maximum so that no other value of recombination can ever result in a higher mean fitness at equilibrium (Lewontin, 1971). These two conditions together do not mean that for intermediate values of R a slight tightening of linkage will necessarily cause an increase in equilibrium fitness. There may be more than one local fitness maximum, and reducing the recombination slightly from one of these values will decrease fitness.

In the long run, the tightest possible linkage will give the highest possible fitness, so we expect, as Fisher suggested, that recombination values will evolve toward zero if there is any genetic variance for recombination. The experiments of Chinnici (1971a, b) and Kidwell (1972) show that there is a lot of variance for selection to operate on. Yet recombination has not been reduced to zero. We are left with the problem so succinctly put by Turner (1967), "Why does the genome not congeal?"

One possible answer is that genetic flexibility for adaptation to a fluctuating environment demands recombination to generate adaptive combinations different from those previously selected. There is a close relationship between this problem and the explanation of inversion homogeneity in marginal populations. Inversions restrict recombinations and result in a few well-defined genetic modes at high frequency. Free recombination allows the generation of a much greater array of adaptive possibilities. We must be very cautious about heuristic and seemingly reasonable arguments on this issue. The effects of linkage on selection in a *fluctuating* environment are complicated and depend strongly on the statistical properties of the environment, especially its serial autocorrelation. Only a beginning has been made in analyzing this problem (Levins, 1968). Some of its complexity is revealed in a study of selection for an intermediate optimum.

INTERMEDIATE OPTIMA

The interest in studying selection for an intermediate optimum phenotype is twofold. First, the intermediate optimum must be a fairly common form of selection, perhaps the most common. For each element of the morphological and behavioral phenotype, there is an appropriate range of sizes that is fittest in a given range of environments. It is seldom the largest or the smallest size that confers the highest fitness, although in the early stages of an evolutionary process the extreme forms that exist in the population are still far below the eventual optimum range. Like all other rules, $\mu\epsilon\delta\epsilon\nu\ \alpha\gamma\alpha\nu$ applies only in appropriate circumstances.

Second, selection for an intermediate optimum generates extreme nonadditivity of fitness, and there will be a great deal of interaction variance if the population is near the optimum. The effect of linkage on selection depends upon the coupling coefficient ϵ/R. Most schemes of selection generate rather low values of ϵ unless fitness differences are very extreme. For example, in the two-locus symmetrical model, if the fitnesses of the double homozygotes are 0.97 and of the single homozygotes 0.99, then $\epsilon = 0.01$ and there will be no effect of linkage on the equilibrium unless $R < 0.0025$, a very tight linkage indeed. On the other hand, because there is not a monotone relationship between gene action and fitness in optimum models, a relatively large amount of epistasis is generated for a given intensity of selection.

Most optimum models of selection do not lead to stable intermediate equilibria of gene frequency, contrary to what might be guessed intuitively. When more than a single locus affects a character,

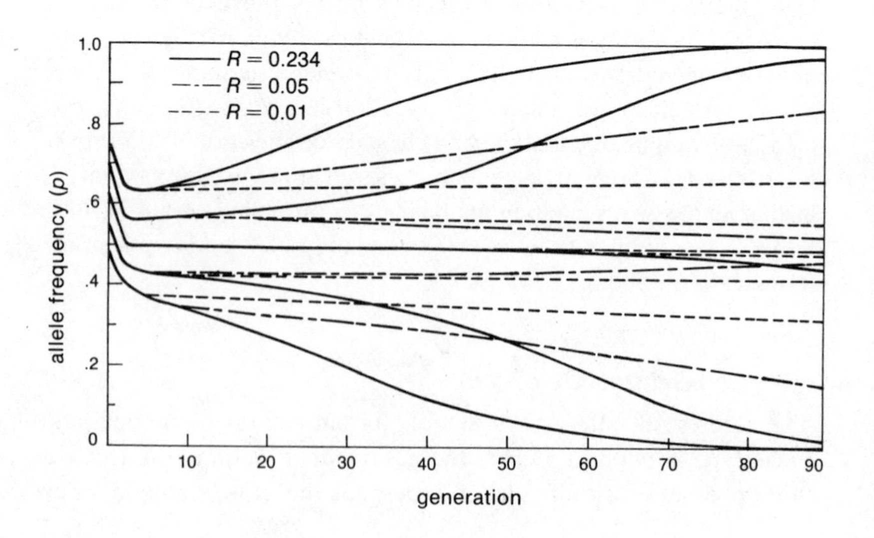

roughly the same intermediacy of phenotype can be obtained by a multiple homozygote $+-/+-$ as by a multiple heterozygote $+-/-+$, so that special circumstances are required to avoid fixation of the mixed homozygous type (Kojima, 1959). Even then a severely limited amount of genetic variance can be preserved at equilibrium, about equal to the squared additive effect of one locus (Lewontin, 1964b). Taking linkage into account makes a drastic difference in the outcome of selection for an intermediate optimum.

Figure 27 shows the evolution of a population in which five loci are affecting a phenotypic character. The loci are perfectly additive in their effect on the phenotypes, but only those genotypes survive whose phenotypes fall within an intermediate range. The solid lines in figure 27a are the paths of gene-frequency change if the loci are loosely linked to each other. Two loci go rapidly toward fixation at $p = 1.0$, two fix at $p = 0.0$, and the fifth remains segregating for a long time but slowly goes toward the lower fixation point. The final composition of the population will be a mixed homozygote $++---/++---$ that will be inside the acceptable range of phenotypes, and the population will have mean fitness of 1.0. The dotted and dashed lines in figure 27a are the paths of evolution when the recombination between adjacent loci is reduced to $R = 0.05$ and $R = 0.01$, respectively. The tight linkage causes a considerable slowing of the fixation process. There is a sort of quasi-equilibrium

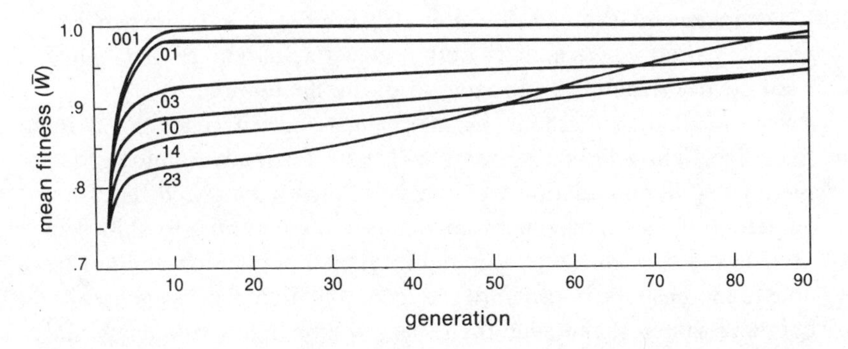

FIGURE 27

Evolutionary changes at five loci selected for an intermediate phenotype optimum. At left, gene frequency at each locus as a function of generation. Above, mean fitness of the population, $\bar{W}$, as a function of generations, for different values of R between adjacent loci.

of gene frequencies, especially when $R = 0.01$, with loci 2 and 4 actually converging *toward* $p = 0.50$ while loci 1 and 5 move slowly toward fixation. If the process were followed long enough, all the loci would eventually fix, but the rate would be exceedingly slow.

In contrast to the apparent lack of evolutionary change shown by the gene frequencies when linkage is tight, there is rapid evolution of mean fitness (figure 27b) as opposed to the slow increase of fitness when recombination is free. The rapid increase in fitness despite the quasi-equilibrium of gene frequencies is the result of the elimination of coupling gametes and the build-up of classes of complementary repulsion gametes like +−+−+/−+−+−. When all the gametes are of this kind, both heterozygotes +−+−+/−+−+− and homozygotes +−+−+/+−+−+ are of an intermediate phenotype, and the fitness of the population is high. The loosely linked system slowly overtakes the tightly linked ones and eventually surpasses them as the loci become fixed, because in the long run the fittest population is the monomorphic one.

Tight and loose linkage are thus alternative adaptive strategies. Tight linkage results in a rapid exhaustion of *actual* variance in fitness by selection of a few gametic types but preserves *potential* variance in fitness by stalling gene frequencies at intermediate values. Loose linkage increases fitness only slowly, leaving actual variance in fitness in the population for longer periods, but in the end will lead to a loss of both actual and potential variance through homozygosity. Tight linkage provides immediate rapid response, loose linkage gives greater long term response, yet at the expense of genetic variance. Which is better in a fluctuating environment? That clearly depends on the period of the fluctuation.

Suppose that an optimum is maintained for ten generations. By that time tightly linked systems will have responded almost completely and the population will consist of mostly balanced repulsion gametes. If the environment now causes the optimum to shift well outside the current range, this tightly linked population will not be able to increase its fitness until the recombination process generates coupling gametes, and even then these will be in low frequency and selection response will be slow. The loosely linked system will have made rather poor response during the ten generations, but it will be able to respond better to the new optimum because it has coupling gametes +++++ and −−−−− in fair abundance.

Alternately, consider a cycle 100 generations long. After 100 generations both tightly and loosely linked systems will have about the same fitness and both will consist almost entirely of repulsion gametes. But the tightly linked system will have complementary classes of repulsion gametes and will, by recombination, be able to produce the extreme combination $+++++$ or $-----$ required by the new environment, whereas the loosely linked system will be at gene fixation for alternate alleles and different loci and will not be able to generate any of the required extreme coupling types. Even this is only a glimpse of the immense variety of possible patterns that results from varying environment in a system of linked genes. An understanding of chromosomal polymorphisms in nature, for example, will demand a much more complete analysis.

MULTIPLE LOCI

The theory of linkage and selection for two loci, if it is taken literally, could not really cause us to revise our ideas about polymorphism. All though the discussion of the relation of theory to fact, I have emphasized the necessity of using realistic values for the intensity of selection at a locus. Selection intensities of 1 or 2 percent at a locus must mean weakly coupled systems unless the two genes are very close together on the linkage map. For example, if loci are multiplicative in their effects in fitness and each locus has as much as 5 percent heterosis (a great deal), the value of ϵ is only 0.0025 and the critical value R is then only 0.0006, an extremely close linkage.

In general, for multiplicative gene action the critical value of R is the product of the genetic loads of the two loci (Bodmer and Felsenstein, 1967). Linkage phenomena would appear to be only a curiosity of major polymorphisms, like mimicry or snail-shell patterns, or else a feature of the microstructure of the genome, not detectable or important for most pairs of interacting loci in the genome. In a book that claims to be asking questions about the real world, why I am so concerned with what appear, quantitatively, to be trivial effects? Because systems that are weakly coupled when considered in isolation may, in fact, be very strongly coupled when embedded in an aggregate of such systems.

The first indication of cumulative effects was in a study of five-locus heterotic models under strong selection (Lewontin, 1964a). Table 59 gives the equilibrium gametic composition, the relative

TABLE 59

Equilibrium gametic frequencies, linkage disequilibria, and fitnesses for a multiplicative cumulative heterotic model. Only 16 gametic types are shown, their complements having exactly equal frequencies

Gametes	*R between adjacent loci*									
	000	.01	.02	.03	.04	.05	.06	.063	.0645	.065
00000	.50000	.46199	.42053	.37444	.32183	.25904	.17488	.13627	.09817	.03125
00001	0	.01083	.02193	.03316	.04418	.05411	.05997	.05874	.05413	.03125
00010	0	.00016	.00074	.00201	.00438	.00863	.01675	.02119	.02567	.03125
00011	0	.00775	.01572	.02384	.03192	.03947	.04495	.04515	.04336	.03125
00100	0	.00010	.00048	.00133	.00299	.00611	.01254	.01642	.02087	.03125
00101	0	.00000	.00003	.00013	.00044	.00135	.00458	.00754	.01213	.03125
00110	0	.00013	.00061	.00166	.00363	.00723	.01443	.01869	.02344	.03125
00111	0	.00775	.01572	.02384	.03192	.03947	.04497	.04515	.04336	.03125
01000	0	.00016	.00074	.00201	.00438	.00863	.01675	.02119	.02567	.03125
01001	0	.00000	.00003	.00015	.00050	.00155	.00524	.00859	.01370	.03125
01010	0	.00000	.00000	.00001	.00006	.00029	.00164	.00341	.00700	.03125
01011	0	.00000	.00003	.00013	.00044	.00135	.00458	.00754	.01213	.03125
01100	0	.00013	.00061	.00166	.00363	.00723	.01443	.01869	.02344	.03125
01101	0	.00000	.00003	.00015	.00050	.00155	.00524	.00859	.01370	.03125
01110	0	.00019	.00088	.00234	.00504	.00985	.01905	.02410	.02912	.03125
01111	0	.01083	.02193	.03316	.04418	.05411	.05997	.05874	.05413	.03125
D'_{12}	1.00000	.95476	.90300	.84164	.76508	.66172	.49236	.39660	.28448	0
D'_{13}	1.00000	.92352	.83888	.74296	.63072	.49236	.29912	.20836	.11928	0
D'_{14}	1.00000	.89284	.77752	.65208	.51380	.35836	.17452	.10408	.04720	0
D'_{15}	1.00000	.85140	.69840	.54184	.38376	.22808	.08192	.03984	.01344	0
D'_{23}	1.00000	.96744	.92944	.88300	.82260	.73604	.58104	.48584	.36680	0
D'_{24}	1.00000	.93572	.86316	.77876	.67656	.54488	.34880	.25104	.15068	0
D'_{25}	1.00000	.89284	.77752	.65208	.51380	.35836	.17452	.10408	.04720	0
D'_{34}	1.00000	.96744	.92944	.88300	.82260	.73604	.58104	.48584	.36680	0
D'_{35}	1.00000	.92352	.83888	.74296	.63072	.49236	.29912	.20836	.11928	0
D'_{45}	1.00000	.95476	.90300	.84164	.76508	.66172	.49236	.39660	.28448	0
W	.49500	.45688	.41927	.38203	.34491	.30738	.26720	.25240	.24021	.22781

Note: From

linkage disequilibria D' between all pairs of genes, and the mean fitness, for various amounts of recombination R between *adjacent* genes in a simple, symmetrical, multiplicative model of heterosis in which each added locus that is homozygous cuts the fitness of the genotype in half. That is, the quintuple heterozygote has a fitness of 1.0, a quadruple heterozygote has a fitness of 0.50, triple heterozygotes 0.25, and so on.

Three important phenomena are seen in this table. First, loci far apart on the chromosome are held out of linkage equilibrium with each other by the loci between them. Locus 1 and locus 5 are 20 crossover units apart when R between adjacent loci is 0.05. If these two loci were considered in isolation there would be no effect of linkage on them at all. The critical linkage distance for this model of selection is $R = 0.0625$, so loci 1 and 5 are much too far apart. Yet $D'_{15} = 0.38$ when they are 20 units apart, held in association with each other by the cumulative correlations between adjacent loci along the length of the chromosome.

Second, the critical distance even for adjacent loci in the five-locus case is slightly larger than predicted from two-locus theory (0.0650 as opposed to 0.0625).

Third, loci in the middle of the chromosome are more out of random association with each other than pairs at the ends of the chromosome. For example, when $R = 0.05$, $D'_{12} = 0.66172$ but $D'_{23} = 0.73604$, even though the amount of recombination and the fitness relations are the same for both pairs.

These last two effects, although small, are the first indication of an *embedding effect*, quite different from the cumulative effect first discussed. It is to be expected that loci at opposite ends of the chromosome may be kept out of linkage equilibrium by a sort of chain of correlation between adjacent loci connecting them. The embedding effect is quite another matter. A pair of adjacent loci have a linkage disequilibrium between them that is magnified by being surrounded by loci that interact with them. This embedding phenomenon is much stronger than the small location effect seen in table 59. If we compare the value of D' calculated for a pair of loci embedded in a five-locus segregation with the value predicted for that pair from two-locus theory, the result is remarkable. This comparison is made in table 60 and figure 28. When they are embedded in a five-locus, interacting segregation, the two loci have a higher linkage disequili-

TABLE 60

Comparison of the relative linkage disequilibrium for two adjacent loci R units apart as predicted from two-locus theory with the disequilibrium observed when they are embedded in a five-locus segregation

R	*Five-locus segregation*	*Two-locus prediction*	*Ratio*
0.00	0.967	0.916	1.06
0.01	0.929	0.825	1.13
0.02	0.883	0.721	1.22
0.03	0.823	0.600	1.37
0.04	0.736	0.447	1.64
0.05	0.581	0.200	2.91
0.063	0.481	0	∞
0.0645	0.367	0	∞

Note: From Franklin and Lewontin (1970).

brium than predicted from two-locus theory, and they lose their correlation much more slowly with increasing recombination.

The observation of these interactive effects of complex systems involving only a few loci led Franklin and Lewontin (1970) to study systems with many more loci and with much weaker selection at each locus, as an approach to a realistic model of the genome. They found that with weak epistatic interactions and long chromosome segments, very close correlation among all the loci in the segment was maintained. Consider a 36-locus model in which each locus is identical, has a 10 percent heterotic effect, and multiplies this effect by that at other loci. The product of genetic loads at a pair of such loci is $(0.05)^2 = 0.0025$, the critical value of R between adjacent loci according to 2-locus theory, above which the loci should be uncorrelated at equilibrium. In fact, the critical value of R in the 36-locus case turns out to be 0.01, four times higher than predicted, which means a total recombination length of 35 percent of the whole segment. In *Drosophila*, in which there is no recombination in males, this is a genetic map length of 70 units, more than the total length of one chromosome arm.

If the actual recombination between adjacent loci is much smaller than 0.01, the entire chromosome segment is "locked up" in one pair of complementary gametic types at equilibrium. For example, if $R = 0.003$, the entire map is 10.5 recombination units long (21 centimorgans in *Drosophila,* about 20 percent of a chromosome), and

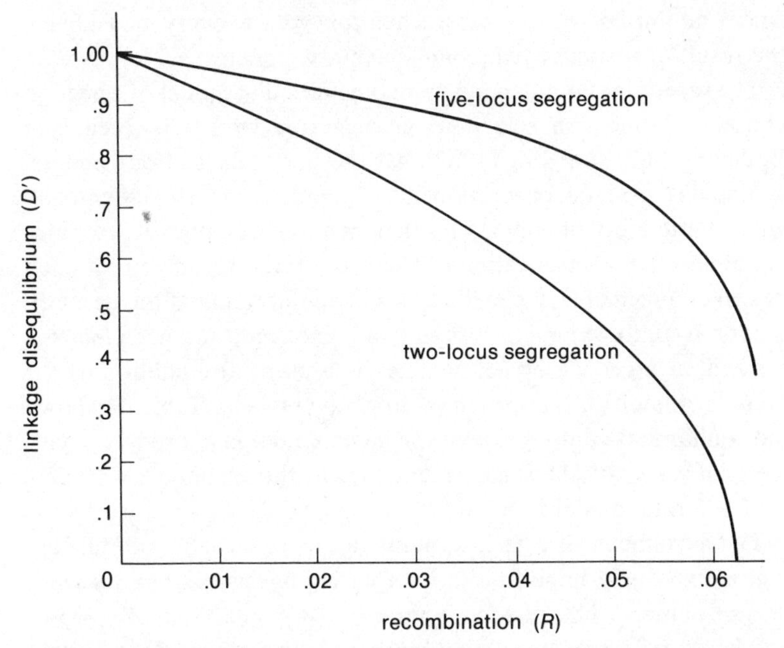

FIGURE 28

Comparison of the observed linkage disequilibrium between two genes R units apart when embedded in a five-locus chromosome with the predicted value from two-locus theory.

85 percent of the population of gametes consists of only two complementary gametic types, for instance:

001 100 100 011 ... 111

and

110 011 011 100 ... 000

the other 15 percent being derivable from these by a few recombinations. The average value of D' between adjacent pairs is 0.96 in these circumstances.

The effect of locking up the genome into only a few gametic types at equilibrium is a drastic increase in fitness over that predicted from independent gene theory. If each of 36 loci is 10 percent heterotic, the fitness of an equilibrium population in which all the loci are distributed at random with respect to each other is only $(0.95)^{36} = 0.158$. The actual fitness of a population in which 85 percent of the gametes are of two complementary classes is about 0.40, since one-third of the zygotes are heterozygous at every locus, being the result of fusion of two complementary gametes.

The result does not depend upon the particular model of selection chosen. Truncation selection, as suggested by Sved, Reed, and Bodmer (1967) and King (1967), has the same effect. For example, setting the surviving proportion of the population to 20 percent, and with a heritability of only 0.1 for the phenotypic character determining fitness, King's truncation model produced 40.5 percent of each of two complementary genetic types at equilibrium. If the assumption of absolute symmetry of the two homozygotes at each locus is relaxed, a larger variety of gametes is present at equilibrium, but there is still a high proportion of only a few types. Table 61 shows the equilibrium composition of the gamete pool in a strongly asymmetrical case with the same genetic load as the symmetrical 10 percent heterotic model.

The asymmetrical case also points up a second effect of linkage that is extremely important in interpreting observed gene frequencies in nature. The one-locus fitnesses $W_{11} = 0.9375$, $W_{10} = 1.0$, and $W_{00} = 0.75$ predict an equilibrium gene frequency of the *1* allele of 0.80 at each locus. But the observed gene frequency, on the average over loci, is only 0.69. It is a general rule in multilocus systems that the equilibrium gene frequencies are closer to 0.5 than is predicted from single-locus kinetics. The reason is easy to see. As linkage tightens, the multilocus chromosome becomes a single "super locus" with "pseudo-alleles." If these are two complementary pseudo-alleles, one with a proportion *k* of its alleles of the *1* type

TABLE 61

Gametic array at equilibrium from a numerical simulation of a 36-locus model with $W_{11} = 0.9375$, $W_{01} = 1.0$, $W_{00} = 0.75$ at each locus and $R = 0.003$ between adjacent loci

Gamete	*Frequency*
111 010 110 101 110 101 001 011 110 101 101 101	0.265
000 111 001 010 101 110 110 111 101 111 010 110	0.134
000 111 001 010 101 110 110 111 101 111 010 111	0.108
111 101 111 101 111 011 111 100 111 010 111 011	0.111
All others	0.392

Note: From Franklin and Lewontin (1970).

and $1 - k$ of the *0* type, while the other has the reverse frequencies, the equilibrium of the two pseudo-alleles from one-locus selection theory is

$$\hat{q} = \frac{1 - W_1}{2 - (W_1 + W_0)} = \frac{1 - (W_{11}^{k}\, W_{00}^{1-k})^n}{2 - (W_{11}^{k}\, W_{00}^{1-k} + W_{11}^{1-k}\, W_{00}^{k})^n}$$

so that

$$\lim_{n \to \infty} \hat{q} = 1/2 \text{ if } 0 \lesseqgtr W_{11}, W_{00} < 1$$

This means that tight linkage and strong linkage disequilibrium increase the degree of heterozygosity at each locus from that predicted by considering the loci individually. Moreover, relative selective values of substitution at a locus cannot be judged from the frequencies of the alleles in nature because *the selection of the chromosome as a whole is the overriding determinant of allelic frequencies.*

THE NUMBER OF EQUILIBRIA

For the simplest, totally symmetrical, two-locus model, Lewontin and Kojima (1960) and Bodmer and Parsons (1962) showed that there are three possible equilibria with both loci segregating, one with $D = 0$ and two complementary equilibria with $D \neq 0$. Which of these equilibria is stable depends upon R. If R is greater than its critical value, then the linkage equilibrium solution is stable, whereas if R is less than its critical value, the pair with $D \neq 0$ is stable. But $D \neq 0$ and $D = 0$ are mutually exclusive and cannot both be stable for a given value of R. Depending upon the amount of recombination, every population will evolve to linkage dependence or linkage independence. There is no uncertainty.

Karlin and Feldman (1970) later showed that for a slightly more general, symmetrical fitness model which includes the Lewontin and Kojima model as a special case, there are four additional stable equilibria possible with both loci segregating. Again, for a fixed value of R either the linkage equilibrium or the linkage disequilibrium solution is stable, but not both. Numerical solutions of multilocus problems have shown that for more than two loci the mutual exclusion of the linkage disequilibrium and linkage equilibrium solutions does not apply (Franklin and Lewontin, 1970; A. Beiles, personal communication, 1972). For some sets of recombination values between loci, a population may evolve to either $D = 0$ or $D \neq 0$, depending upon its initial conditions. When R is large, only $D = 0$ is stable; when R is very small, only $D \neq 0$ is stable; and when R is intermediate, both equilibria are stable. This result has been obtained analytically for three loci by Feldman, Franklin, and Thomson (1973).

The total number of possible equilibria for which all loci are segregating is remarkably large, and if fitnesses are asymmetrical these alternative stable states will correspond to different population compositions with different mean fitnesses. Whether they are detectably different depends upon the total variance in fitness, but, at least as far as future evolution is concerned, they represent different potential histories.

I. Franklin (personal communication, 1972) has derived an iterative procedure for finding the total number of equilibria with all loci segregating, with extraordinary results. While two-locus systems have 7 potential equilibria, three-locus systems have 193, four-locus systems have 63,775, and five-locus systems have 4,294,321,153! Unfortunately we do not yet know the conditions for stability of all these equilibria, but our analysis of numerical cases for five and more loci suggests that a large proportion are stable for reasonably small values of recombination (Franklin and Lewontin, 1970; A. Beiles, personal communication, 1972). The sorting out of this immense complexity is a major task of theoretical work in the future.

PASSING TO THE LIMIT

It is unrealistic to postulate larger and larger numbers of loci affecting fitness, while holding the effect of each locus constant. There might be as many as 36 loci on a chromosome with effects as large as 10 percent on fitness, but there cannot be 3600. Yet there are hundreds of loci segregating in fairly short sections of the genome. If 40 percent of all loci are segregating in Drosophila, a minimum estimate, and if there are only 10,000 loci, a conservative estimate, there are 4000 segregating loci distributed along about 300 recombination map units, or about 13 segregating loci per centimorgan, making the recombination distance between adjacent loci

0.0008. A model of the genome that includes so many segregating loci must assume that the marginal fitness effect of a gene substitution is small, if for no other reason than to arrive at a realistic figure for inbreeding depression.

A reasonable way to investigate the effect of increasing the number of loci is to fix the total effect of homozygosity of a chromosome, its total inbreeding depression, and to ask what happens as more and more loci are packed into this chromosome of fixed homozygous fitness and fixed recombination length. Two competing effects result from this process. As more and more loci are assigned to the chromosome, the recombination between adjacent loci, R, grows smaller linearly with n, the number of loci. In addition, the marginal fitness effect of each locus decreases. If $1 - s$ is the marginal effect of making one locus homozygous and K is the fitness of a totally homozygous chromosome, then

$$K = (1 - s)^n$$

in a multiplicative model, so that for small s,

$$s \cong -\frac{\ln K}{n}$$

and the selection intensity decreases linearly with n. The epistatic deviation ϵ is equal to s^2, so ϵ decreases as n^2. The effect of linkage and selection depends roughly on the coupling coefficient ϵ/R, which in this case is proportional to $(1/n^2)/(1/n) \sim 1/n$. On this basis we would predict that as more and more loci are packed into a fixed chromosome length, with a fixed total inbreeding depression, the loci will interact more and more weakly with each other and their correlation will become increasingly small as n grows. We have also shown, however, that there is an embedding effect that causes an intensification of the coupling between two loci from interaction with all other loci. The number of these interactions grows factorially with n, and this might be sufficient to counterbalance the predicted decrease of the coupling coefficient. What actually happens?

By a combination of algebraic, numerical, and simulation techniques, Franklin and Lewontin (1970) calculated the equilibrium compositions of populations in which 2 to 360 loci were placed on chromosome segments of fixed lengths and with fixed inbreeding depressions. Table 62 shows the equivalence between K, the

TABLE 62
Fitness of a homozygote at a single locus among *n* loci when the fitness of the *n*-tuple homozygote equals *K*

	K	
n	*0.0225*	*0.4832*
2	0.1501	0.6951
5	0.4683	0.8646
18	0.8100	0.9604
36	0.9000	0.9800
360	0.9895	0.9980

Note: From Franklin and Lewontin (1970).

homozygous fitness of the whole chromosome segment, and $1 - s$, the marginal fitness of a single-locus homozygote. For large numbers of loci the effect of a single gene substitution is extremely small and the epistatic effect, being proportional to s^2, is vanishingly small. The equilibrium compositions corresponding to these different cases are shown in figures 29a and 29b, drawn so that the 2-locus relation is linear. The measure of linkage disequilibrium is $\bar{D}'$, the average relative disequilibrium over *all* pairs of loci, including those close together and those far apart on the chromosome. It is thus an underestimate of the intensity of correlation between loci that are in short, contiguous sections.

As it should, the disequilibrium between loci falls off as the total map length over which the loci are distributed increases. A rise from 2 to 5 loci results in some decrease in correlation between loci at each total map length. Dividing the total chromosome among more loci has had a weakening effect on the interaction. A further subdivision of chromosomes into 18 loci again decreases the interactions. But when the number of loci is doubled to 36, no further change occurs, and even 360 loci are indistinguishable from 18 and 36. As the number of loci increases beyond a mere handful, *the average correlation structure of the genome at equilibrium is totally independent of the number of genes or their individual effects and depends solely on the total map length of the chromosome segment and the total inbreeding depression.*

This limiting invariance of the equilibrium structure has been confirmed by an approximate analytic treatment of the problem (Slatkin, 1972) that also allows the evaluation of the relative linkage disequilibrium $D'(x, y)$ between two loci at points x and y along the chromosome length. The correlation falls off exponentially with distance between the loci

$$D'(x, y) = e^{-(y - x)/l_c(x, y)}$$

where $l_c(x, y)$ is the *characteristic length* familiar in the treatment of

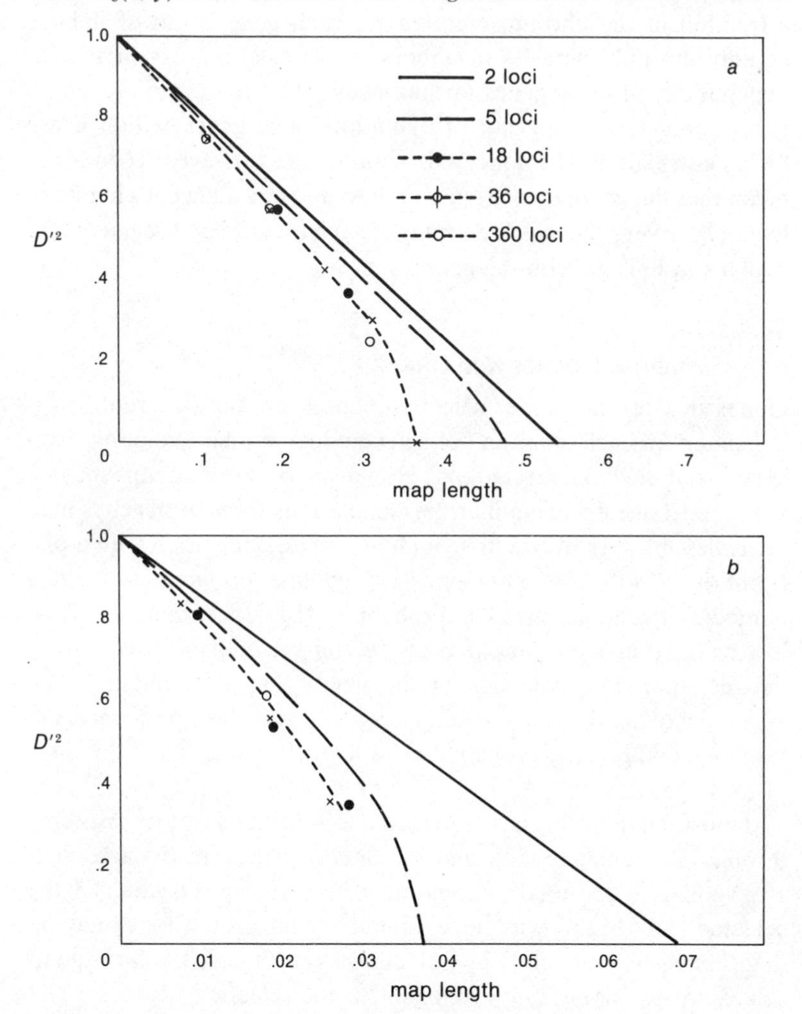

FIGURE 29

The relation between the average relative linkage disequilibrium D' between *all* pairs of genes, and the total map length for different numbers of loci making up the map. *a*: strong selection, K=0.0225. *b*: weak selection, K = 0.4832. From Franklin and Lewontin (1970).

coupled systems. It is the distance over which the coupling is effective and can be calculated from the inbreeding depression of the total chromosome segment and the map length. When the characteristic distance is equal to or greater than the entire length of the chromosome segment being considered, then all the loci are highly correlated and only a few gametic types are present. When l_c is only a fraction of the chromosome length, each gene is out of linkage equilibrium only with its neighbors but is assorted essentially independently of other genes farther away. The characteristic length is, in some sense, the unit of evolution since genes within it are highly correlated. The concept is a subtle one, however. It does not mean that the genome is broken up into discrete adjacent chunks of length l_c. Every locus is the center of such a correlated segment and evolves in linkage with the genes near it.

NEUTRAL GENES AND LINKAGE

Genes that are not under selection should, in the long run, be independent of each other and of selected loci in their evolution, irrespective of the linkage relations. However, "in the long run" may be very long indeed, and equilibrium calculations for neutral genes may be irrelevant. Mutations arise and, in a finite population with a distribution of offspring numbers, will be lost or fixed in varying numbers of generations. The probability that a new mutant will be lost or fixed and the time it takes to run the course of its history depend upon the dynamics of the genes that surround it. The steady-state distributions of heterozygosity and allele frequencies in turn depend upon the probabilities of fixation and loss and the rates of these processes.

If most polymorphisms are indeed neutral, but a few loci sprinkled through the genome are under selection, the steady-state heterozygosity of the total genome may be a strong reflection of the selected loci. In this way the evolutionary fate of most loci may be determined or influenced by selection even though the loci themselves are physiologically irrelevant to the selection process. The phenomenon of linkage forces us to define much more exactly what we mean when we say that a locus is "controlled" by selection.

First, a locus may appear to be under direct selection just because it is currently out of linkage equilibrium with a selected locus. If the selected locus is overdominant, the the value of this "associative overdominance" (Frydenberg, 1963) will be sD'^2 if both loci are near 0.5 in gene frequency, where s is the selection intensity at the overdominant locus (Ohta and Kimura, 1970; Franklin and Lewontin, 1970). It is not likely that D'^2 will be large for a neutral locus segregating at intermediate frequency in a population that has been at a large size for a long time, because such a neutral allele during its random rise to intermediate frequency will have become uncorrelated with the selected locus recombination. But if a population has been founded recently from a small colony, a considerable linkage disequilibrium may be induced by random sampling. Laboratory experiments that pretend to measure selection at an allozyme locus by watching gene frequency changes in newly formed laboratory populations are completely misleading unless there is some assurance that the locus in question does not have a high D' with some selected locus or gene bloc. That is why a large number of independently derived lines of each allele from nature must be used to initiate laboratory experiments.

Second, associative overdominance will be a permanent feature of neutral loci in a population of restricted size. Ohta and Kimura showed that if overdominant loci are widely enough spaced through the genome so that coupling between them is too weak to produce a supergene, there will nevertheless be associative overdominance of neutral genes located between them. The intensity of the associative overdominance s' for a locus embedded in a group of truly overdominant genes is approximately

$$s' = \frac{s}{4NR}(1 + \ln n_1 n_2)$$

where s is the overdominance at the selected loci, R is the distance between adjacent genes, n_1 and n_2 are the number of overdominant genes to the right and left of the neutral locus, and N is the effective population size. So, for example, under the classical hypothesis a chromosome that is 50 units long might have 20 overdominant loci. In a population of 1000, a neutral gene in the middle of the chromosome would have an apparent selection of 0.056s. In populations much larger than this the effect would be negligible.

Third, there is the Hill-Robertson effect (Hill and Robertson, 1966). The rate of fixation of a neutral gene and the probability of eventual fixation of a favorable gene are both a function of the effective population size. A selected gene has a probability of eventual fixation of $N_e s$, whereas a neutral gene drifts at a rate of $1/2N_e$. But effective population size is itself sensitive to the variance in number of offspring. A large variance means that many genes are descended from a few ancestral genes, so drift is more effective. But selection causes an increase in variance of offspring number, some alleles leaving more on the average than others. Any locus that is out of linkage equilibrium with a selected locus will then also have an increase in the variance of its offspring number, just because of the correlation with the first selected locus. Genetic drift will thus be more important for the second locus than it would have been had the locus been independent. If an allele at the second locus is advantageous, it will then have a *reduced* probability of eventual fixation because $N_e s$ will be smaller. If the second locus is neutral, it will drift more rapidly for the same reason.

OBSERVATIONS OF LINKAGE DISEQUILIBRIUM

There is an increasing body of direct evidence for the correlated states of loci. The most direct is the analysis of the three third-chromosome loci in *Drosophila pseudoobscura* by Prakash and Lewontin (1968, 1971). Particular alleles at these loci are almost constantly restricted to particular inversions so that the loci are strongly out of linkage equilibrium with each other (p. 135). Inversion heterozygosity strongly suppresses recombination and in *Drosophila pseudoobscura* does so even outside the limits of the inversion (Dobzhansky and Epling, 1948). The characteristic correlation length of the third chromosome is then likely to include the entire element if there is any selection at all of the loci.

The observed associations cannot be explained by a random deviation from linkage equilibrium, because the same inversions carry the same alleles no matter from what part of the species range they are drawn. Moreover the association between gene and inversion transcends the species boundary and so must antedate the separation of *D. pseudoobscura* and *D. persimilis*. These observations are the direct genetic confirmation of Dobzhansky's theory of the coadaptation of the allelic contents of inversions. Our finding that three randomly chosen allozyme polymorphisms are strongly out of linkage equilibrium, and that the disequilibrium is identical in all populations of the species, is *prima facie* evidence that the loci are under natural selection. They need not be under strong natural selection as individual loci, as I have shown, because if recombination is low the coupling coefficient between genes may be very large without large differences in marginal fitness at its separate loci. *The observation of significant linkage disequilibrium that is consistent between populations is a very sensitive detector of natural selection.*

There is an inversion polymorphism on the X chromosome of *D. pseudoobscura*, the "sex-ratio" polymorphism. Prakash and Merritt (1973) have found strong association of two sex-linked loci, *esterase-5* and *acid phosphatase-6*, with the inversion polymorphism. In both California and Colorado the sex-ratio inversion is 85 to 90 percent $est\text{-}5^{1.04}$ and 100 percent *aph-6*(−), but the standard chromosome is only 1 percent $est\text{-}5^{1.04}$ and 60 to 70 percent *aph-6*(−).

Consistency between populations is, unfortunately, necessary if the interpretation is to be unambiguous. Different intensities and different signs of D in different populations may easily be the result of selection since there are so many alternative equilibria possible, but they may also be a result of random drift, which gives rise to linkage disequilibrium from sampling of gametes. For gene frequencies near 0.50, Sved (1968) showed that

$$D'^2 = \frac{1}{4NR + 1}$$

on the average. D' will sometimes be positive and sometimes negative. So, for genes recombining 1 percent of the time in a population of 1000, $D'^2 = 0.024$ and $D' = \pm 0.156$. If two loci are in a "coupling" disequilibrium in some populations and a "repulsion" equilibrium in others, one would need to show that they are of too large an effective size for the observed disequilibrium to be explained by drift.

Drift cannot explain disequilibrium of the same sign in all populations, because it would be necessary to invoke migration between populations that is sufficient to make them all one large population. But if that were the case, N would be too large for linkage disequilibrium to occur by genetic drift. The populations cannot be both small enough to cause random linkage disequilibrium and large enough to exchange migrants to prevent differentiation in the nature of the linkage association.

Several other examples of associations between allozyme polymorphisms and inversions are known. M. Loukas and C. B. Krimbas (personal communication, 1972), found an association between esterase alleles and inversions on chromosome E of two Greek populations of *Drosophila subobscura*. In *Drosophila pavani* the standard arrangement of chromosome arm IV-L is 100 percent of one allele at an alkaline phosphatase locus, while an inverted arrangement is segregating at about 0.5:0.5 for two other alleles at this locus (Nair and Brncic, 1971). Four genes on the second chromosome of *Drosophila melanogaster* with about 20 map units between adjacent loci were found to be in linkage equilibrium in a natural population in two successive years, but the alleles at two of the loci, *alcohol dehydrogenase* and *amylase*, were significantly associated with short inversions in both years (Mukai, Mettler, and Chigusa, 1971).

Evidence for significant and consistent linkage disequilibrium between loci when inversions are not involved has proved harder to come by. Charlesworth and Charlesworth (1973) have made a large-scale attempt in *Drosophila melanogaster*, with some success. They studied five allozyme loci from the middle of chromosome III, with the following genetic map:

est-6—6.6—*pgm*—2.9—*aph*—5.7—*xdh*—4.6—*ao*

Since males in *Drosophila* have no recombination, these distances should really be cut in half to estimate the actual recombination in the population. Three population samples were taken, two in successive years from a wild population and one from a population cage established from several hundred lines from a different locality. Of the 30 possible comparisons, 4 showed significant linkage disequilibrium. If a pair was significant in one sample, then its *D* value in the other two samples was greater than the avearge of all *D* values, so there is consistency in the data. Nevertheless, the results are not strong enough to make an airtight case. Over 400 gametes were scored with respect to all five loci, a considerable effort. Clearly such attempts must be repeated because their outcome bears directly and critically on the classical and balanced hypotheses of population structure.

RELATIONSHIP OF THEORY TO OBSERVATIONS

Multilocus theory has led us in a curious circle. We began with a theory of population genetics that dealt with alternative alleles at single loci, and we now have considerable evidence about the frequency of alleles at various loci in natural populations. To bring together the theory of single loci with observations about single loci requires the measurement of parameters of single loci, especially the fitnesses of single-locus genotypes. But it is unlikely that these parameters are measurable to the degree of accuracy that theory requires.

A consideration of two or more loci, without the complication of linkage, made the problem even worse. Because of interactions, the number of dimensions rose to n, the number of loci, and the number of parameters to be measured rose exponentially with n. Moreover, since all relative fitnesses are somehow to be crammed into the interval 0–1, the more fitnesses to be measured, the more difficult it is to distinguish one from another even in their relative orders. It might be possible to order the 3 fitnesses of the three genotypes at one locus with two alleles, although the problems of estimation are bad enough, but what faith do we have in the ordering between 0 and 1 of the 27 fitnesses for a three-locus segregation?

Bringing in linkage created a problem of dimensionality beyond any ability to deal with it. Interactions of loci force the dimensionality to 2^n, but at least the number of fitnesses to be measured is not made greater by the complication of linkage. How can we ever hope to understand the evolution of a whole genome if for just five loci there are 4,294,321,153 possible equilibrium conditions in 32 dimensions? The introduction of linkage, while terribly interesting and exciting from a theoretical standpoint, made a predictive theory an absurdity. Yet suddenly, when pushed to its extreme, the dimensionality collapsed into a new space of description with only a few variables. *Moreover, these variables are the very ones that can be measured with reasonably high accuracy.*

The equilibrium structure (we cannot yet tell about the dynamical structure) of complex genetic systems is a function only of the effect of homozygosity of whole chromosome segments and of the genetic map length of these segments. These are precisely the parameters that we can estimate. Measuring map length is trivial. The effect of inbreeding chromosome segments of various lengths is measurable because whole chromosome segments can be made homozygous, unlike single loci, and the inbreeding depressions are large enough to be measured, unlike the situation for single loci. The variable that is to be related to these parameters, the genetic correlation between loci at various points on the chromosome, can be directly measured by counting *gametic combinations* for pairs of loci that are studied by electrophoresis.

The relationship between the astronomical dimensionality of the gametic ensemble and the severely reduced dimensionality of the limiting description, in terms of measurable variables, is analogous to the situation in thermodynamics. A complete description of a gas in phase space is utterly hopeless because the dimensionality of the phase space is too great, but the macroscopic law

$$PV = nRT$$

collapses the dimensionality of the phenomenon into three eminently measurable dimensions—pressure, temperature, and volume.

The significance of the transformation of our problem from a micro- to a macro-description does not lie only in what can be measured, but also in what exists. Even if it were possible to randomize the alleles at a single locus with respect to the rest of the genome and then to measure the marginal fitnesses of the alternative genotypes at that locus to an arbitrary level of accuracy, it would be a useless occupation. Genes in populations do not exist in random combinations with other genes. The alleles at a locus are segregating in a context that includes a great deal of correlation with the segregation of other genes at nearby loci. The fitness at a single locus ripped from its interactive context is about as relevant to real problems of evolutionary genetics as the study of the psychology of individuals isolated from their social context is to an understanding of man's sociopolitical evolution. In both cases context and interaction are not simply second-order effects to be superimposed on a primary monadic analysis. Context and interaction are of the essence.

BIBLIOGRAPHY

Alexander, M. L. 1949. Note on genetic variability in natural populations of Drosophila. Univ. Texas Publ. 4920:63–69.

Alexander, M. L. 1952. Gene variability in the *americana-texana-novamexicana* complex of the *virilis* group of Drosophila. Univ. Texas Publ. 5204:73–105.

Allison, A. C. 1955. Aspects of polymorphism in man. Cold Spring Harbor Symp. Quant. Biol. 20:239–55.

Altukhov, Yu. P., E. A. Salmenkova, V. T. Omelchenko, G. D. Satchko, and V. I. Slynko. 1972. (The number of monomorphic and polymorphic loci in populations of the salmon, *Oncorynchus keta*—one of the tetraploid species [in Russian]). Genetika 8:67–75.

Anderson, W. W., C. Oshima, T. Watanabe, Th. Dobzhansky, and O. Pavlovsky. 1968. Genetics of natural populations. XXXIX. A test of the possible influence of two insecticides on the chromosomal polymorphism in *Drosophila pseudoobscura*. Genetics 58:423–34.

Ayala, F. J., and J. R. Powell. 1972. Allozymes as diagnostic characters of sibling species of *Drosophila*. Proc. Nat. Acad. Sci. U.S. 69: 1094–96.

Ayala, F. J., J. R. Powell, and Th. Dobzhansky. 1971. Polymorphisms in continental and island populations of *Drosophila willistoni*. Proc. Nat. Acad. Sci. U. S. 68:2480–83.

Ayala, F. J., C. A. Mourão, S. Perez-Salas, R. Richmond, and Th. Dobzhansky. 1970. Enzyme variability in the *Drosophila willistoni* group. I. Genetic differentiation among sibling species. Proc. Nat. Acad. Sci. U.S. 67:225–32.

Ayala, F. J., J. R. Powell, M. L. Tracey, C. A. Mourão, and S. Perez-Salas. 1972. Enzyme variability in the *D. willistoni* group. IV. Genic variation in natural populations of *D. willistoni*. Genetics 71:113–39.

Baker, C. M. A. 1968. Molecular genetics of avian proteins. IX. Interspecific and intraspecific variation of egg white proteins of the genus *Gallus*. Genetics 58:211–26.

Baker, C. M. A., and C. Manwell. 1967. Molecular genetics of avian proteins. VIII. Egg white proteins of the migratory quail *Coturnix coturnix*—new concepts of "hybrid vigour." Comp. Biochem. Physiol. 23:21–42.

Baker, C. M. A., C. Manwell, R. F. Labisky, and J. A. Harper. 1966. Molecular genetics of avian proteins. V. Egg, blood and tissue proteins of the ring-neck pheasant, *Phasianus colchicus* L. Comp. Biochem. Physiol. 17:467–99.
Barlow, G. H. and E. Margoliash. 1966. Electrophoretic behavior of mammalian-type cytochrome-c. J. Biol. Chem. 241:1473–77.
Bell, A. E., C. H. Moore, and D. C. Warren. 1955. The evaluation of new methods for the improvement of quantitative characteristics. Cold Spring Harbor Symp. Quant. Biol. 20:197–212.
Bennett, J. 1960. A comparison of selective methods and a test of the preadaptation hypothesis. Heredity 15:65–77.
Berger, E. M. 1970. A comparison of gene-enzyme variation between *D. melanogaster* and *D. simulans*. Genetics 66:672–83.
Berger, E. M. 1971. A temporal survey of allelic variation in natural and laboratory populations of *D. melanogaster*. Genetics 67:121–36.
Bodmer, W. F. 1968. Demographic approaches to the measurement of differential selection in human populations. Proc. Nat. Acad. Sci. U.S. 59:690–99.
Bodmer, W. F., and J. Felsenstein. 1967. Linkage and selection: Theoretical analysis of the deterministic two locus random mating model. Genetics 57:237–65.
Bodmer, W. F., and P. A. Parsons. 1962. Linkage and recombination in evolution. Advan. Genet. 11:1–100.
Breese, E. L., and K. Mather. 1960. The organization of polygenic activity within a chromosome in Drosophila. II. Viability. Heredity 14:375–99.
Brewbaker, J. 1964. *Agricultural Genetics*. Prentice-Hall, Englewood Cliffs, N.J.
Brown, W. L., and E. O. Wilson. 1956. Character displacement. Syst. Zool. 5:49–64.
Brues, A. M. 1954. Selection and polymorphism in the ABO blood groups. Amer. J. Phys. Anthropol. 12:559–97.
Bulmer, M. G. 1971. Protein polymorphism. Nature 234:410–11.
Bundgaard, J., and F. B. Christiansen. 1972. Dynamics of polymorphisms: I. Selection components in an experimental population of *Drosophila melanogaster*. Genetics 71:439–60.
Burla, H., A. B. da Cunha, A. R. Cordeiro, Th. Dobzhansky, C. Malagolowkin, and C. Pavan. 1949. The *willistoni* group of sibling species of Drosophila. Evolution 3:300–314.
Cain, A. J., and P. M. Sheppard. 1950. Selection in the polymorphic land snail *Cepaea nemoralis*. Heredity 4:275–94.
Cain, A. J., and P. M. Sheppard. 1954. Natural selection in *Cepaea*. Genetics 39:89–116.
Carson, H. L. 1958. The population genetics of *Drosophila robusta*. Advan. Genet. 9:1–40.
Carson, H. L. 1959. Genetic conditions which promote or retard the formation of species. Cold Spring Harbor Symp. Quant. Biol. 24:87–105.
Carson, H. L. 1967. Select on for parthenogenesis in *D. mercatorum*. Genetics 55:157–71.
Cavalli-Sforza, L. L. 1966. Population structure in human evolution. Proc. Roy. Soc. Ser. B 164:362–79.
Cavalli-Sforza, L. L., and W. F. Bodmer. 1971. *The Genetics of Human Populations*. Freeman, San Francisco.
Charlesworth, B. 1970. Selection in populations with overlapping genera-

tions. I. The use of Malthusian parameters in population genetics. Theor. Pop. Biol. 1:352–70.

Charlesworth, B., and D. Charlesworth. 1973. A study of linkage disequilibrium in populations of *Drosophila melanogaster.* Genetics 73:351–59.

Charlesworth, B., and J. T. Giesel. 1972. Selection in populations with overlapping generations. II. Relations between gene frequency and demographic variables. Amer. Natur. 106:388–401.

Chetverikov, S. S. 1926. On certain features of the evolutionary process from the viewpoint of modern genetics. Translated from Russian, 1961. Proc. Amer. Phil. Soc. 105:167–95.

Chinnici, J. P. 1971a. Modification of recombination frequency in Drosophila. I. Selection for increased and decreased crossing-over. Genetics 69:71–83.

Chinnici, J. P. 1971b. Modification of recombination frequency in Drosophila. II. The polygenic control of crossing-over. Genetics 69:85–96.

Chovnick, A. 1966. Genetic organization in higher organisms. Proc. Roy. Soc. Ser. B 164:198–208.

Christiansen, F. B., and O. Frydenberg. 1973. Selection component analysis of natural polymorphisms using population samples including mother-child combinations. Theor. Pop. Biol., in press.

Clarke, B. 1970. Selective constraints on amino acid substitutions during the evolution of proteins. Nature 228:159–60.

Clarke, J. M., J. M. Smith, and K. Sondhi. 1961. Asymmetrical responses to selection for rate of development in *D. subobscura.* Genet. Res. 2:70–81.

Clayton, G., and A. Robertson. 1955. Mutation and quantitative variation. Amer. Natur. 89:151–58.

Clegg, M. T., and R. W. Allard. 1972. Patterns of genetic differentiation in the slender wild oat species *Avena barbata.* Proc. Nat. Acad. Sci. U.S. 69:1820–24.

Cordeiro, A. R. 1952. Experiments on the effects in heterozygous condition of second chromosomes for natural populations of *Drosophila willistoni.* Proc. Nat. Acad. Sci. U.S. 38:471–78.

Creed, R., ed. 1971. *Ecological Genetics and Evolution.* Blackwell, Oxford and Edinburgh.

Crow, J. F. 1958. Some possibilities for measuring selection intensities in man. Hum. Biol. 30:1–13.

Crow, J. F., and M. Kimura. 1965. Evolution in sexual and asexual populations. Amer. Natur. 99:439–50.

da Cunha, A. B., H. Burla, and Th. Dobzhansky. 1950. Adaptive chromosomal polymorphism in *Drosophila willistoni.* Evolution 4:212–35.

da Cunha, A. B., and Th. Dobzhansky. 1954. A further study of chromosomal polymorphism in *Drosophila willistoni* in its relation to the environment. Evolution 8:119–34.

da Cunha, A. B., Th. Dobzhansky, O. Pavlovsky, and B. Spassky. 1959. Genetics of Natural Populations. XXVIII. Supplementary data on the chromosomal polymorphism in *Drosophila willistoni* in its relation to the environment. Evolution 13:389–404.

Dempster, E. 1955. Maintenance of genetic heterogeneity. Cold Spring Harbor Symp. Quant. Biol. 20:25–32.

Dessauer, H. C., and E. Nevo. 1969. Geographic variation of blood and liver proteins in cricket frogs. Biochem. Genet. 3:171–88.

de Wit, C. T. 1960. On competition. Versl. Landb. Onderzoek 66:8–82.

Dickerson, G. 1955. Genetic slippage in response to selection for multiple objectives. Cold Spring Harbor Symp. Quant. Biol. 20:213–24.

Dickerson, R. E. 1971. The structure of cytochrome C and the rates of molecular evolution. J. Mol. Evol. 1:26–45.

Dixon, M., and E. C. Webb. 1964. *Enzymes*, 2d ed. Longmans, London.

Dobzhansky, Th. 1936. Studies on hybrid sterility. II. Localization of sterility factors in *Drosophila pseudoobscura* hybrids. Genetics 21:113–35.

Dobzhansky, Th. 1939. Genetics of natural populations. IV. Mexican and Guatemalan populations of *D. pseudoobscura*. Genetics 24:391–412.

Dobzhansky, Th. 1943. Genetics of natural populations. IX. Temporal changes in the composition of populations of *Drosophila pseudoobscura*. Genetics 28:162–86.

Dobzhansky, Th. 1946. Genetics of natural populations. XIII. Recombination and variability in populations of *Drosophila pseudoobscura*. Genetics 31:269–90.

Dobzhansky, Th. 1951. *Genetics and the Origin of Species*, 3d ed., rev. Columbia, New York.

Dobzhansky, Th. 1955. A review of some fundamental concepts and problems of population genetics. Cold Spring Harbor Symp. Quant. Biol. 20:1–15.

Dobzhansky, Th. 1957. Genetics of natural populations. XXVI. Chromosomal variability in island and continental populations of *Drosophila willistoni* from Central America and the West Indies. Evolution 11:280–93.

Dobzhansky, Th. 1963. Genetics of natural populations. XXXIII. A progress report on genetic changes in populations of *Drosophila pseudoobscura* and *Drosophila persimilis* in a locality in California. Evolution 17:333–39.

Dobzhansky, Th. 1970. *Genetics of the Evolutionary Process*. Columbia, New York.

Dobzhansky, Th., and C. Epling. 1944. Contributions to the genetics, taxonomy and ecology of *Drosophila pseudoobscura* and its relatives. Carnegie Inst. Wash. Publ. 554:1–183.

Dobzhansky, Th., and C. Epling. 1948. The suppression of crossing over in inversion heterozygotes of *Drosophila pseudoobscura*. Proc. Nat. Acad. Sci. U.S. 34:137–41.

Dobzhansky, Th., A. M. Holz, and B. Spassky. 1942. Genetics of natural populations. VIII. Concealed variability in the second and the fourth chromosomes of *D. pseudoobscura* and its bearing on the problem of heterosis. Genetics 27:463–90.

Dobzhansky, Th., C. Krimbas, and M. G. Krimbas. 1960. Genetics of natural populations. XXX. Is the genetic load in *D. pseudoobscura* a mutational or balanced load? Genetics 45:741–53.

Dobzhansky, Th., and H. Levene. 1948. Genetics of natural populations. XVII. Proof of the operation of natural selection in wild populations of *Drosophila pseudoobscura*. Genetics 33:537–47.

Dobzhansky, Th., R. C. Lewontin, and O. Pavlovsky. 1964. The capacity for increase in chromosomally polymorphic and monomorphic populations of *Drosophila pseudoobscura*. Heredity 19:597–614.

Dobzhansky, Th., and O. Pavlovsky. 1953. Indeterminate outcome of certain experiments on Drosophila populations. Evolution 7:198–210.

Dobzhansky, Th., and O. Pavlovsky. 1957. An experimental study of the interaction between genetic drift and natural selection. Evolution 11:311–19.

Dobzhansky, Th., and M. L. Queal. 1938. Genetics of natural populations. II. Genetic variation in populations of *D. pseudoobscura* inhabiting isolated mountain ranges. Genetics 23:463–84.

Dobzhansky, Th., and B. Spassky. 1953. Genetics of natural populations. XXI. Concealed variability in two sympatric species of Drosophila. Genetics 38:471–84.

Dobzhansky, Th., and B. Spassky. 1954. Genetics of natural populations. XXII. A comparison of the concealed variability in *Drosophila prosaltans* with that in other species. Genetics 39:472–87.

Dobzhansky, Th., and B. Spassky. 1963. Genetics of natural populations. XXXIV. Adaptive norm, genetic load, and genetic elite in *D. pseudoobscura*. Genetics 48:1467–85.

Dobzhansky, Th., and B. Spassky. 1967. Geotactic and phototactic behavior in Drosophila. I. Proc. Roy. Soc. Ser. B 168:27–47.

Dobzhansky, Th., and B. Spassky. 1968. Genetics of natural populations. XL. Heterotic and deleterious effects of recessive lethals in populations of *Drosophila pseudoobscura*. Genetics 59:411–25.

Dobzhansky, Th., B. Spassky, and T. Tidwell. 1963. Genetics of natural populations. XXXII. Inbreeding and the mutational and balanced loads in natural populations of *D. pseudoobscura*. Genetics 48:361–73.

Dobzhansky, Th., and S. Wright. 1941. Genetics of natural populations. V. Relations between mutation rate and accumulation of lethals in populations of *Drosophila pseudoobscura*. Genetics 26:23–51.

Dobzhansky, Th., O. Pavlovsky, B. Spassky, and N. Spassky. 1955. Genetics of natural populations. XXIII. Biological role of deleterious recessives in populations of *D. pseudoobscura*. Genetics 40:781–808.

Dobzhansky, Th., H. Levene, B. Spassky, and N. Spassky. 1959. Release of genetic variability through recombination. III. *D. prosaltans*. Genetics 44:75–92.

Dobzhansky, Th., A. S. Hunter, O. Pavlovsky, B. Spassky, and B. Wallace. 1963. Genetics of natural populations. XXXI. Genetics of an isolated marginal population of *D. pseudoobscura*. Genetics 48:91–103.

Dubinin, N. P., D. D. Romashov, M. A. Heptner, and Z. A. Demidova. 1937. (Aberrant polymorphism in *Drosophila fasciata* Meig. [in Russian]). Biol. Zh. 6:311–54.

Dunn, L. C. 1956. Analysis of a complex gene in the house mouse. Cold Spring Harbor Symp. Quant. Biol. 21:187–95.

Ehrman, L. 1967. Further studies on genotype frequency and mating success in *Drosophila*. Amer. Natur. 101:415–24.

Epstein, C. J. 1967. Nonrandomness of amino acid change in the evolution of homologous protein. Nature 215:355–59.

Erlenmeyer-Kimling, L., and L. F. Jarvik. 1963. Genetics and intelligence: A review. Science 142:1477–79.

Eshel, I., and M. Feldman. 1970. On the evolutionary effect of recombination. Theor. Pop. Biol. 1:88–100.

Ewens, W. J. 1968. A genetic model having complex linkage behavior. Theor. Appl. Genet. 38:140–43.

Ewens, W. J. 1972. The sampling theory of selectively neutral alleles. Theor. Pop. Biol. 3:87–112.

Falconer, D. S. 1964. *Introduction to Quantitative Genetics.* Oliver and Boyd, Edinburgh and London.

Falk, R. 1961. Are induced mutations in Drosophila overdominant? II. Experimental results. Genetics 46:737–57.

Feldman, M., I. Franklin, and G. Thomson. 1973. Selection in complex genetic systems. I: The symmetric equilibria of the three locus symmetric viability model. Theor. Pop. Biol., in press.

Felsenstein, J. 1965. The effect of linkage on directional selection. Genetics 52:349–63.

Felsenstein, J. 1971. On the biological significance of the cost of gene substitution. Amer. Natur. 105:1–11.

Fisher, R. A. 1930. *The Genetical Theory of Natural Selection.* Clarendon, Oxford.

Fisher, R. A. 1958. Polymorphism and natural selection. J. Ecol. 46:289–93.

Fitch, W. M., and E. Markowitz. 1970. An improved method for determining code variability in a gene and its application to the rate of fixation of mutations in evolution. Biochem. Genet. 4:579–93.

Ford, E. B. 1953. The genetics of polymorphism in the Lepidoptera. Advan. Genet. 5:43–88.

Ford, E. B. 1971. *Ecological Genetics*, 3d ed. Chapman and Hall, London.

Franklin, I., and R. C. Lewontin. 1970. Is the gene the unit of selection? Genetics 65:707–34.

Frydenberg, O. 1963. Population studies of a lethal mutant in *D. melanogaster.* I. Behavior in populations with discrete generations. Heredity 50:89–116.

Frydenberg, O., D. Møller, G. Naendal, and K. Sick. 1965. Haemoglobin polymorphism in Norwegian cod population. Heredity 53:257–71.

Geiringer, H. 1944. On the probability theory of linkage in Mendelian heredity. Ann. Math. Statist. 15:25–57.

Gibson, J. 1970. Enzyme flexibility in *D. melanogaster.* Nature 227:959–60.

Gillespie, J., and K. Kojima. 1968. The degree of polymorphism in enzymes involved in energy production compared to that in nonspecific enzymes in two *D. ananassae* populations. Proc. Nat. Acad. Sci. U.S. 61:582–85.

Greenberg, R., and J. F. Crow. 1960. Comparison of effects of lethal and detrimental chromosomes from Drosophila populations. Genetics 45:1153–68.

Grossman, A. I., L. G. Koreneva, and L. E. Ulitskaya. 1969. (Variation of the alcohol dehydrogenase locus in natural populations of *Drosophila melanogaster* [in Russian]). Genetika 6:91–96.

Gustafsson, A. 1953. The cooperation of genotypes in barley. Hereditas 39:1–18.

Haigh, J., and J. M. Smith. 1972. Population size and protein variation in man. Genet. Res. 19:73–89.

Haldane, J. B. S. 1922. Sex-ratio and unisexual sterility in hybrid animals. J. Genet. 12:101–9.

Haldane, J. B. S. 1927. A mathematical theory of natural and artificial selection. Part V. Selection and mutation. Proc. Cambridge Phil. Soc. 23:838–44.

Haldane, J. B. S. 1957. The cost of natural selection. J. Genet. 55:511–24.

Haldane, J. B. S., and S. D. Jayakar. 1963. Polymorphism due to selection of varying direction. J. Genet. 58:237–42.

Hamrick, J. L., and R. W. Allard. 1972. Microgeographical variation in allozyme frequencies in *Avena barbata*. Proc. Nat. Acad. Sci. U.S. 69:2100–2104.

Harper, J. L. 1968. The regulation of numbers and mass in plant populations. In: *Population Biology and Evolution*, ed. R. C. Lewontin, pp. 131–58. Syracuse, Syracuse, N.Y.

Harris, H. 1966. Enzyme polymorphisms in man. Proc. Roy. Soc. Ser. B 164:298–310.

Harris, H. 1970. *The Principles of Human Biochemical Genetics*. North Holland, Amsterdam and London.

Harris, H. 1971. Annotation: Polymorphism and protein evolution: the neutral mutation-random drift hypothesis. J. Med. Genet. 8:444–52.

Harris, H., and D. A. Hopkinson. 1972. Average heterozygosity in man. J. Hum. Genet. 36:9–20.

Hartley, B. S. 1970. Homologies in serine proteinases. Phil. Trans. Roy. Soc. London Ser. B 257:77–87.

Hildreth, P. E. 1956. The problem of synthetic lethals in *Drosophila melanogaster*. Genetics 41:729–42.

Hill, W. G., and A. Robertson. 1966. The effect of linkage on limits to artificial selection. Genet. Res. 8:269–94.

Hiorns, R. W., and G. A. Harrison. 1970. Sampling for the detection of natural selection by age group genetic differences. Hum. Biol. 42:53–64.

Hiraizumi, Y., and J. F. Crow. 1960. Heterozygous effects on variability, fertility, rate of development and longevity of Drosophila chromosomes that are lethal when homozygous. Genetics 45:1071–83.

Hochman, B. 1971. Analysis of chromosome 4 in *D. melanogaster*. II. Ethyl methanesulfonate induced lethals. Genetics 67:235–52.

Huang, S. L., M. Singh, and K. Kojima. 1971. A study of frequency-dependent selection observed in the esterase-6 locus of *Drosophila melanogaster* using a conditioned medium method. Genetics 68:97–104.

Hubby, J. L., and R. C. Lewontin. 1966. A molecular approach to the study of genic heterozygosity in natural populations. I. The number of alleles at different loci in *Drosophila pseudoobscura*. Genetics 54:577–94.

Hubby, J. L., and S. Narise. 1967. Protein differences in Drosophila. III. Allelic differences and species differences in *in vitro* hybrid enzyme formation. Genetics 57:291–300.

Hubby, J. L., and L. H. Throckmorton. 1965. Protein differences in Drosophila. II. Comparative species genetics and evolutionary problems. Genetics 52:203–15.

Hubby, J. L., and L. H. Throckmorton. 1968. Protein differences in Drosophila. IV. A study of sibling species. Amer. Natur. 102:193–205.

Ingram, V. M. 1963. *The Hemoglobins in Genetics and Evolution.* Columbia, New York.

Irwin, M. R. 1953. Evolutionary patterns of antigenic substances of the blood corpuscles in Columbidae. Evolution 7:31–50.

Ives, P. T. 1945. The genetic structure of American populations of *Drosophila melanogaster.* Genetics 30:167–96.

Jacquard, A. 1970. *Structures Génétiques des Populations.* Masson, Paris.

Jepsen, G. L., E. Mayr, and G. G. Simpson, eds. 1949. *Genetics, Paleontology and Evolution.* Princeton, Princeton, N.J.

Johnson, F. M. 1971. Isozyme polymorphisms in *Drosophila ananassae.* Genetic diversity among island populations of the South Pacific. Genetics 68:77–95.

Johnson, F. M., C. G. Kanapi, R. H. Richardson, M. R. Wheeler, and W. S. Stone. 1966. An analysis of polymorphisms among isozyme loci in dark and light *D. ananassae* strains from American and Western Samoa. Proc. Nat. Acad. Sci. U.S. 56:119–25.

Johnson, F. M., H. E. Schaffer, J. E. Gillespy, and E. S. Rockwood. 1969. Isozyme genotype-environment relationships in natural populations of the harvester ant, *Pogonomyrmex barbatus,* from Texas. Biochem. Genet. 3:429–50.

Johnson, G. B. 1971. Analysis of enzyme variation in natural populations of the butterfly, *Colias eurytheme.* Proc. Nat. Acad. Sci. U.S. 68:997–1001.

Johnson, M. S. 1971. Adaptive lactate dehydrogenase variation in the crested blenny, *Anoplorchus.* Heredity 27:205–26.

Johnson, W. E., and R. K. Selander. 1971. Protein variation and systematics in kangaroo rats (genus *Dipodomys*). Syst. Zool. 20:377–405.

Judd, B. H., M. W. Shen, and T. C. Kaufman. 1972. The anatomy and function of a segment of the X-chromosome of *Drosophila melanogaster.* Genetics 71:139–56.

Karlin, S., and M. W. Feldman. 1970, Linkage and selection: Two locus symmetric viability models. Theor. Pop. Biol. 1:39–71.

Karlin, S., and J. McGregor. 1972. Polymorphisms for genetic and ecological systems with weak coupling. Theor. Pop. Biol. 3:210–38.

Kenyon, A. 1967. Comparison of frequency distribution of viability of second and fourth chromosomes from caged *Drosophila melanogaster.* Genetics 55:123–30.

Kidwell, M. G. 1972. Genetic change of recombination value in *Drosophila melanogaster.* I. Artificial selection for high and low recombination and some properties of recombination-modifying genes. Genetics 70:419 :32.

Kimura, M. 1955. Stochastic processes and distribution of gene frequencies under natural selection. Cold Spring Harbor Symp. Quant. Biol. 20:33–53.

Kimura, M. 1956. A model of a genetic system which tends to closer linkage by natural selection. Evolution 10:278–87.

Kimura, M. 1962. On the probability of fixation of mutant genes in a population. Genetics 47:713–19.

Kimura, M. 1968. Evolutionary rate at the molecular level. Nature 217:624–26.

Kimura, M., and J. F. Crow. 1964. The number of alleles that can be maintained in a finite population. Genetics 49:725–38.

Kimura, M., and T. Ohta. 1971a. *Theoretical Aspects of Population Genetics.* Princeton, Princeton, N.J.

Kimura, M., and T. Ohta. 1971b. Protein polymorphism as a phase of molecular evolution. Nature 229:467–69.

Kimura, M., and T. Ohta. 1971c. On the rate of molecular evolution. J. Mol. Evol. 1:1–17.

Kimura, M., and T. Ohta. 1972. Population genetics, molecular biometry and evolution. In: *VI Berkeley Symposium on Probability and Statistics,* ed. L. Le Cam, J. Neyman, and E. L. Scott, pp. 43–68. U. of California, Berkeley, Calif.

Kimura, M., and G. H. Weiss. 1964. The stepping stone model of population structure and the decrease of genetic correlation with distance. Genetics 49:561–76.

King, J. L. 1967. Continuously distributed factors affecting fitness. Genetics 55:483–92.

King, J. L., and T. H. Jukes. 1969. Non-Darwinian evolution: Random fixation of selectively neutral mutations. Science 164:788–98.

Knight, G. R., A. Robertson, and C. H. Waddington. 1956. Selection for sexual isolation within a species. Evolution 10:14–22.

Koehn, R. K. 1969. Esterase heterogeneity: Dynamics of a polymorphism. Science 163:943–44.

Koehn, R. K., and D. J. Rasmussen. 1967. Polymorphic and monomorphic serum esterase heterogeneity in catastomid fish populations. Biochem. Genet. 1:131–44.

Kojima, K. 1959. Stable equilibria for the optimum model. Proc. Nat. Acad. Sci. U.S. 45:989–93.

Kojima, K. 1971. Is there a constant fitness for a given genotype? No! Evolution 25:281–85.

Kojima, K., J. Gillespie, and Y. N. Tobari. 1970. A profile of *Drosophila* species enzymes assayed by electrophoresis. I. Number of alleles, heterozygosities and linkage disequilibrium in glucose-metabolizing systems and some other enzymes. Biochem. Genet. 4:627–37.

Kojima, K., and Y. N. Tobari. 1969. The pattern of viability changes associated with genotype frequency at the alcohol dehydrogenase locus in a population of *D. melanogaster.* Genetics 61:201–9.

Kojima, K., and K. M. Yarbrough. 1967. Frequency dependent selection at the Esterase-6 locus in *D. melanogaster.* Proc. Nat. Acad. Sci. U.S. 57:645–49.

Krimbas, C. B. 1961. Release of genetic variability through recombination. IV. *D. willistoni.* Genetics 46:1323–34.

Lakovaara, S., and A. Saura. 1971a. Genetic variation in natural populations of *D. obscura.* Genetics 69:377–84.

Lakovaara, S., and A. Saura. 1971b. Genic variation in marginal populations of *D. subobscura.* Hereditas 69:77–82.

Lamotte, M. 1951. Recherches sur la structure génétique des populations naturelles de *Cepaea nemoralis* (L.). Bull. Biol. Fr. Belg., suppl. 35:1–238.

Lamotte, M. 1959. Polymorphism of natural populations of *Cepaea nemoralis*. Cold Spring Harbor Symp. Quant. Biol. 24:65–86.

Latter, B. D. H. 1972. Selection in finite populations with multiple alleles. III. Genetic divergence with centripetal selection and mutation. Genetics 70:475–90.

Levene, H. 1953. Genetic equilibrium when more than one ecological niche is available. Amer. Natur. 87:131–33.

Levene, H., O. Pavlovsky, and Th. Dobzhansky. 1954. Interaction of the adaptive values in polymorphic experimental populations of *Drosophila pseudoobscura*. Evolution 8:335–49.

Levin, D. A., G. P. Howland, and E. Steiner. 1972. Protein polymorphism and genic heterozygosity in a population of the permanent translocation heterozygote *Oenothera biennis*. Proc. Nat. Acad. Sci. U.S. 69:1475–77.

Levine, P. 1958. The influence of the ABO system on Rh hemolytic disease. Hum. Biol, 20:14–28.

Levins, R. 1968. *Evolution in Changing Environments*. Princeton, Princeton, N.J.

Lewontin, R. C. 1955. The effects of population density and competition on viability in *Drosophila melanogaster*. Evolution 9:27–41.

Lewontin, R. C. 1957. The adaptations of populations to varying environments. Cold Spring Harbor Symp. Quant. Biol. 22:395–408.

Lewontin, R. C. 1958. A general method for investigating the equilibrium of gene frequency in a population. Genetics 43:419–34.

Lewontin, R. C. 1964a. The interaction of selection and linkage. I. General considerations. Heterotic models. Genetics 49:49–67.

Lewontin, R. C. 1964b. The interaction of selection and linkage. II. Optimum models. Genetics 50:757–82.

Lewontin, R. C. 1965. Selection for colonizing ability. In: *The Genetics of Colonizing Species*, ed. H. Baker, pp. 77–94. Academic, New York.

Lewontin, R. C. 1967a. An estimate of average heterozygosity in man. Amer. J. Hum. Genet. 19:681–85.

Lewontin, R. C. 1967b. The principle of historicity in evolution. In: *Mathematical Challenges to the Neo-Darwinian Interpretation of Evolution*. Wistar Symp. Monogr. no. 5, pp. 81–94.

Lewontin, R. C. 1968. The concept of evolution. In *International Encyclopedia of Social Science*, ed. D. L. Sills, vol. 5, pp. 202–10. Macmillan, Free Press, New York.

Lewontin, R. C. 1971. The effect of genetic linkage on the mean fitness of a population. Proc. Nat. Acad. Sci. U.S. 68:984–86.

Lewontin, R. C. 1972a. The apportionment of human diversity. Evol. Biol. 6:381 :98.

Lewontin, R. C. 1972b. Testing the theory of natural selection. Nature 236:181–82.

Lewontin, R. C. 1973. Darwin and Mendel: The triumph of materialism. In: *The Copernican Revolution*, ed., J. Neyman. In press.

Lewontin, R. C., and L. C. Birch. 1966. Hybridization as a source of variation for adaptation to new environments. Evolution 20:315–36.

Lewontin, R. C. and C. C. Cockerham. 1959. The goodness-of-fit test for detecting natural selection in random mating populations. Evolution 13:561–64.

Lewontin, R. C., and L. C. Dunn. 1960. The evolutionary dynamics of a polymorphism in the house mouse. Genetics 45:705–22.

Lewontin, R. C., and J. L. Hubby. 1966. A molecular approach to the study of genic heterozygosity in natural populations. II. Amount of variation and degree of heterozygosity in natural populations of *Drosophila pseudoobscura*. Genetics 54:595–609.

Lewontin, R. C., and K. Kojima. 1960. The evolutionary dynamics of complex polymorphisms. Evolution 14:458–72.

Lewontin, R. C., and J. Krakauer. 1973. Distribution of gene frequency as a test of the theory of the selective neutrality of polymorphisms. Genetics 74:175–95.

Lewontin, R. C., and Y. Matsuo. 1963. Interaction of genotypes determining viability in *Drosophila busckii*. Proc. Nat. Acad. Sci. U.S. 49:270–78.

Lewontin, R. C., and T. Prout. 1956. Estimation of the number of different classes in a population. Biometrics 12:211–23.

Lewontin, R. C., and M. J. D. White. 1960. Interaction between inversion polymorphisms of two chromosome pairs in the grasshopper, *Moraba scurra*. Evolution 14:116–29.

L'Heritier, Ph., and G. Teissier. 1933. Étude d'une population de Drosophiles en equilibre. C. R. Acad. Sci. 198:770–72.

Lindsley, D. L., and E. H. Grell. 1967. Genetic variations of *Drosophila melanogaster*. Carnegie Inst. Wash. Publ. 627.

Luffman, J. E., and H. Harris. 1967. A comparison of some properties of human red cell acid phosphatases in different phenotypes. Ann. Hum. Genet. 30:387–401.

MacArthur, R., and R. Levins. 1967. The limiting similarity, convergence and divergence of coexisting species. Amer. Natur. 101:377–85.

Malécot, G. 1959. Les modèles stochastiques en génétique de population. Publ. Inst. Statist. Univ. Paris 8:173–210.

Manwell, C., and C. M. A. Baker. 1968. Genetic variation of isocitrate, malate and 6-phosphogluconate dehydrogenases in snails of the genus *Cepaea*. Introgressive hybridization, polymorphism and pollution? Comp. Biochem. Physiol. 26:195–209.

Marinkovic, D. 1967. Genetic loads affecting fecundity in natural populations of *D. pseudoobscura*. Genetics 56:61–71.

Markert, C. L., and F. Møller. 1959. Multiple forms of enzymes: Tissue, ontogenetic and species specific patterns. Proc. Nat. Acad. Sci. U.S. 45:753–63.

Marshall, D. R., and R. W. Allard. 1970. Isozyme polymorphisms in natural populations of *Avena fatua* and *Avena barbata*. Heredity 29:373–82.

Mather, K., and B. S. Harrison. 1949. The manifold effect of selection. Heredity 3:1–52.

Mayhew, S. H., S. K. Kato, F. M. Ball, and C. Epling. 1966. Comparative studies of arrangements within and between populations of *Drosophila pseudoobscura*. Evolution 28:646–62.

Maynard Smith, J. 1968. Evolution in sexual and asexual populations. Amer. Natur. 102:469–73.

Maynard Smith, J., and K. C. Sondhi. 1960. The genetics of a pattern. Genetics 45:1039–50.

Mayr, E. 1963. *Animal Species and Evolution*. Harvard, Belknap Press, Cambridge, Mass.

Merrell, D. J. 1949. Selective mating in *Drosophila melanogaster*. Genetics 34:370–89.

Merritt, R. B. 1972. Geographic distribution and enzymatic properties of lactate dehydrogenase in the fathead minnow, *Pimephales promelas*. Amer. Natur. 106:173–84.

Milkman, R. D. 1964. The genetic basis of natural variation. V. Selection for crossveinless polygenes in new wild strains of *Drosophila melanogaster*. Genetics 50:625–32.

Milkman, R. D. 1967. Heterosis as a major cause of heterozygosity in nature. Genetics 55:493–95.

Modiano, G., R. Scozzari, F. Gigliani, C. Santolamazza, G. Spennati, and P. Saini. 1970. Enzyme activity in two red cell adenylate kinase phenotypes. Amer. J. Hum. Genet. 22:292–98.

Monod, J. 1971. *Chance and Necessity*. Knopf, New York.

Moran, P. A. P. 1962. *The Statistical Processes of Evolutionary Theory*. Clarendon, Oxford.

Morton, N., J. Crow, and H. J. Muller. 1956. An estimate of the mutational damage in man from data on consanguinous marriages. Proc. Nat. Acad. Sci. U.S. 42:855–63.

Mukai, T. 1964. The genetic structure of natural populations of *Drosophila melanogaster*. I. Spontaneous mutation rates of polygenes controlling variation. Genetics 50:1–19.

Mukai, T. 1970. Spontaneous mutation rates of isozyme genes in *Drosophila melanogaster*. Drosophila Inform. Serv. 45:99.

Mukai, T., S. Chigusa, and I. Yoshikawa. 1964. The genetic structure of natural populations of *Drosophila melanogaster*. II. Overdominance of spontaneous mutant polygenes controlling viability in homozygous genetic background. Genetics 50:711–15.

Mukai, T., S. Chigusa, and I. Yoshikawa. 1965. The genetic structure of natural populations of *Drosophila melanogaster*. III. Dominance effect of spontaneous mutant polygenes controlling viability in heterozygous genetic background. Genetics 52:493–501.

Mukai, T., L. E. Mettler, and S. Chigusa. 1971. Linkage disequilibrium in a local population of *Drosophila melanogaster*. Proc. Nat. Acad. Sci. U.S. 68:1065–69.

Mukai, T., I. Yoshikawa, and K. Sano. 1966. The genetic structure of natural populations of *Drosophila melanogaster*. IV. Heterozygous effects of radiation induced mutations on viability in various genetic backgrounds. Genetics 53:513–27.

Muller, H. J. 1928. The problem of gene modification. Verhandlung des V Internationalen Kongresses für Vererbungswissenschaft 1:234–260. (Suppl. I, Z. indukt. Abstamm. Vererbungsl.)

Muller, H. J. 1932. Some genetic aspects of sex. Amer. Natur. 66:118–38.

Muller, H. J. 1949. The Darwinian and modern conceptions of natural selection. Proc. Amer. Phil. Soc. 93:459–70.

Muller, H. J. 1950. Our load of mutations. Amer. J. Hum. Genet. 2:111–76.

Muller, H. J., and R. Falk. 1961. Are induced mutations in Drosophila overdominant? I. Experimental design. Genetics 46:727–35.

Nair, P. S., and D. Brncic. 1971. Allelic variations within identical chromosomal environs. Amer. Natur. 105:291–94.

Neel, J. V., and W. J. Schull. 1962. The effect of inbreeding on mortality and morbidity in two Japanese cities. Proc. Nat. Acad. Sci. U.S. 48:573–82.

Nevo, E., and C. R. Shaw. 1972. Genetic variation in the subterranean mammal, *Spalax ehrenbergi.* Biochem. Genet. 7:235–41.

O'Brien, S. J., and R. J. MacIntyre. 1969. An analysis of gene-enzyme variability in natural populations of *Drosophila melanogaster* and *D. simulans.* Amer. Natur. 103:97–113.

O'Gower, A. K., and P. I. Nicol. 1968. A latitudinal cline of hemoglobins in a bivalve mollusc. Heredity 23:485–92.

Ohta, T., and M. Kimura. 1970. Development of associative overdominance through linkage disequilibrium in finite populations. Genet. Res. 16:165–77.

Ohta, T., and M. Kimura. 1971a. Behavior of neutral mutants influenced by associated overdominant loci in finite populations. Genetics 69:247–60.

Ohta, T., and M. Kimura. 1971b. On the constancy of the evolutionary rate of cistrons. J. Mol. Evol. 1:18–25.

Patterson, J. T., and H. Muller. 1930. Are "progressive" mutations produced by X-rays? Genetics 15:495–577.

Perutz, M. F., and H. Lehman. 1968. Molecular pathology of human hemoglobin. Nature 219:902–9.

Petit, C. 1958. Le déterminisme génétique et psycho-physiologique de la compétition sexuelle chez *Drosophila melanogaster.* Bull. Biol. Fr. Belg. 92:248–329.

Polivanov, S. 1964. Selection in experimental populations of *Drosophila melanogaster* with different genetic backgrounds. Genetics 50:81–100.

Prakash, S. 1969. Genic variation in a natural population of *Drosophila persimilis.* Proc. Nat. Acad. Sci. U.S. 62:778–84.

Prakash, S. 1972. Origin of reproductive isolation in the absence of apparent genic differentiation in a geographic isolate of *Drosophila pseudoobscura.* Genetics 72:143–55.

Prakash, S. 1973. Patterns of gene variation in central and marginal populations of *Drosophila robusta.* Genetics, in press.

Prakash, S., and R. C. Lewontin. 1968. A molecular approach to the study of genic heterozygosity. III. Direct evidence of co-adaptation in gene arrangements of *Drosophila.* Proc. Nat. Acad. Sci. U.S. 59:398–405.

Prakash, S., and R. C. Lewontin. 1971. A molecular approach to the study of genic heterozygosity. V. Further direct evidence of co-adaptation in inversions of Drosophila. Genetics 69:405–8.

Prakash, S., R. C. Lewontin, and D. W. Crumpacker. 1973. A molecular approach to the study of genic heterozygosity in natural populations. VI. The remarkable genetic similarity of geographical populations of *D. pseudoobscura.* Genetics, in press.

Prakash, S., R. C. Lewontin, and J. L. Hubby. 1969. A molecular approach to the study of genic heterozygosity in natural populations. IV. Patterns of genic variation in central, marginal, and isolated populations of *Drosophila pseudoobscura.* Genetics 61:841–58,

Prakash, S., and R. B. Merritt. 1972. Direct evidence of genic differentiation between sex ratio and standard arrangements of X chromosome in *Drosophila pseudoobscura.* Genetics 72:169–75.

Prout, T. 1952. Selection against heterozygotes for autosomal lethals in natural populations of *Drosophila willistoni.* Proc. Nat. Acad. Sci. U.S. 38:478–81.

Prout, T. 1954. Genetic drift in irradiated experimental populations of *Drosophila melanogaster.* Genetics 39:529–45.

Prout, T. 1965. The estimation of fitness from genotypic frequencies. Evolution 19:546–51.

Prout, T. 1969. The estimation of fitness from population data. Genetics 63:949–67.

Prout, T. 1971a. The relation between fitness components and population prediction in Drosophila. I. The estimation of fitness components. Genetics 68:127–49.

Prout, T. 1971b. The relation between fitness components and population prediction in Drosophila. II. Population prediction. Genetics 68:151–67.

Rendel, J. M. 1959. Canalization of the scute phenotype of Drosophila. Evolution 13:425–39.

Rendel, J. M. 1967. *Canalization and Gene Control.* Academic, Logos Press, London.

Rendel, J. M., and B. L. Sheldon. 1960. Selection for canalization of scute phenotype in *Drosophila melanogaster.* Aust. J. Biol. Sci, 13:36–47.

Richmond, R. C. 1972. Enzyme variability in the *Drosphila willistoni* group. III. Amounts of variability in the superspecies *D. paulistorum.* Genetics 71:87–112.

Richmond, R. C., and J. R. Powell. 1970. Evidence of heterosis associated with an enzyme locus in a natural population of *Drosophila.* Proc. Nat. Acad. Sci. U.S. 67:1264–67.

Robertson, F. W. 1959. Studies in quantitative inheritance. XII. Cell size and number in relation to genetic and environmental variation of body size in Drosophila. Genetics 44:869–96.

Robertson, F. W., and E. Reeve. 1952. Studies in quantitative inheritance. I. The effects of selection of wing and thorax length in *Drosophila melanogaster.* J. Genet. 50:414–48.

Rockwood, E. S. 1969. Enzyme variation in natural populations of *Drosophila mimica.* Studies in Genetics V. Tex. Univ. Publ. 6918:111–25.

Roderick, T. H., F. H. Ruddle, V. M. Chapman, and T. B. Shows. 1972. Biochemical polymorphism in field and inbred mice (*Mus musculus*). Biochem. Genet. 5:457–66.

Ruddle, F. H., T. H. Roderick, T. B. Shows, P. G. Weigl, R. K. Chipman, and P. K. Anderson. 1969. Measurement of genetic heterogeneity by means of enzyme polymorphism. J. Hered. 60:321–22.

Salthe, S. N. 1969. Geographic variation of the lactate dehydrogenase of *Rana pipiens* and *Rana palustris.* Biochem. Genet. 2:271–304.

Sang, J. H., and J. M. McDonald. 1954. Production of phenocopies in Drosophila using salts, particularly sodium metaborate. J. Genet. 52:392–412.

Scharloo, W. 1964. The effect of disruptive and stabilizing selection on the expression of cubitus interruptus in Drosophila. Genetics 50:553–62.

Scharloo, W. 1971. Reproductive isolation by disruptive selection: Did it occur? Amer. Natur. 105:83–86.

Schlager, G., and M. M. Dickie. 1971. Natural mutation rates in the house mouse. Estimates for five specific loci and dominant mutations. Mutat. Res. 11:89–96.

Schopf, T. J., and J. L. Gooch. 1971. Gene frequencies in a marine ectoproct. A cline in natural populations related to sea temperature. Evolution 25:286–89.

Selander, R. K. 1970. Behavior and genetic variations in natural populations. Amer. Zool. 10:53–66.

Selander, R. K., W. G. Hunt, and S. Y. Yang. 1969. Protein polymorphism and genetic heterozygosity in two European subspecies of the house mouse. Evolution 23:379–90.

Selander, R. K., and S. Y. Yang. 1969. Protein polymorphism and genic heterozygosity in a wild population of the house mouse (*Mus musculus*). Genetics 63:653–67.

Selander, R. K., S. Y. Yang, and W. G. Hunt. 1969. Polymorphism in esterases and hemoglobin in wild populations of the house mouse (*Mus musculus*). Studies in Genetics V. Tex. Univ. Publ. 6918:271–338.

Selander, R. K., S. Y. Yang, R. C. Lewontin, and W. E. Johnson. 1970. Genetic variation in the horseshoe crab (*Limulus polyphemus*), a phylogenetic "relic." Evolution 24:402–14.

Selander, R. K., M. H. Smith, S. Y. Yang, W. E. Johnson, and J. B. Gentry. 1971. Biochemical polymorphisms and systematics in the genus *Peromyscus*. I. Variation in the old-field mouse. Studies in Genetics VI. Tex. Univ. Publ. 7103:49–90.

Sen, B. K., and A. Robertson. 1964. An experimental examination of methods for the simultaneous selection of two characters using *Drosophila melanogaster*. Genetics 50:199–209.

Serov, O. L. 1972. Monomorphism of some structural genes in populations of wild and laboratory rats. Isozyme Bull. 5:38.

Shaw, C. R. 1965. Electrophoretic variation in enzymes. Science 149:936–43.

Simpson, G. G. 1944. *Tempo and Mode in Evolution*. Columbia, New York.

Slatkin, M. 1970. Selection and polygenic characters. Proc. Nat. Acad. Sci. U.S. 66:87–93.

Slatkin, M. 1972. On treating the chromosome as the unit of selection. Genetics 72:157–68.

Solbrig, O. 1971. The population biology of dandelions. Amer. Sci. 59:686–96.

Spassky, B., Th. Dobzhansky, and W. W. Anderson. 1965. Genetics of natural populations. XXXVI. Epistatic interaction of the components of the genetic load in *Drosophila pseudoobscura*. Genetics 52:653–64.

Spassky, B., N. Spassky, H. Levene, and Th. Dobzhansky. 1958. Release of genetic variability through recombination., I. *Drosphila pseudoobscura*. Genetics 43:844–67.

Spencer, W. P. 1946. High mutant gene frequencies in a population of *Drosophila immigrans*. Ohio J. Sci. 46:143–51.

Spencer, W. P. 1947. Mutations in wild populations of Drosophila. Advan. Genet. 1:359–402.

Spencer, W. P. 1957. Genetic studies on *Drosophila mulleri*. I. Genetic analysis of a population. Tex. Univ. Publ. 5721:186–205.

Sperlich, D., and H. Feuerbach. 1969. Austausch-Ungleichgewicht zwischen unabhängigen Inversionen in natürlichen Populationen von *Drosophila subobscura.* Theor. Appl. Genet. 39:104–12.

Spiess, E. B. 1957. Relation between frequency and adaptive values of chromosomal arrangements in *Drosophila persimilis.* Evolution 11:84–93.

Spiess, E. B. 1959. Release of genetic variability through recombination. II. *Drosophila persimilis.* Genetics 44:43–58.

Spiess, E. B. 1968. Low frequency advantage in mating of *Drosophila pseudoobscura* karyotypes. Amer. Natur. 102:363–79.

Spiess, E. B., and A. C. Allen. 1961. Release of genetic variability through recombination. VII. Second and third chromosomes of *Drosophila melanogaster.* Genetics 46:1531–53.

Stadler, L. J. 1942. Some observations on gene variability and spontaneous mutation. Spragg Mem. Lect. 3:3–15.

Stebbins, G. L., and R. C. Lewontin. 1972. Comparative evolution at the levels of molecules, organisms and populations. In: *Proceedings of VI Berkeley Symposium on Mathematical Statistics and Probability*, vol. 5, pp. 23–42.

Stern, C., G. Carson, M. Kinst, E. Novitski, and D. Uphoff. 1952. The viability of heterozygotes for lethals. Genetics 37:413–49.

Stone, W. S., M. R. Wheeler, F. M. Johnson, and K. Kojima. 1968. Genetic variation in natural island populations of members of the *Drosophila nasuta* and *D. ananassae* subgroups. Proc. Nat. Acad. Sci. U.S. 59:102–9.

Strickberger, M. W., and C. J. Wills. 1966. Monthly frequency changes of *Drosophila pseudoobscura* third chromosome gene arrangements in a California locality. Evolution 20:592–602.

Subak-Sharpe, J. H. 1969. The doublet pattern of nucleic acid in relation to the origin of viruses. In: *Handbook of Molecular Cytology*, ed. A. Lima de Faria, pp. 68–87. North-Holland, Amsterdam.

Sved, J. A. 1968. The stability of linked systems of loci with small population size. Genetics 59:543–63.

Sved, J. A. 1971. An estimate of heterosis in *Drosophila melanogaster.* Genet. Res. 18:97–105.

Sved, J. A., and F. J. Ayala. 1970. A population cage test for heterosis in *Drosophila pseudoobscura.* Genetics 66:97–113.

Sved, J. A., T. E. Reed, and W. F. Bodmer. 1967. The number of balanced polymorphisms that can be maintained in a natural population. Genetics 55:469–81.

Temin, R. G. 1966. Homozygous variability and fertility loads in *Drosophila melanogaster.* Genetics 53:27–46.

Temin, R. G., H. V. Meyer, P. S. Dawson, and J. F. Crow. 1969. The influence of epistasis on homozygous viability depression in *Drosophila melanogaster.* Genetics 61:497–519.

Thoday, J. M., and T. B. Boam. 1959. Effects of disruptive selection. II. Polymorphism and divergence without isolation. Heredity 13:205–18.

Thoday, J. M., and T. B. Boam. 1961. Regular responses to selection. I. Description of responses. Genet. Res. 2:161–76.

Thoday, J. M., and J. B. Gibson. 1962. Isolation by disruptive selection. Nature 193:1164–66.

Thomas, D. M., and H. Harris. 1972. Comparison of thermo-stabilities of different human placental alkaline phosphatase phenotypes. Ann. Hum. Genet. 35:221–24.

Timofeeff-Ressovsky, N. W. 1940. Zur Analyse des Polymorphismus bei *Adalia bipunctata*. Biol. Zentralbl. 60:130–37.

Tobari, I. 1966. Effects of temperature on the viability of homozygotes and heterozygotes for several chromosomes of *Drosophila melanogaster*. Genetics 54:783–91.

Tobari, Y. N., and K. Kojima. 1972. A study of spontaneous mutation rates at ten loci detectable by starch gel electrophoresis in *Drosophila melanogaster*. Genetics 70:397–403.

Tsakas, S., and C. B. Krimbas. 1970. The genetics of *Dacus oleae*. IV. Relation between adult esterase genotypes and survival to organophosphate insecticides. Evolution 24:807–15.

Turner, J. R. 1967. Why does the genome not congeal? Evolution 21:645–56.

Vogel, F. 1964. A preliminary estimate of the number of human genes. Nature 201:847.

Vogel, F. 1970. ABO blood groups and disease. Amer. J. Hum. Genet. 22:464–75.

Waddington, C. H. 1953. Genetic assimilation of an acquired character. Evolution 7:118–26.

Waddington, C. H. 1960. Experiments on canalizing selection. Genet. Res. 1:140–50.

Wallace, B. 1950. Allelism of second chromosome lethals in *Drosophila melanogaster*. Proc. Nat. Acad. Sci. U.S. 36:654–57.

Wallace, B. 1954. Genetic divergence of isolated populations of *Drosophila melanogaster*. Proc. Int. Congr. Genet, 9th., 761–64.

Wallace, B. 1957. The effect of heterozygosity for new mutations on viability in *Drosophila melanogaster*: A preliminary report. Proc. Nat. Acad. Sci. U.S. 43:404–7.

Wallace, B. 1958a. The average effect of radiation-induced mutations on viability in *D. melanogaster*. Evolution 12:532–52.

Wallace, B. 1958b. The role of heterozygosity in Drosophila populations. Proc. Int. Congr. Genet. 10th, 1:408–19.

Wallace, B. 1963. Further data on the overdominance of induced mutations. Genetics 48:633–51.

Wallace, B. 1968. Polymorphism, population size and genetic load. In: *Population Biology and Evolution*, ed. R. C. Lewontin, pp. 87–108. Syracuse, Syracuse, N.Y.

Wallace, B., and Th. Dobzhansky. 1962. Experimental proof of balanced genetic loads in *Drosophila*. Genetics 47:1027–42.

Wallace, B., and C. Madden. 1953. The frequencies of sub- and supervitals in experimental populations of *Drosophila melanogaster*. Genetics 38:456–70.

Wallace, B., J. C. King, C. V. Madden, B. Kaufman, and E. C. McGunnigle. 1953. An analysis of variability arising through recombination. Genetics 38:272–307.

Weiner, A. S. 1970. Blood groups and disease. Amer. J. Hum. Genet. 22:476–83.

Weiss, M. L., and M. Goodman. 1972. Frequency and maintenance of genetic variability in natural populations of *Macaca fascicularis*. J. Hum. Evol. 1:41–48.

Wills, C. 1966. The mutational load in two natural populations of *Drosophila pseudoobscura*. Genetics 53:281–94.

Wills, C., and L. Nichols. 1971. Single gene heterosis in Drosophila revealed by inbreeding. Nature 233:123–25.

Wills, C., and L. Nichols. 1972. How genetic background masks single gene heterosis in Drosophila. Proc. Nat. Acad. Sci. U.S. 69:323–25.

Wilson, J. 1968. Experimental determination of fitness interactions in *Drosophila melanogaster* by the method of marginal populations. Genetics 59:501–11.

Wilson, J. 1970. Experimental design in fitness estimation. Genetics 66:555–67.

Wilson, J. 1972. Experimentally determined two locus fitnesses of *Drosophila melanogaster* males. Genetics 70:445–55.

Wright, S. 1937. The distribution of gene frequencies in populations. Proc. Nat. Acad. Sci. U.S. 23:307–20.

Wright, S. 1948. On the roles of directed and random changes in gene frequency in the genetics of populations. Evolution 2:279–94.

Wright, S. 1949. Adaptation and selection. *Genetics, Paleontology and Evolution*, ed. G. L. Jepsen, pp. 365–89. Princeton, Princeton, N.J.

Wright, S. 1969. *Evolution and the Genetics of Populations*, vol. 2. U. of Chicago, Chicago.

Wright, S., and Th. Dobzhansky. 1946. Genetics of natural populations. XII. Experimental reproduction of some of the changes caused by natural selection in certain populations of *Drosophila pseudoobscura*. Genetics 31:125–56.

Wright, S., Th. Dobzhansky, and W. Hovanitz. 1942. Genetics of natural populations. VII. The allelism of lethals in the third chromosome of *Drosophila pseudoobscura*. Genetics 27:363–94.

Yarbrough, K. M., and K. Kojima. 1967. The mode of selection at the polymorphic esterase-6 locus in cage populations of *Drosophila melanogaster*. Genetics 47:677–89.

Yamazaki, T. 1971. Measurement of fitness at the esterase-5 locus in *Drosophila pseudoobscura*. Genetics 67:579–603.

INDEX

Accuracy: of prediction, 8, 9; of measurement, 9
Acris crepitans, 122, 147, 193
Adaptation, 4, 30
Adaptive values, 11
Alexander, M. L., 34
Allard, R. W., 120, 248
Alleles: activity, 261; fixation rates, 221; frequencies, 95, 212, 213, 216, 243–45, 248, 257, 267, 270, 275, 281–85, 289, 306, 307, 312; lethal, 11, 38; mobility of, 263; mutation rates, 221; probability of fixation, 225; rare deleterious, 24; rate of substitution, 200, 265
Allen, A. C., 46
Allison, A. C., 37, 98
Allozymes, 107, 240, 261; fitness, 256; frequencies, 253; heterozygosity, 202; inversions, 316; *in vitro,* 263; polymorphism, 259, 282; variation, 120, 246
Altukhov, Yu. P., 121
Ambystoma maculatum, 120
Amino acids, 36, 99, 101; evolution of sequences, 218; rate of substitution, 199, 200, 222; substitutions, 35, 198, 199, 227, 231, 263, 267
Andara trapezia, 249
Anderson, W. W., 11, 46, 78, 205
Animal breeding, 15, 87
Anoplorchus purpurescens, 248; *A. insignis,* 248
Avena barbata, 120, 248; *A. fatua,* 120
Ayala, F. J., 62, 64, 65, 115, 142, 143, 175–77, 205, 246, 258

Baker, C. M. A., 114, 120
Balance hypothesis, 23, 25, 27, 53, 194, 196, 200, 202, 204, 206, 207, 217, 235, 265, 267, 317
B/A ratio, 74
Barlow, G. H., 263
Behavior, 90
Beiles, A., 308
Bell, A. E., 90
Bennett, J., 92
Berger, E. M., 118, 247, 251, 252
Biochemical genetics, 24
Biometrical genetics, 13, 15
Birch, L. C., 190
Boam, T. B., 161
Bodmer, W. F., 152, 202, 206, 236, 240, 241, 301, 306, 307
Breeding system, 11, 19
Breese, E. L., 65
Bristle number in Drosophila, 89
Brncic, D., 316
Brown, W. L., 162
Brues, A. M., 154
Bryozoa, 247
Bulmer, M. G., 263
Bundgaard, J., 255
Burla, H., 142, 177

Cain, A. J., 234
Carson, H. L., 92, 150, 151
Catastomus clarkii, 249

Cavalli-Sforza, L. L., 152, 206, 215, 240, 241
C/B technique, 10
Cepaea nemoralis, 120, 233–35; *C. hortensis*, 120
Charlesworth, B., 236, 316
Charlesworth, D., 316
Chetverikov, S. S., 23
Chicken, polymorphism, 114, 120
Chigusa, S., 85, 316
Chinnici, J. P., 92, 297
Chovnik, A., 35
Christiansen, F. B., 237, 255
Chromosomes: heterozygotes, 11; homozygotes, 10, replication, 48, 71, 98; salivary, 35; semilethals, 42; viabilities, 53–55
Clarke, B. C., 90, 227, 228
Classical hypothesis, 23, 26, 34, 53, 194–96, 207, 208, 263, 313, 317
Clayton, G., 67
Clegg, M. T., 120, 248
Clines: geographical, 245; topographical, 244
Coadaptation, 314
Cockerham, C. C., 241
Columbidae, 165, 166
Competition, 220, 258
Cordeiro, A. R., 44
Coturnix coturnix, 120
Coupling coefficient, 294, 298
Creed, R., 233
Crow, J. F., 26, 27, 44, 74, 76, 82, 196, 199, 201
Crumpacker, D. W., 138
Cytochrome *c*, 36, 263
Cytogenetics, 29

da Cunha, A. B., 142, 149
Dacus olea, 256
Dandelion, 120
Darwin, C., 3, 28, 30, 194, 196
Darwinism, 3, 4, 6, 159, 195, 198
Dempster, E., 192, 196
Dessauer, H. C., 122, 147
Deterministic theory, 267
Developmental sensitivity, selection for, in Drosophila, 91
Development time, 47; correlated with fitness, 62; selection for, in Drosophila, 90
de Wit, C. T., 258
Dickerson, G., 36, 88, 230, 260
Dickie, M. M., 36
Diderot, D., 157
Differentiation: genetic, ix, 4, 25, 86, 213, 267; random, 216
Diffusion processes, 265
Dimensionality, 9
Dipodomys merriami, 122; *D. ordii*, 122
Disequilibrium, 303, 310; linkage, 291, 293, 311–16
Dixon, M., 221
D/L ratio, 81, 82
DNA, 99, 123, 220, 221, 226
Dobzhansky, Th., 10, 11, 19, 27, 30, 31, 42, 44, 45, 46, 47, 49, 54, 55, 61, 72, 77, 78, 90, 113, 115, 130, 135, 140, 142, 143, 149, 150, 151, 159, 164, 174, 182, 203, 239, 242, 243, 244, 246, 258, 268, 281, 314
Domain of explanation, 9
Dominance, 15, 264, 265, 276; in population sampling, 96; relation to regression, 73
Doves, 165, 166
Drosophila, 10, 11, 48, 51, 55, 60, 79, 82, 89, 91, 92, 103, 201, 202, 222, 305, 308; cistrons, 35; classical mutants, 32, 37; controlled model selection experiments, 89; eye-facet number, 265; fitness variation, 65; lethal frequency in populations, 42; micro-chromosomes, 58; minority mating effect, 258; mutations, 29
Drosophila species: *D. affinis*, 120, 212; *D. ananassae*, 114, 120, 147, 245; *D. athabasca*, 120; *D. azteca*, 140; *D. equinoxialis*, 176, 178, 218; *D. flavomontana*, 167; *D. hydei*, 34; *D. immigrans*, 34; *D. lacicola*, 167; *D. lebanonensis*, 170; *D. melanogaster*, 32, 34, 44, 46, 47, 57, 58, 63, 66, 73, 78, 118, 123, 124, 172, 175, 178, 179, 182, 204, 205, 247, 248, 253, 256, 258, 259, 260, 262, 316, —bristle number, 67, —forked locus, 36, —"norm," 26, —rosy locus, 35; *D. mercatorum*, 92; *D. mimica*, 120; *D. mulleri*, 33, 34, 37; *D. nasuta*, 120, 147; *D. novamexicana*, 34; *D. obscura*, 144; *D. paulistorum*, 171, 172, 176, 178, 180, 182, 183, 185, 218; *D.*

Drosophila species (*Continued*)
pavani, 316; *D. persimilis*, 135, 136, 164, 172, 173, 174, 175, 178, 182, 184, 217, 231, 258, 281, 314; *D. prosaltans*, 42, 55, 57; *D. pseudoobscura*, 22, 44, 45, 46, 47, 49, 61, 72, 73, 77, 78, 90, 104, 105, 107, 108, 113, 114, 118, 128, 134, 135, 136, 140, 142, 144, 148, 149, 150, 164, 172, 173, 174, 175, 178, 179, 182, 184, 185, 212, 214, 216, 217, 226, 231, 242, 243, 247, 253, 256, 257, 258, 259, 263, 281, 282, 314, 315; *D. robusta*, 149, 150, 246; *D. simulans*, 118, 120, 172, 175, 178, 182, 210, 212; *D. subobscura*, 144, 316; *D. texana*, 167; *D. tropicalis*, 176, 178, 218; *D. victoria*, 170; *D. virilis*, 167, 168, 169, 171; *D. willistoni*, 57, 68, 143, 149, 150, 171, 172, 176, 178, 180, 182, 183, 185, 210, 213, 217, 218, 231, 246, —inversion polymorphisms, 142
Dubinin, N. P., 32
Dunn, L. C., 240
Dynamic sufficiency, 6, 7, 8, 11

Egg size, selection for, in Drosophila populations, 90
Ehrman, L., 258
Electrophoresis, 102, 198, 318; detection of variation, 197; studies of populations, 105, 167–70
Electrophoretic mobility classes, 218
Electrophoretic variation, 108, 197, 217, 249
Elm, 149
Embedding effect, 303, 309
Empirical sufficiency, 9–11
Environment, 248, 250, 256; fluctuating, 297, 300; marginal, 151; niche, 260; relationship with gene frequency, 248; spatial differences, 242; statistical properties, 297; temporal differences, 242; temporal sequence, 271; variables, 244; variation, 247
Enzymes, 24, 198; activity, 261, 265; concentration, 265; functional properties, 267; genes coding for, 123; glycolytic, 125; impairment, 35
Epistasis, 276
Epling, C., 130, 135, 314
Epstein, C. J., 228
Equilibrium, 258, 274, 294, 301, 303, 305, 306, 309–17; compositions of populations, 296; frequency, 244; genetic composition of an equilibrium population, 295; linkage, 289, 291, 293, 314, 315; stable, 192, 279; states, 269, 295; systems, 7; theory, 269; unstable, 279
Erlenmeyer-Kimling, L., 22
Eshel, I., 196
Estimation techniques, 11
Evolution, 3, 6, 8, 9, 31; adaptive, x, 196, 199, 221, 226; allele substitution, 200; allozyme frequency, 253; comparative, 226; gene replacement, 201, 281–82; inversion system, 275; laws, 21; molecular, 267; nonadaptive, x; non-Darwinian, 4; phyletic, 12, 229; progressive, 30; protein, 226; rates, 220, 229, 230; rates of amino acid substitution, 199, 200; whole genome, 317
Ewens, W. J., 210, 294
Extinction, 12
Eye facets in Drosophila, 91

Falk, R., 84, 85, 86
Fecundity, 10, 65; selection for in Drosophila populations, 90
Feldman, M., 196, 210, 289, 295, 307, 308
Felsenstein, J., 220, 291, 301
Fertility: component of fitness, 60; differential, 238; variance in, 60
Feuerbach, H., 144
Fish, 237–38, 248–49, 261
Fisher, R. A., 257, 288, 289, 297
Fitch, W. M., 227
Fitness, x, 15, 220, 256, 258, 263, 264, 276, 279, 289, 290, 293, 295–305, 308, 310, 317, 318; allozyme, 256; component, 10; constants, 278; differences, 252; differential, 239; distribution, 289; effect of linkage, 295; estimates, 11, 236, 237, 254, 259; genetic variance, 206; genotypes, 10; heterozygotes, 37, 71, 72; homozygotes, 37, 71, 72, 257; marginal, 277–78; mean, 201; measurement, 10, 65, 235, 236, 242, 250, 260; model,

Fitness (*Continued*)
291; modifiers, 47; multiplicative model, 202, 203; net, 65, 235, 238, 241; nonadditivity, 294; one-locus, 306; overdominance, 191, 199; partial, 239, 241; relations, 278; standard error, 11; symmetrical model, 307; values, 235; variance, 195, 206
Fixation probability, 225, 226
Ford, E. B., 30, 97, 233
Franklin, I., 289, 304, 305, 308, 309, 310, 311, 313
Frequencies: allozyme, 253, equilibrium, 244; gametic, 282–85, 288, 289; gene, 95, 212, 213, 216, 243–45, 248, 267, 270, 275, 281–83, 285, 289, 306, 307, 312; Hardy-Weinberg, 238, 241, 242; independent, 278; inversion, 274, 281; ratios, 268; zygotic, 282
Frogs, 122, 147, 193
Frydenberg, O., 237, 249, 313
$\hat{f}$ values, heterogeneity of, 216

Galton, F., 5, 6
Geiringer, H., 284
Gene: coding specific enzymes, 123; deleterious, 77; detection of invariant, 100; fixation, 225; multiplicative action, 294, 301; neutral, 312; nonrandom association, 290; replacement cost, 220; replacement in evolution, 201; species-differentiating, 174; substitution, 223, 225
Gene frequencies, 275, 281–83, 285, 289, 307, 312; between populations, 212; characterization, 95; cyclic fluctuation, 243; distribution between species, 216, 267; historicity of change, 270; homogeneity, 213; in nature, 306; mosaic distribution, 245; relationship with environment, 248; spatial variation, 244; stable equilibrium, 257
Genetic divergence, 4, 25, 86, 213, 267
Genetic elite, 30
Genetic load, 15, 74, 79, 202, 218–20, 257, 260, 290, 305, 306; balanced polymorphism, 199; defined as loss of fitness, 80; gene substitution, 199; populations at equilibrium, 195
Genetic map length, 310–12, 318
Genetic revolution, 200
Genetics: biochemical, 24; ecological, 233, 235; evolutionary, x, 6, 8, 9; molecular, 198
Genetic variance, 10, 15; additive, 14; fitness, 206; short chromosomal segments, 58
Genetic variation, ix, x, 12, 16, 30, 31, 48, 235; assessment, 35; description, 19; hidden, 23; natural populations, 92; populations, 232; reduction, 193; removed from populations, 192
Genome: as unit selection, 273; constitution, 123; haploid model, 195
Genotype: distribution in a population, 28, 189; enumeration, 21, 22, 96; fecundity, 236; fitness, constant, 15; fitness of combination, 278; fitness with respect to single loci, 10; fittest, hypothetical, 202; frequency, 11, 96; ideal, 220; probability of survival, 236; relation to phenotype, 19; statistical distribution, 20; superior, 31; variations in time and space, 96
Geographical races, 4, 212
Gibson, J., 161, 253, 262
Giesel, J. T., 236
Gillespie, J., 118, 120, 124, 175
Gooch, J. L., 247
Goodman, M., 121
Grasshoppers, 242, 275, 279, 280
Greenberg, R., 82; *see also* Temin, R. G.
Grell, E. H., 34
Grossman, A. I., 249
Gustafsson, A., 258

Haigh, J., 270
Haldane, J. B. S., 28, 184, 192, 199, 224
Hamrick, J. L., 248
Haploids, 195
Hardy-Weinberg equilibrium, 238, 241, 242, 276, 282
Harper, J. L., 258
Harris, H., 108, 112, 115, 116, 124, 126, 152, 198, 262
Harrison, B. S., 89
Harrison, G. A., 241
Hartley, B. S., 260

Hemoglobin, 100, 198; evolution, 270; variants in man, 270
Heritability, 14, 15; of I.Q. in Caucasians, 22
Hershey, A. D., 116
Heterosis, 37, 196, 204, 205, 239, 258, 262, 276; detection, 74; multiplicative model, 302, 303; single-locus, 191; viability, 242
Heterozygosity, 27; allozyme, 202; amount maintained by mutation, 199; as allelic variation, 207; at different loci, 119–27; classical hypothesis, 26; human blood groups, 126–27; in populations, 200; in various species, 116–28; neutral mutations, 213; neutral theory, 208; optimum level, 86; relationship with polymorphism, 210; steady state, 208, 312; total, 201
Heterozygotes: chromosomal, 11, 53; correlations with homozygotes, 71, 72, 74; fitness, 27, 37, 71, 72; multiple, 202, 204
Hill, W. G., 313
Hiorns, R. W., 241
Hiraizumi, Y., 44
Hochman, B., 44, 47
Homo sapiens, 241; "type" of, 26
Homozygosity, classical hypothesis, 26
Homozygotes: chromosomal, 10, 53; correlations with heterozygotes, 71, 72, 74; fitness, 37, 71, 72; multiple, 204, quasi-normal, 54; viability depression, 68
Hopkinson, D. A., 115, 124, 126
Hovanitz, W., 268
Howland, G. P., 120
Huang, S. L., 253, 259
Hubby, J. L., 107, 108, 115, 116, 130, 167, 168, 169, 171, 173, 185, 202
Human blood groups, 125; frequencies, 269; polymorphisms, 98, 234, 269
Hunt, W. G., 145, 148, 180, 245
Husbandry, technology of, 87
Hybridization, 190

Ideal: Aristotelian, 5; Platonic, 5
Idealized quantities, 80
Immunology, 198; techniques for measuring variation, 100
Inbreeding, 41; depression, 74, 204, 205, 207, 251, 257, 258, 267, 309, 310, 312, 318
Independent assortment, 295
Ingram, V. M., 38
Insecticide resistance: selection for, in Drosophila populations, 91
Invariant patterns: selection for, in Drosophila populations, 90
Inversions, 297, 316; evolution, 275; frequency, 274, 281; in grasshoppers, 275, 279, 280; polymorphism, 150, 273, 315
Irwin, M. R., 165, 166
Isozymes, 107
Ives, P. T., 27

Jacquard, A., 154
Jarvik, L. F., 22
Jepsen, G. L., 3
Johnson, F. M., 114, 120, 147, 248
Johnson, G. B., 109, 210
Johnson, M. S., 248
Johnson, W. E., 122
Judd, B. H., 35
Jukes, T. H., 199, 219, 222, 226, 227

Kangaroo rat, 120
Karlin, S., 194, 295, 307
Kaufman, T. C., 35
Kenyon, A., 58, 73
Kidwell, M. G., 92, 297
Kimura, M., 4, 26, 27, 30, 192, 193, 196, 199, 200, 220, 221, 224, 227, 228, 245, 288, 295, 313
Kinetics, *in vitro,* 261, 263
King, J. L., 199, 202, 203, 204, 205, 219, 222, 226, 227, 306
Knight, G. R., 90, 161
Koehn, R. K., 249
Kojima, K., 118, 120, 124, 175, 222, 239, 253, 259, 279, 284, 288, 291, 295, 299, 307
Koreneva, L. G., 249
Krakauer, J., 215
Krimbas, C. B., 44, 72, 256, 316

Lactate dehydrogenase, 100
Ladybirds, pattern polymorphism, 97
Lakovaara, S., 144

Lamotte, M., 97, 233, 234
Latter, B. D. H., 263
Lehman, H., 198
Lepidoptera, pattern polymorphism, 97
Lethals: allelism, 45; average dominance, 44; detection of recessive, 40; frequency in Drosophila, 42; heterozygous fitness, 44; mapping, 46; Poisson distribution, 42; recombination, 46; synthetic, 46
Levene, H., 192, 242, 258
Levin, D. A., 120
Levine, P., 234
Levins, R., 162, 297
Lewontin, R. C., 44, 62, 65, 107, 108, 115, 116, 122–24, 126, 130, 135, 138, 151, 155, 173, 185, 190, 202, 215, 233, 239, 240, 241, 242, 257, 258, 259, 271, 275, 279, 282, 284, 288, 291, 294, 295, 297, 299, 301, 304, 305, 307, 308, 309, 310, 311, 313, 314
L'Heritier, Ph., 11
Limulus polyphemus, 122, 124
Lindsley, D. L., 34
Linkage, 287, 290, 292, 294, 297–312, 316, 317; disequilibrium, 284, 291, 293, 311, 313–16; effect on fitness, 295; equilibrium, 289, 291, 293, 314, 315; interaction with selection, 290
Litke, A., 137
Load ratios, 79; estimated in birds, 88
Loukas, M., 316
Luffman, J. E., 262

Macacus fascicularis, 121
MacArthur, R., 162
McDonald, J. M., 265
McGregor, J., 194
MacIntyre, R. J., 118
Madden, C. V., 10
Maize, 27
Malaria, 37
Malécot, G., 245
Man: differential activity of enzymes, 261; heterozygosities, 118, 119; racial variation, 152
Manwell, C., 120
Margoliash, E., 263
Marinkovic, D., 62; study of fecundity, 61, 73
Marker effect, 48
Markert, C. L., 107
Markowitz, E., 227
Marshall, D. R., 120
Mather, K., 89
Mating: assortive, by genotype, 282; laws, 13; minority advantage, 259; random, 251; scheme, 41; selection for, in Drosophila, 90
Matsuo, Y., 258, 259
Maynard Smith, J., 90–92, 270
Mayr, E., 3, 4, 148, 159, 160, 163, 164, 171, 190, 200
Mendel, G., 6, 13, 25, 96
Mendelism, 6, 13, 31, 159, 194, 195, 196; genetic system, 23, 24, 47; mechanism, 193; principles, 48
Merrell, D. J., 37
Merritt, R. B., 261, 315
Mettler, L. E., 316
Mice, 115, 245; see also *Mus musculus* and *Peromyscus polionotus*
Migration, 214, 216, 225, 234, 245, 246, 267, 268; between populations, 213, 315; effect on random differentiation, 213, 215; laws, 13
Milkman, R. D., 90, 202
Modiano, G., 262
Mole rat, 108, 122
Møller, F., 107
Mollusks, 249
Monod, J., 157
Moore, C. H., 90
Moraba scurra, 242, 275, 279, 380
Moran, P. A. P., 225
Morgan, T. H., 13, 30
Morphogenetic disturbance, 36
Mortality: environmental, 76; pre-adult in human population, 76
Morton, N., 74, 76
Mukai, T., 66, 85, 86, 95, 316
Muller, H. J., 4, 26, 29, 30, 31, 36, 38, 69, 73, 74, 76, 82, 84, 85, 195, 196; C/B technique, 10
Multi-locus theory, 317
Mus musculus, 144, 180, 195, 217; *M. m. musculus*, 148, 180, 190; *M. m. domesticus*, 148, 180, 190
Mutant expression, selection for in Drosophila, 91

Mutants: "leaky", 24; rate of introduction of new, 222; revertants of morphological, 36; visible, 34
Mutations, 28, 234; adaptive, 223, 224, 267; chain-terminating, 99; defined, classical school of formal genetics, 30; deleterious, 27, 69, 195, 197, 264; favorable, 198, 199; forward and reverse, 36; heterozygous effect of new, 82; induction, 29; lethal, 97; maintenance of heterozygosity, 199; manipulating, 29; neutral, 197, 199, 213, 224; overdominant, 71; product of variability, 66; random in homozygous stock, 83; rates, 26, 27, 222, 267, 268; redundant, 197–98; visible, 35, 36, 97

Nair, P. S., 316
Narise, S., 173
Natural selection, x, 3, 4, 19, 22, 191, 194–96, 224, 227, 230, 232, 288, 289, 315; direction and intensity, 27; laws, 13
Neel, J. V., 77
Neoclassical theory, 197–200, 207–10, 217, 218, 223–31, 250, 264; *see also* Neutralist theory
Neutralist theory, 197, 218, 250; explanation of observed heterozygosity, 208; prediction of random variation, 213; *see also* Neoclassical theory
Nevo, E., 122, 147
Newton, I., 5, 6
Nicol, P. J., 249, 256
Nucleotides, 99

Oats, 120, 248
O'Brien, S. J., 118
Oenothera biennis, 120
O'Gower, A. K., 249
Ohta, T., 4, 30, 199, 220, 221, 224, 227, 228, 313
Oncorhyncus keta, 121
Oshima, C., 11
Overdominance, 70, 207; associative, 313; fitness, 191, 199; mutations, 71; unconditional, 192

Paradox, epistomological, 10
Parsons, P. A., 307
Pattern, selection for in natural populations, 90
Patterson, J. T., 36
Pavlovsky, O., 11, 239, 258, 281
Peromyscus polionotus, 124, 145, 146
Perturbation technique, 273
Perutz, M. F., 198
Petit, C., 258
Phasianus calchicus, 120
Pheasant, 120
Phenotype: distribution, 289; enumeration, 22, 23; manifestation, 19, 21; relation with genotype, 19; selection for intermediate optimum, 297; variance, 14, 15, 21
Phyletic evolution, 12, 229
Pimephales promelas, 261
Plant breeding, 15, 87
Plants, allozyme variation, 120
Pogonomyrmex barbatus, 120, 248
Polivanov, S., 37
Polymorphisms, 232, 233, 239, 266, 301; allozyme, 316; amino acid substitution, 263; balanced, 287; Batesian mimicry, 288; biochemical, 97, 98; electrophoretic variants, 197, 249; evolutionary dynamics of complex, 288; human blood group, 98, 234; human electrophoresis, 263; inversion, 150, 273, 315; single-gene balanced, 37; substitutions, 267
Polypeptide chain, 37, 99, 101
Population cages, 11
Population genetics, 10, 12, 267, 268; theory, 12, 266, 268, 288
Populations: biological future, 86; breeding size, 224; differences between, 25; evolution, 286, 289; genetic description, 281; marginal, 148; variation in outbred, 92
Population size, 268
Population structure, theory of, 23, 24
Powell, J. R., 142, 143, 175, 176, 177, 242, 246
Prakash, S., 107, 130, 135, 138, 150, 172, 173, 175, 184, 185, 282, 314, 315
Prediction: accuracy, 8; quantitative, 28; sensitivity, 10
Progress, notion of, 269

Proteins, 35; amino acid sequence, 99; electrophoretic mobility, 105, 167; enzymatic, 105; evolution, 226; genetic differences between species, 167–70; nonenzymatic, 125; physicochemical properties, 101
Prout, T., 11, 44, 65, 66, 238, 255, 256

Quail, 120

Races: ecological, 4; genetic differences among, 25; geographical, 4
Random drift, 195, 233, 245, 246, 314, 315
Randomization, 10; dynamics, 284
Random processes, 3, 4
Rare-genotype advantage, 257
Rasmussen, D. J., 249
Rat, 121
Rattus rattus, 121
Recombination, 190, 196, 249, 283, 288–91, 294–97, 300, 303, 305, 308, 313, 314; creation of lethals, 46; generation of genetic variants, 67; production of variability, 66
Reed, T. E., 202, 306
Reeve, E., 89
Regression, 6, 72; relation to dominance, 73
Rendel, J. M., 90, 91
Richmond, R., 180, 182, 242
Robertson, A., 93, 313
Robertson, F. W., 67, 89, 90, 161
Roderick, T. H., 115
Ruddle, F. H., 115

Salamanders, 120
Salmon, Pacific, 121
Salthe, S. N., 101
Sang, J. H., 265
Sano, K., 85
Saura, A., 144
Scharloo, W., 91, 161
Schizoporella unicornis, 247
Schlager, G., 36
Schopf, T. J., 246
Schull, W. J., 77
Segregation, allozyme variants, 240
Selander, R. K., 122, 124, 145, 148, 180, 245
Selection, 68, 215, 234, 242, 245, 249, 250, 254, 256, 267–69, 273, 288, 290, 295, 298, 300, 301; additive genetic variation, 87; against heterozygotes, 193; balancing, 27, 191, 199, 200, 201, 212, 232, 233, 235, 239, 241; directional, 195; disruptive, 91; experiments to measure, 251; fertility component, 255; frequency-dependent, 239, 256–60, 274, 276; in the laboratory, 250; independent, 93; index, 93; interaction with linkage, 290; intermediate optimum phenotype, 297; limit, 16; one-locus theory, 307; plateau models, 204; progress under, 16; stabilizing, 91; tandem, 93; truncation, 306
Selective changes, 287
Selective constraints, 227
Semispecies comparisons, 183
Sen, B. K., 93
Sensitivity of predictions, 10
Serov, O. L., 121
Sexual reproduction, importance in evolution, 196
Shaw, C. R., 108, 122
Sheldon, B. L., 91
Shen, M. W., 35
Sheppard, P. M., 234
Sickle-cell anemia, 37, 44, 98, 198, 199, 262
Simpson, G. G., 3, 339
Singh, M., 253, 259
Singh, R., 169
Single-gene: effects, 54; heterosis, 70; substitution, 55
Single-locus substitution, 82, 310
Single-locus theory, 317
Size, selection for in Drosophila, 89
Slatkin, M., 15, 310
Snails, 97, 120, 233–35
Socio-biological theories, 30
Solbrig, O., 120
Sondhi, K. C., 90, 91, 92
Spalax ehrenbergi, 122
Spassky, B., 10, 11, 42, 44, 45, 46, 61, 72, 77, 78, 90, 205
Speciation, 4, 12, 27, 159, 161, 179, 183, 185, 186, 196, 200, 201, 225; genetic variation and, 28; geographic, 160,

Speciation (*Continued*)
162; paradox of, 28
Species: antigenic differences, 165; biological future, 86; border of, 148; breeding size, 224; closely related, 216; differentiating genes, 174; divergence, 160–62; formation of new, 225; genetic difference between, 163, 164; incipient, 4, 216; *in statu nascendi*, 179; internal margin, 149; isolated, 4; Mendelizing differences, 165; morphological differences, 165; self-fertilizing, 282; sibling, 171; sterility barrier, 165
Spencer, H., 30
Spencer, W. P., 35, 37
Sperlich, D., 144
Spiess, E.B., 46, 258, 281
Stability, 269; measures, 263
Stabler, L. J., 27
Stable states, alternative, 308
Standard errors of fitnesses, 11
State space, 8
State variables, 8, 16
Steady-state distributions, 269
Stebbins, G. L., 231
Steiner, E., 120
Stern, C., 44
Stochastic theory, 223, 267–69
Stone, W. S., 120, 147
Streptopelia chinensis, 165, 166; *S. resoria*, 165, 166; *S. senegalensis*, 166
Stress, environmental, 265
Subak-Sharpe, J. H., 227
Substitution: neutral, 228; polymorphic, 267; probabilities, 227, 228; segregating, in populations, 261; rates over geological time, 228
Subvitals, 54
Sufficient dimensionality, 9, 15
Survival probabilities, 10, 22, 41
Sved, J. A., 62, 63, 64, 65, 79, 202, 204, 205, 258, 306

Taraxacum officinale, 120
Teissier, G., 11
Temin, R. G., 44, 46, 78, 205; *see also* Greenberg, R.
Thalassemia, 98
Theory construction, 9
Thoday, J. M., 161
Thomas, D. M., 262
Thomson, G., 289, 308
Throckmorton, L. H., 162, 168, 169, 171
Tidwell, T., 77
Timofeeff-Ressovsky, N. W., 97
Tobari, Y. N., 73, 118, 120, 124, 175, 222, 259
Tolerance set, 8, 9
Tsakas, S., 256
Turner, J. R., 297

Ulitskaya, L. E., 249
Ulmus americanus, 149
Unmeasurability, 9
Uzzell, T., 120

Variance, 268, 298; additive genetic, 14; analysis of, 285; binomial, 268; developmental, 60; environmental, 51, 60; fitness, 195, 206; genetic, 10, 15, 58, 206; maintenance, 192; phenotypic, 14, 15; sampling error, 51; selection for in Drosophila, 90
Variation, 68; allozyme, 120, 246; between groups, 28; characterization, 23; conversion, 12; electrophoretic, 108, 197, 217; environmental, 35, 247; geographical, 212; genetic, ix, x, 12, 16, 19, 23, 32, 35, 192, 193, 232, 235; in populations, 27, 32, 190, 200; in viabilities, 51; individual, 4; intrapopulation, 207; materialistic view, 4; measurement, 95, 100, 101; metaphysical mutation and recombination, 66; stabilization, 196; standing, 267; under domestication, 5; within populations, 201, within species, 28
Viabilities, 10, 49, 255; bimodality, 54; differences, 241; heterozygous and homozygous, 74; interactions, 78; larval, 258; unimodal distribution, 58
Vogel, F., 219, 235

Waddington, C. H., 90, 91, 161
Wallace, B., 10, 25, 42, 46, 47, 72, 83, 84, 85, 86, 95, 161, 220
Warren, D. C., 90

Watanabe, T., 11
Webb, E. C., 221
Weiner, A. S., 235
Weiss, M. L., 121, 245
White, M. J. D., 242, 275
Wild-type concept, basis of racism, 26
Wills, C., 73, 256
Wilson, J., 66, 162, 255
Wright, S., 11, 27, 45, 47, 192, 193, 233, 257, 268

Yamazaki, T., 253, 256, 257, 259
Yang, S. Y., 122, 124, 145, 148, 180, 245
Yarbrough, K. M., 239, 253, 259
Yoshikawa, I., 85

Zoarces viviparus, 237–38